Physical Gels from Biological and Synthetic Polymers

Presenting a unique perspective on the state-of-the-art of physical gels, this interdisciplinary guide provides a complete, critical analysis of the field and highlights recent developments. It shows the interconnections between the key aspects of gels, from molecules and structure through to rheological and functional properties, with each chapter focusing on a different class of gel. There is also a final chapter covering innovative systems and applications, providing the information needed to understand current and future practical applications of gels in the pharmaceutical, agricultural, cosmetic, chemical and food industries. Many research teams are involved in the field of gels, including theoreticians, experimentalists and chemical engineers, but this interdisciplinary book collates and rationalizes the many different points of view to provide a clear understanding of these complex systems for researchers and graduate students.

Madeleine Djabourov is Professor at the Ecole Supérieure de Physique et Chimie Industrielles de la Ville de Paris (ESPCI-ParisTech), in charge of thermodynamics and rheology courses. She was involved in pioneering studies on gelation and percolation initiated by Pierre-Gilles de Gennes.

Katsuyoshi Nishinari is a Special Appointment Professor at the Graduate School of Human Life Science, Osaka City University. He has investigated gels and gelling processes as part of a collaborative research group on gellan gum affiliated with the Research Group on Polymer Gels in the Society of Polymer Science, Japan.

Simon B. Ross-Murphy is currently a Visiting Professor at the University of Manchester (Materials Science) and the University of Nottingham (Biopolymer Science and Technology). He has previously been a Professor in Biopolymer Science at King's College London, and before that was at Unilever Research.

'Many innovative functional gel materials have been developed over the past decade, and our understanding of physical gels and their functionalities is advancing at a rapid pace, but so far there has been a lack of comprehensive textbooks suited to introduce graduate students, teachers and research workers into the science of physical gels. This wonderful book perfectly fills this need. Written in an elegant and accessible style with lucid concepts, plenty of examples, and spectacular figures, including the authors' original scientific works on rheology and phase transitions, the book takes the readers gently from the most elementary concepts of physical gels to the forefront of current research. The book can therefore be warmly recommended as a textbook or reference work for both undergraduate and graduate courses whether or not the readers are familiar with the subject.'

Fumihiko Tanaka, Kyoto University, Japan

'As a class of fascinating materials, physical gels hold realized and potential application in many fields. This interdisciplinary book provides basic approaches to rationally designing and fabricating a physical gel along with molecular level understanding of the gelation mechanism.'

Hongbin Zhang, Shanghai Jiao Tong University, China

'This book is a very important and original one; with 351 pages, it is devoted to a good and up to date review on physical gels. The 3 authors are well-known in this field and they are able to cover it extensively: the different mechanisms of physical gel formation, especially important in the domain of natural polysaccharides are discussed. The techniques able to characterize the gels at different scales are given in the same time as method for the gel point (sol-gel transition) determination. The cases of proteins, synthetic polymers and polysaccharides are well documented. I can recommend this book for people starting in the field of physical gels and those yet involved in the study of physical gels.'

Marguerite Rinaudo, European Synchrotron Radiation Facility (ESRF)

Physical Gels from Biological and Synthetic Polymers

MADELEINE DJABOUROV

Ecole Supérieure de Physique et Chimie Industrielles de la Ville de Paris (ESPCI-ParisTech)

KATSUYOSHI NISHINARI

Osaka City University, Japan

SIMON B. ROSS-MURPHY

University of Manchester

CAMBRIDGE
UNIVERSITY PRESS

CAMBRIDGE UNIVERSITY PRESS
Cambridge, New York, Melbourne, Madrid, Cape Town,
Singapore, São Paulo, Delhi, Mexico City

Cambridge University Press
The Edinburgh Building, Cambridge CB2 8RU, UK

Published in the United States of America by Cambridge University Press, New York

www.cambridge.org
Information on this title: www.cambridge.org/9780521769648

First published 2013

Printed and bound in the United Kingdom by the MPG Books Group

A catalogue record for this publication is available from the British Library

Library of Congress Cataloguing in Publication data
Djabourov, Madeleine, 1949–
Physical gels from biological and synthetic polymers / Madeleine Djabourov, Ecole Superieure de Physique et de Chimie Industrielles de la Ville de Paris, Katsuyoshi Nishinari, Osaka City University, Japan, Simon B. Ross-Murphy, University of Manchester.
pages cm
Includes bibliographical references.
ISBN 978-0-521-76964-8 (hardback)
1. Polymer colloids. I. Nishinari, Katsuyoshi. II. Ross-Murphy, S. B. III. Title.
QD549.2.P64D56 2013
541′.345–dc23

2012038172

ISBN 978-0-521-76964-8 Hardback

Contents

Preface

This book is the result of both long personal friendships and a series of scientific collaborations, between the three authors. Although we have been working in this field essentially independently, as is usual we have also intersected with one another on many occasions in various parts of the globe, and visited each other's facilities. From all of this, we have come to realize that our thoughts have evolved through various aspects of this complex topic. The idea of writing a book represents some culmination of these intersections, both geographical and ideological. This, together with our will to put together our personal views and experiences to try to reflect the specific ways in which the topic has been appreciated in our original cultures and environments, has resulted in this volume.

Writing the book, which is necessarily then the synthesis of our views, has been a matter of considerable debate, because of the large area that the topic obviously covers. Even though this book reflects the overall complexity of this still-developing subject imperfectly, we feel that we have produced an appropriate survey of the present state of art. We hope that the final result could be an introduction to a larger public, but more particularly for physical chemists, condensed matter physicists and all disciplines in between, and that it will encourage other workers to adopt these topics and so to introduce their own ideas and hypotheses.

We wish to acknowledge the fruitful discussions with, and advice from, many distinguished scientists and colleagues. Among these the thoughts and encouragement of Dr Kawthar Bouchemal, Professor Walther Burchard, Professor Allan Clark, Professor Lucilla de Arcangelis, Professor Emanuela Del Gado, Professor Masao Doi, Professor Jacques Leblond, Professor Edwin Morris, Professor Kunio Nakamura, Dr Klaas te Nijenhuis, Professor Marguerite Rinaudo, Professor Fumihiko Tanaka, Professor Masayuki Tokita and Professor Peter A. Williams are especially appreciated.

We thank Mrs Trang Le Viet for her precious contribution in realizing the artwork of the book with great talent and patience, and SRM thanks Dr Adam Corrigan, University of Cambridge, for kindly providing a copy of his PhD thesis.

Madeleine Djabourov
Katsuyoshi Nishinari
Simon B. Ross-Murphy

1 Introduction

A number of high polymer systems can assemble to form networks and gels and this assembly can occur by a variety of mechanisms. Some of these involve chemical treatments and processes, others are largely physical in nature – it is on the latter that this volume is focused. In this introductory chapter we consider the nature of polymer gels and networks, the processes of assembly, their characterization by physical methods and the coverage of succeeding chapters.

1.1 Gels from colloidal and polymer networks: a brief survey

Traditional colloidal gels were first investigated by Thomas Graham in the Philosophical Transactions of the Royal Society (Graham, 1861). Graham proposed a definition of substances according to their diffusive power. Colloidal substances (from the Greek κολλα, 'glue') are slowly diffusing substances, held together in solution by what he termed 'feeble forces'. In the classification of these substances Graham included hydrated silicic acid, hydrated alumina, starch, gelatin, albumen and gum.

Graham continued his discourse, saying 'opposed to the colloidal, is the crystalline condition, the distinction is no doubt of one of intimate molecular constitution', although the means of characterizing the latter were not available at that time. This very profound introduction to the colloidal state also includes this following intuitive statement: 'another eminently characteristic quality of colloids is their mutability', that is to say their ability to change. He writes: 'Their existence is a continued metastasis', i.e. again capable of change, which immediately points towards the current fully accepted difficulties in characterizing this particular state of the matter. He was also the first to recognize the specificity of the colloidal state and even went as far as suggesting that 'it is the source of vitality' revealing the mystery of life!

Efforts to analyse the colloidal state appear again in 1899, in a paper by W. B. Hardy with the subtitle 'The Structure of Colloidal Matter and the Mechanism of Setting and of Coagulation'. He addressed in particular the question of the effect of the fixing reagents used in cell observation:

> I would start the discussion with no statement as to the nature of cell-protoplasm other than that it is, as Dujardin described it, 'glutinous'. Now this glutinous character is a special characteristic of that state of matter to which Graham applied the word 'colloidal'. This statement holds without modification whether the initial stage, that is the soluble colloid, be entirely fluid (colloidal

solution), entirely solid (jelly), or a mixture of the two; or whether the physical change is or is not accompanied by chemical change.

He recalls also that:

Graham's nomenclature is as follows: The fluid state, colloidal solution, is the 'sol', the solid state the 'gel'. The fluid constituent is indicated by a prefix. Thus an aqueous solution of gelatin is a 'hydrosol', and on setting it becomes a 'hydrogel'.

This nomenclature is still largely valid nowadays, other extensions having been introduced. Gelatin, in some ways the paradigm hydrogel, and which derives its name directly from 'gel', is produced by hydrolysis of collagen, so gelatin gels were probably recognized by early man during cooking, much before their properties were understood.

Gelation was initially regarded in the same light for supersaturated solutions of inorganic compounds or natural polymers, although now the two are regarded very differently. Natural rubber gave the first estimate of molecular mass (referred to informally, and in the early literature, as 'molecular weight') in the thousands of daltons, but these were then largely dismissed and so posed the question of the nature of such molecules. Paul J. Flory wrote: 'The gap between molecules of ordinary size and those of hundreds or thousands times as large was too great to be bridged in a single leap' (Flory, 1953). Indeed, recognition of polymer molecules (macromolecules) came much later, when Hermann Staudinger, against the views of many of his contemporaries, asserted the existence of covalently bound long chains (Mülhaupt, 2004). Rubber, gelatin and cellulose were all considered to belong to this category. The gels described in this volume mostly involve such polymeric chains. Covalently bound polymer networks became very important in polymer science, since, depending upon conditions, they can form 'rubber-like' elastomers – they often show high elasticity – rigid resins or foams. According to Treloar (1975) such elastomer systems, for example, have been studied for many years, going back to the work of Gough and Joule in the nineteenth century. In excess of solvent such a polymer network no longer dissolves; instead it simply swells – often, but not exclusively, since this depends on the polymer–solvent system – to fill the volume of its container. The solvent can be replaced by another fluid inside the network, giving rise to macroscopic changes of gel volume either by swelling or by shrinking

Covalently linked networks are permanent, since the junctions can be assumed to be formed irreversibly; in physical networks the junctions are sometimes irreversible even though they do not involve covalent linkages (Ca alginates or, heat-set protein gels), but they can also sometimes associate or dissociate reversibly under thermodynamic (temperature, pH, ionic strength) or mechanical action (shear, elongation), depending on their species and mode of formation. In fact any definition of physical gels which simply implies shear or temperature reversibility is far too narrow. As we will see below, some of the systems of interest to us *are* reversible when they are subject to, for example, stirring ('shear'), others are temperature reversible, but some are neither of these. Many papers, particularly some theoretical papers, seem not to have grasped this essential facet.

That said, the widely cited definition by Dorothy Jordan Lloyd that 'the colloidal condition, the gel, is one which is easier to recognise than to define' (Jordan Lloyd, 1926)

still reflects the unclear description. An early, more formal classification of the gel state was noted by Hermans (1949), but still stresses the traditional colloidal viewpoint. According to Hermans' definition:

- Gels should be coherent two-component systems formed by solid substances finely dispersed or dissolved in a solvent.
- They should exhibit solid-like behaviour.
- The dispersed component and the solvent should extend continuously throughout the whole system (bicontinuous systems).

In an earlier review (Djabourov, 1988), four different structures were illustrated: aggregates of colloidal particles, with crystalline or amorphous structures; frameworks of rod-like particles as in supersaturated solutions; polymer gels with long linear chains held together by physical bonds such as 'crystalline junction zones'; and covalent bonds.

1.2 Structural characteristics and their study

At a Royal Society of Chemistry Faraday Division Meeting, Flory (1974) proposed a structural classification of gels which has some features in common with Hermans' definition, but with more emphasis on specifically polymeric systems:

- Well-ordered lamellar structure, including gel mesophases
- Covalent polymeric networks, completely disordered
- Polymer networks formed through physical aggregation, predominantly disordered, but with regions of local order
- Particular, disordered structures.

In the present volume we are concerned with aspects of all of these classes, although, as we have already stated, the second of these, the so-called chemical gels, although very important, falls naturally outside the definition of a physical gel. Instead we will be concerned with the other three types. Incidentally, the term 'physical gel' itself has not been in existence for that long: although it is difficult to establish quite who first coined the term, it seems to go back to the early 1970s, to Pierre-Gilles de Gennes.

At a later Faraday Division Meeting, Andrew Keller, another distinguished polymer scientist, noted (Keller, 1995) that

> there is no simple and unique definition of the gel state. Its main constituent is a fluid, yet it retains its shape, which is a characteristic feature of the solid state matter. It can support large strains to a high elastic limit, which, however, a real solid cannot do. Everyday one can experience that a jelly wobbles! The retention of shape implies a connectedness of the system which, in view of the fact that the major component is a fluid, means that there is a connective pathway along the non-fluid component which is the existence of a 'network'.

Consequently he coined a further definition in which gels were 'fluid containing self supporting disperse systems where the non-fluid connecting elements are not confined to individual chain molecules but can also be larger assemblies of molecules, finely divided struts or even membranes of the appropriate solid phase'.

Despite the above definitions, we need to clarify that many of the systems described as 'gels', such as shower gels and pain relief or topical gels, are often not gels even in terms of the various descriptions above. For example, in the cosmetic and pharmaceutical area, the term 'hydrogel' is widely employed. The 'hydro' part of this again shows that the solvent is water or electrolyte, but the 'gel' part, by contrast, does not always require that the underlying structures are of the types of network defined above. Instead, in some of these application areas, this term has been applied rather more broadly to just 'simple' viscous solutions. The importance of this distinction is introduced below, and detailed in Chapters 2, 3 and 4. Noting how widely the term is applied in the external literature, we will still use the term 'hydrogel' in this volume, but underline the necessary caveat that we are using it implicitly in a more restricted sense than some workers in these application areas.

1.2.1 Solids versus liquids

What should be clear from almost all of the definitions given above is that a gel has some solid or solid-like characteristics, but the field is not made easier by the number of different terms employed. These include hydrogel, as already noted, microgel, nanogel, strong gel, weak gel, fluid gel, topical gel and liquid gel.

A further question must be considered: if a gel is solid-like, then what constitutes a solid? This is not as straightforward a distinction as might at first be thought. For example, some materials, such as pitch or bitumen, appear to be solids, but are actually flowing, although their rate of flow is obviously extremely small. In the famous pitch drop experiment, for example, one drop falls approximately every decade.[1] At the same time, some unfortunate jumping off a very high river bridge dies not of drowning but of multiple fractures, since over very short times water responds like a solid. This underlies the reality that almost all materials are actually 'viscoelastic' – they combine both liquid (viscous) and solid (elastic) properties, and which of these dominates tends to depend on factors such as temperature (and pressure) and the duration of observation. In practice most conscious human activities correspond to time scales from, say, 0.1 s up to tens of seconds, so specialist instruments that sample over much shorter or much longer time scales (Chapter 2) tend to be much better at making distinctions such as 'solid' versus 'liquid'.

In fact the transition from liquid to solid – the sol–gel transition – has been quite extensively studied. This is examined in Chapter 3, including a discussion of the best way to determine the characteristics of the transition. In food applications, a very traditional area of use for gels, there are many methods recommended for beginners when preparing jams, marmalades or creamy desserts, for example, to produce soft creams or wobbling solids rather than a viscous, sticky, unpleasant, lumpy texture. Every country has its own recipes, because gel textures are appreciated in food preparation everywhere.

For a long while it was thought that simply inverting a tube was an appropriate laboratory technique. Indeed, even today this approach is sometimes still used, not

[1] http://en.wikipedia.org/wiki/Pitch_drop_experiment

least because it is both cheap and easy. Slightly more sophisticated methods, which involve capillary tubes or falling ball methods, have subsequently been proposed. However, except for quick screening of systems, these approaches are little exploited nowadays, and the science of rheology – now understood as the study of the mechanical behaviour of liquids and solids – has been exploited instead.

In fact there have been a number of attempts to define the nature of a gel in more formal 'rheological' terms, including by the late John Ferry (1980) and more recently by Burchard and Ross-Murphy (1990), Almdal *et al.* (1993) and Nishinari (2009). All of these definitions are in terms of the mechanical (rheological) properties of the materials in question, and the response of a gel to deformation (for example shear) and time, as detailed in later chapters.

However, the easiest of these defnitions to explain is that of Ferry, who requires that a gel is not able to sustain a steady-state flow. In other words, when it is subjected to a steady flow-rate experiment, for example stirring at a constant rate, it will tend to fracture or rupture, as we would expect for a solid, rather than to flow like a liquid. Unfortunately this definition excludes some systems of interest to us, and so appears to be too narrow. Two later approaches (Burchard and Ross-Murphy, 1990; Almdal *et al.*, 1993) try to represent gel characteristics in terms of response of the material over time.

The recent article by Nishinari (2009) re-addresses the rheological definition of a gel and its various arguments and counter-arguments. In the end he cites te Nijenhuis (1997), who was obliged to conclude, along the lines of Jordan Lloyd, that 'a gel is a gel, as long as one cannot prove that it is not a gel'. Nevertheless he makes the important point that a gel can be defined both by its mechanical behaviour and by its structural features. Succeeding sections will try to maintain this structure–property linkage.

1.2.2 Multidisciplinary nature of gel studies

The implicit structural complexity in all these definitions should already have established to the reader's satisfaction that the study of gels is a highly multidisciplinary activity. Indeed we can enumerate the disciplines of direct relevance to the study of physical gels and of their formation, gelation, as follows:

- Gels are an aspect of macromolecular science: polymeric chains are needed to create a network.
- Gels are an aspect of colloid science: aggregation and association of small particles (or macromolecules such as globular proteins) can lead to physical and colloidal gels.
- Mathematical treatments of networks are needed to describe their topology. The network itself may become a mathematical abstraction (like self-avoiding walks), while issues of criticality (gelation and the gel point) are related to other problems in statistical physics.
- Mechanical behaviour: rubber elasticity is expected for networks from flexible chains. When the network includes rod-like chain elements or more rigid connecting elements, things become more complicated. Although the matrix is a fluid, we may consider an analogy with composite materials, where the network is the filler.

- Thermodynamics: in chemical networks, swelling behaviour is related to polymer–solvent interaction properties. In physical gels, partial crystallization, phase separation, glass transition, conformational changes and solubility limitations are also related to thermodynamics.
- A range of spectroscopic and other characterization techniques must be used to measure both static and dynamic characteristics of networks: structure, size of the junctions, their spatial distribution, connectedness and time-dependent properties.
- The wide range of technological applications gives rise to specific investigations related to the functional properties of gels: their texture in food materials, controlled-release drug delivery capacity for biomedical applications, adhesion on various surfaces and the production of new materials.

Overall, then, this book describes structure–property relationships for the class of materials known as physical gels. These are formed by a sequence of local processes or mechanisms that create a partial and/or localized aggregation of the polymers in solution. For reasons of relevance (and space) we discard from our scrutiny other types of polymeric gels, including the chemical gels already mentioned, hybrid gels containing both organic polymers and inorganic particles, and purely inorganic gels. However, since some of these were also included in the Flory classification, we need to mention, albeit briefly, their characteristics and relations with the systems of direct interest here.

1.3 Non-physical gels

This somewhat humorous title reflects not that the excluded gel types do not exist, but simply that they lie outside the scope of this volume. Many have been covered in depth elsewhere, and for the interested reader we include at the end of this chapter a brief bibliography, in addition to the more specific list of references.

1.3.1 Chemical gels

Chemical networks can form either in the course of polymerization (for example, in the formation of a branched macromolecule from small molecule precursors during step-addition polymerization) or by the cross- or end-linking of previously formed macromolecules, either in solution or in the melt ('curing' or vulcanization in the nomenclature of rubber technology). Cross-linking methodology and technological practice have been fully reported in the literature and justify the numerous developments of synthetic polymers from organic chemistry.

Chemical networks have very characteristic properties, the most relevant being rubber elasticity (provided the system is above its glass transition temperature) and swelling in 'solvents'. These properties are influenced both by the number of chemical junctions and their spatial distribution, which can be uniform, random or clustered. The contribution of cross-links is sometimes difficult to establish, however, both because of problems controlling the chemical reaction(s) and because of the natural entanglements of chains in solution. In addition, so-called 'network defects' such as chain ends or closed loops do not generally contribute to connectivity and so to macroscopic properties. Network topology is regarded as a major factor influencing the elasticity of networks, since the number of *elastically effective* junctions of the network is often significantly different from the *total number* of junctions. The determination of the number of cross-links active in the network relies on models, whereas the stoichiometric number of cross-links can, in principle, be measured by analytical means (such as spectroscopy or calorimetry). In physical gels, these difficulties are even more pronounced: such factors as the course of the thermal process, the non-equilibrium state and the slow kinetics of reorganization of labile networks induce substantial changes in macroscopic properties. For this reason, the basic properties of chemically cross-linked networks are outlined in Chapter 4 since they provide some of the necessary background for understanding physical gels.

1.3.2 Hybrid organic–inorganic materials

A new area has opened up with the field of hybrid organic–inorganic materials. The formation of chemical bonds between organic and inorganic components allows molecular composites with novel properties to be produced. We take as an example hybrid networks containing clay particles. Clays are layered aluminosilicates (typically silica tetrahedra bonded to alumina octahedra) present in sheet-like structures with charge compensating counterions (such as Li^+ , Na^+ , K^+ or Ca^{++}) located in the inter-layer spaces. One important consequence of this charged nature is that clays are generally hydrophilic. The compatibility between organic polymers and inorganic hosts results in systems exhibiting so-called 'intercalated' or 'exfoliated' morphologies. In intercalated structures, the organic component is inserted between the layers of the clay in such a way that the inter-layer spacing is expanded. In exfoliated structures, the layers of the clay have been completely separated and are randomly distributed throughout the organic matrix; the type of layer 'delamination' determines the properties of the clay nanocomposites.

For example, Haraguchi and Takehisa (2002) were able to prepare a new type of hydrogel based on poly(N-isopropylacrylamide) (poly(NIPAm)), polymerized *in situ* in the presence of exfoliated, uniformly dispersed clay particles. The free radical polymerization was initiated from the clay surface, without the use of an organic cross-linker, so the clay sheet itself acts as a large cross-linker. Here it was thought that initiator was adsorbed on to the surface of the clay particles, and the monomer and catalyst were in the surrounding liquid. The properties of these gels are very different from those of the polymer cross-linked with a conventional organic cross-linker. Compared to the latter, the change of mechanical properties of the hybrid gel is impressive. For example, in the stress–strain curve, the hybrid gel can reach a maximum elongation of 1000% before fracture, and its behaviour is almost completely reversible, whereas some unswollen chemical gels are relatively brittle. Other hybrids have been developed by intercalation of biopolymers in clays, providing new nanocomposite materials, although this topic again falls outside the scope of this volume.

1.3.3 Inorganic gels

The synthesis of solid materials via 'soft chemistry' has been widely developed over the last two decades. These syntheses involve sol–gel chemistry based on inorganic polymerization of molecular precursors. The sol–gel process is a wet-chemical technique for the fabrication of metal oxide materials, starting either from a chemical solution ('sol' is short for 'solution') or from colloidal particles, to produce an integrated network (a 'gel'). Typical precursors are alkoxides $M(OR)_Z$, where M is a metal with valency z (Si, Ti, Zr, Al, Sn, . . .) and OR is an alkoxide corresponding to a deprotonated alcohol which undergoes hydrolysis and polycondensation (step-addition) reactions to form a system composed of solid particles. The sol evolves towards a continuous inorganic network, composed of particles with sizes ranging from 1 nm to 1 μm, dispersed in a solvent containing a liquid phase.

Sol–gel syntheses developed mainly from 1980 but are nowadays very widely used. The approach is interesting in that it is a cheap, low-temperature technique that allows for fine control of a product's chemical composition. The process can be used for producing monolithic ceramics, glasses, fibres, membranes, aerogels or powders (e.g. microspheres or nanospheres), and it can be fabricated as very thin films of metal oxides for various purposes. Making thin films requires an advanced knowledge of the rheology of the sol–gel transition in order to control the thickness and regularity of the film with precision. Several rheological studies can be found involving the sol–gel transition in silica gels from tetraethoxysilane (TEOS) (Devreux *et al.*, 1993) or from tetramethoxysilane (TMOS) (Martin *et al.*, 1987). Sol–gel transitions exhibiting similarities with the polymerization of organic molecules have been explored in the context of a unified description of gelation phenomena (see Chapter 3). Other than this, such inorganic gels will not be discussed further in this book.

However, physical gels of a wide variety of other types will be covered. As outlined above, these gels can be classified according to their mechanism of network formation. Recent experimental studies not only allow a much better understanding of gel structures at a very local scale (~1 nm) but enable rheological measurements of gels to be carried out under defined small- and large-deformation regimes. Because the number of gelling systems has increased following a series of innovations, it is necessary to both compare and categorize results for current systems. However, it is now possible to reveal the unique 'fingerprint' of some of these systems to rationalize the origin of their properties. This remains the overall objective of this book.

1.4 Physical gels

Throughout this book we identify several mechanisms leading to physical gelation, depending on the polymer and on the solvent. In volume terms, the major component in physical gels is the solvent, and in most of the physical gels presented in this book the solvent is aqueous. However, in a few other systems the choice of appropriate organic solvent plays a very important role in gelation and often helps modify the structure of polymer aggregates (Chapter 8).

Water and electrolyte solutions are often good solvents for biopolymers. However, after chemical modification via inclusion, grafting or substitution of non-polar (hydrophobic) groups, the solubility of such modified polymers may be limited, so that the aqueous solvent becomes poorer, especially with increasing temperature. With electrically charged polymers (polyelectrolytes), solubility is also reduced when water contains large amounts of salt (high ionic strength). The overall mechanism of physical gelation often lies in this subtle interplay between solubility and aggregation. The 'sol state' is a true solution. Then, with a combination of several factors including temperature, pH, polymer concentration, polymer molecular mass and ionic strength, there is a reduction in this ability to solubilize the polymer. This alone would normally produce a precipitate, a two-phase liquid-droplet morphology or even crystallization. However, in gels there is still connectivity within polymer-rich domains or phases. The mechanisms reported below have been identified as being among the features that are characteristic and give rise to the formation of physical gels.

Some of these structures at different distance scales are shown in Figures 1.1 and 1.2. Figure 1.1 introduces the differences between chemical, physical and colloidal networks, drawn to the same scales. Figure 1.2 illustrates the various mechanisms leading to physical gel structures:

1. Conformational changes of the polymer (e.g. the coil–helix transition) which give rise to more rigid domains. In this category we find the coil–triple helix transition in gelatin (Figure 1.2a), the aggregation of helices in certain carrageenans (Figure 1.2b) and the double helices in agarose, and the 'egg-box' structure in alginate gels (Figure 1.2c). The first two cases are driven by lowering the temperature; the last two are induced by temperature changes and/or specific ionic content.
2. Denaturation of globular proteins under conditions where the protein remains essentially globular induces aggregation, so producing colloidal type networks formed with branched structures (Figure 1.2d) or linear ones (Figure 1.2e). These consist of particulate networks including the casein networks formed in milk clotting or cheese-making. Such aggregation is usually irreversible, but shares some characteristics with the phase separation of model colloidal systems. Since the assembly can involve mutual hiding of hydrophobic groups exposed during denaturation, it also has features in common with the systems immediately below.

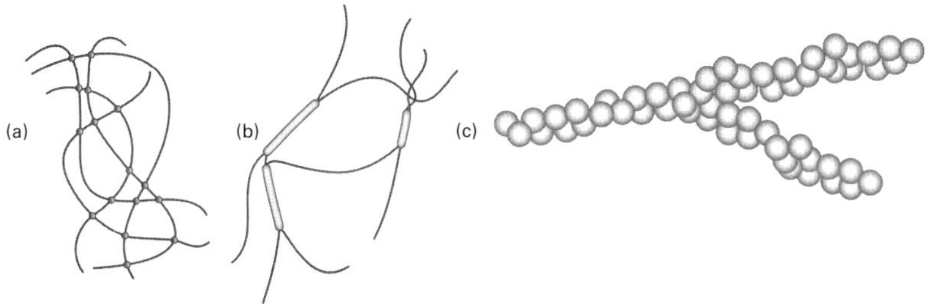

Figure 1.1 Types of network structure: (a) point junctions (chemically cross-linked); (b) junction zone systems; (c) colloidal network strands. These are not drawn to scale, since in (a) and (b) the strands are typically <0.3 nm in thickness, whereas in (c) this dimension can be >2 nm.

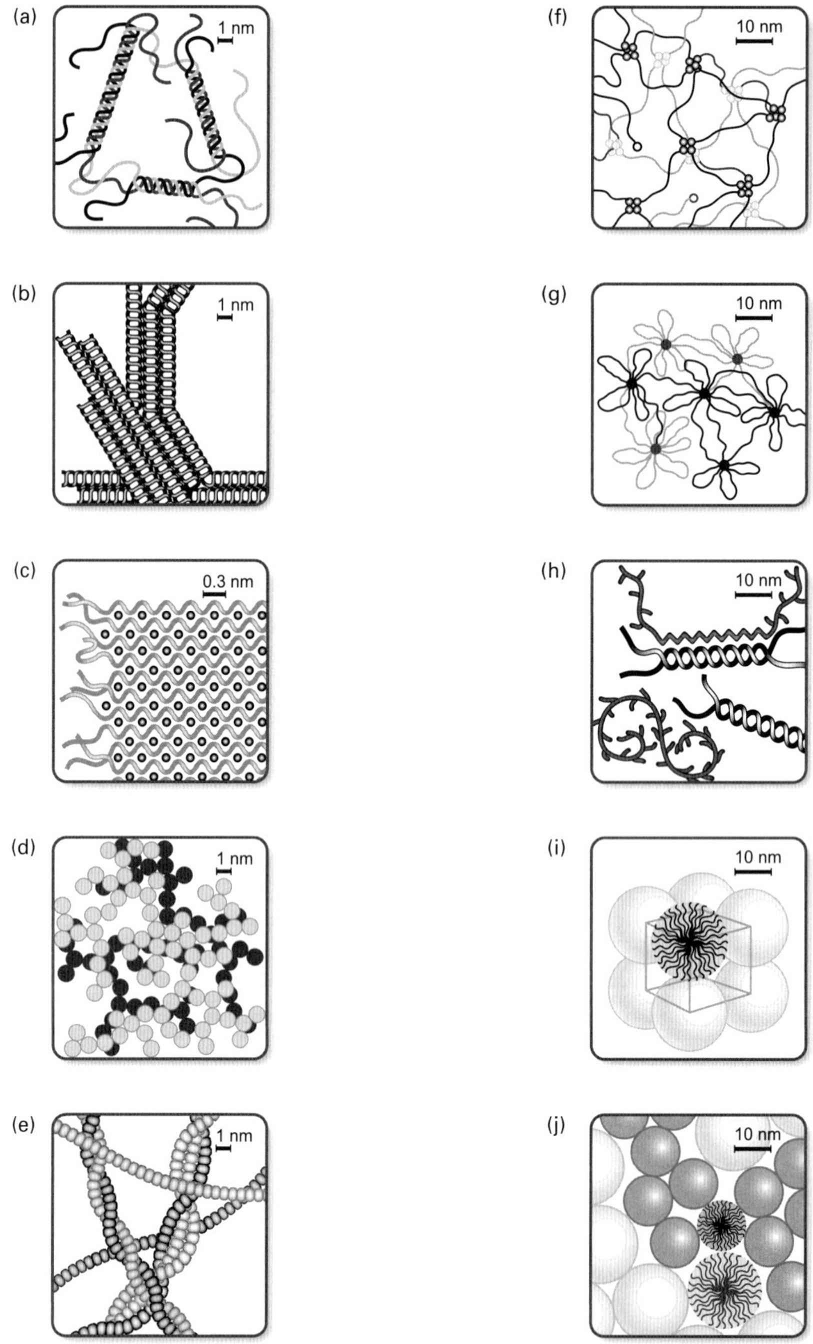

Figure 1.2 Mechanisms of network formation, and the observed structures in physical gel systems: (a) helical association (gelatin, agarose); (b) helical formation in charged polysaccharides such as carrageenans; (c) egg-box complexation in Ca^{2+} alginate gels; (d) colloidal aggregation of proteins close to the isoelectric point; (e) fibrillar aggregation of proteins in amyloid type gels at low pH; (f) hydrophobic effect in telechelic polymer aggregation; (g) flowers in amphiphilic block copolymer aggregation; (h) synergetic interactions between two polysaccharides; (i) colloidal crystal with spherical micelles; (j) mixtures of similar spherical micelles in ternary systems (colloidal glass).

3. The hydrophobic effect occurring with amphiphilic polymers. The insertion of hydrophobic functional groups into a water-soluble (hydrophilic) polymer generates this effect, in which the hydrophobic entities tend to self-associate into micelle-like domains. This invariably gives rise to a variety of structures depending on the architecture of the copolymer: telechelic (Figure 1.2f), random-block (Figure 1.2g), triblock copolymers (Figure 1.2i), linear or branched, hydrophobically modified synthetic and bio-polymers.
4. Synergy between two different polymers (Figure 1.2h). This occurs in some mixtures of polysaccharides, and some specific synthetic polymers, which do not gel as single components. Such a mechanism has features of both category 1 (conformational change) and category 5 (immiscibility). Mixtures of similar triblock copolymers with the same block length leads to immiscibility (Figure 1.2j).
5. Immiscibility or phase separation. This occurs mainly in binary polymer mixtures or in single-polymer solutions when the solvent becomes poor, either at lower temperatures or with selected organic solvents. Sometimes liquid–liquid phase separation is inhibited by the high viscosity of the mixtures, but this is not the only mechanism, since either a glass transition or crystallization may interfere with the phase separation. Under these circumstances the overall morphology of the mixture can be changed from interconnected paths (Figure 1.3a and b) to droplet-like inclusions (Figure 1.3c).

As mentioned initially, the major function of gels in applications is often related to their unique mechanical (rheological) properties. The aim in formulating gels is to try and tune these properties from a molecular starting point, and this is the motive for instituting chemical modifications, new syntheses or new biotechnological processes. Unfortunately, even now the relation between molecular composition (in protein terms, the primary structure), the secondary and tertiary structure (at a larger scale) and rheological properties is far from straightforward. As mentioned before, a priori it is difficult to predict the elastic properties for chemical networks, where the junctions are only point-like assemblies. In physical gels, understanding is only at its beginning and the reader may feel disappointed in not finding here the key to controlling rheological properties. However, some such cases can be illuminated and we hope that this will become apparent after reading this book.

In this volume we analyse the theoretical bases and the subtle experimental procedures that allow us to explore the processes that lead to network formation and to determine this special state called the 'gel point'. Here again, the debate is not yet closed, but we report on some substantial improvements in analysing sol–gel transitions in a fundamental way without disturbing the underlying processes, so helping to establish crucial 'critical' parameters.

Innovative applications are mentioned right at the end of this book, and make up the last chapter. This is a rapidly growing area and there is a broad range of disciplines where gels have found new 'smart' applications. The growing areas particularly include the biomedical area. We try to clarify not only how these new applications work, but also how they can be better controlled or improved. Here, it is fair to point out that the developers of such devices do not always display a comprehensive understanding of the properties of their systems.

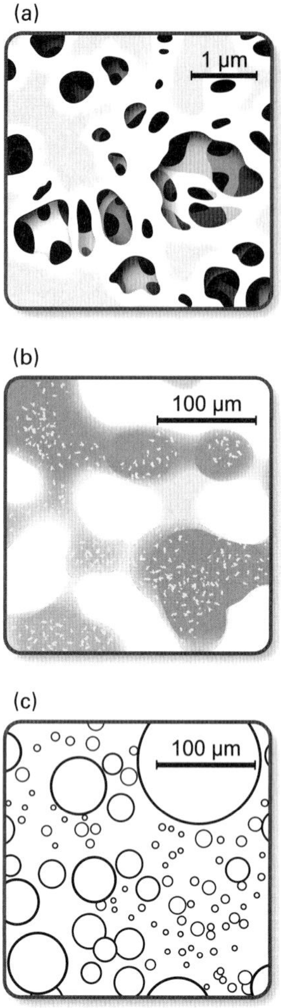

Figure 1.3 Structure of phase separated networks in ternary systems: (a) interconnected paths in the early stages of phase separation; (b) two-phase system making connected paths with small inclusions within the phases; (c) one continuous phase with large droplet inclusions of the second phase (i.e. well-developed phase separation).

1.5 Outline of the book

The coverage and scope of succeeding material follows the pattern delineated below.

Chapter 2 Techniques for the characterization of physical gels

Chapter 2 introduces most of the techniques normally used in the characterization of physical gels, which include some already well known in the physics and chemistry of polymer solutions, in colloid science or in biology for tissue imaging. It is also necessary to perform appropriate molecular characterization (such as molecular mass and composition) of molecular precursors before investigating macroscopic properties. Most of the techniques we discuss here are macromolecular or supramolecular (nano- or micro-scale) in nature. The techniques we concentrate upon include scattering, ultramicroscopy (cryomicroscopy and atomic force microscopy), differential scanning calorimetry (DSC), rheology and simulations. The order in which these are discussed follows, at least roughly, the so-called distance scale approach (Clark and Ross-Murphy, 1987). Because of the convolution of factors, we assert that no single technique should be used in isolation. In other words, in physical gel systems several parallel investigative approaches are needed. Numerical simulations should provide strong support for experimentalists to understand the origin of the effects observed and should also help to predict what may happen when experimental conditions are changed.

Chapter 3 The sol–gel transition

In Chapter 3 we concentrate on a particular aspect that still occupies many chemists and physicists. This is the very special limit when the solution switches from a liquid state, where polymers are free to diffuse, to a solid-like state, where the polymers entrap the solvent. This transition may happen quite suddenly, and this is what is defined as the sol–gel transition. Despite the large number of theoretical and experimental publications dealing with the sol–gel transition, there is still a debate on the best way to determine the precise 'gel-point'. In other words, can there really be a unique definition of the transition from sol to gel? Can we have a single method for all systems or is everyone free to choose the most convenient way to describe their own particular system? In this chapter we present the various ways of defining the gel point from both the theoretical and the experimental viewpoints, and put particular stress on both the problems and potential resolutions.

Chapter 4 General properties of polymer networks

Chapter 4 recalls some general properties ascribed to polymer networks. In 'chemical gels' or covalently cross-linked networks, the junctions are formed by chemical bonds and entanglements; they are simply localized topological constraints. In 'physical gels' or physical networks, the junctions have complex structures, stabilized by secondary forces (such as hydrogen bonding, hydrophobic interactions or labile ionic complexes). Because in both cases macromolecules represent the backbone or the connecting paths of the networks, there are many common features between chemical and physical gels, including their elastomeric behaviour, which is fundamentally different from the classical elasticity of crystals. In chemical gels at least, elastic constraints follow from their three-dimensional network structure, and in turn determine that a polymer network can swell, to a limited extent, but not dissolve. Here it is necessary to highlight the differences between chemical and physical networks, since the latter have (more or less) labile junctions. Another interesting feature appears in the rheology of physical gels, and is related to the dynamics of junction formation and dissociation.

The succeeding chapters analyse various classes of physical gels according to their respective formation mechanisms.

Chapter 5 Ionic gels

In this chapter we consider the physical gels formed from ionic polymers, and particularly ionic polysaccharides, because of their industrial importance. These polyelectrolytes have attached anionic or cationic groups, and sufficient associated low molecular mass counter-ions to ensure electroneutrality. The term 'ionic gels' is well known from those chemically cross-linked polyelectrolytes which form superabsorbent gels (particularly based on acrylamides or acrylic acid) and the swelling of these gels. However, anionic polysaccharide gels formed, respectively, from the carrageenans, gellans, alginates and pectins are well known in a number of other applications (food, cosmetics, personal products, biomedical etc.). They also exhibit considerable variability, and hence allow design flexibility in terms of their properties, arising particularly from the ionic nature of the aqueous solution.

For example, the selection of ions can facilitate changes in the mechanism and the structure of the network, so allowing the gel 'stiffness' and/or temperature of gelation to be 'tuned'. In these systems, not only hydrogen bonds and hydrophobic interactions but electrostatic forces and ordered secondary structures play an important role in gelation. An anionic non-gelling but 'structuring' system, xanthan, and a cationic polyelectrolyte chitosan are also briefly introduced. Both have a number of important applications. For all such naturally occurring polymers, depending on the origin of the raw material, changes in the molecular structure and composition introduce extra complications into their study.

Chapter 6 Hydrophobically associated networks

In this chapter we consider macromolecules which have a dual nature, with one portion soluble in water (hydrophilic) and the other, non-polar, part tending to be excluded from water (hydrophobic); such molecules are sometimes known as amphiphilic. Molecules of this kind are forced to adopt unique orientations with respect to the aqueous medium and this sometimes initiates the formation of organized structures. For example, such molecules were first recognized as playing an important role in living matter, specifically in its organization, and cell membranes are the best examples. Amphiphilic macromolecules can be synthesized with a variety of architectures: by random-block copolymerization, as telechelic copolymers, as triblock copolymers with hydrophobic cores or as modified polysaccharides containing hydrophobic groups. The rheological properties and the microstructures of the gels show some specific features revealed in this chapter. Some rather unusual heat-set gels also belong to this category of macromolecules.

Chapter 7 Helical structures from neutral biopolymers

This chapter is devoted almost exclusively to gelatin and agarose, two systems which form 'cold-set' gels provided the concentration is high enough, via structural transition from a disordered and relatively flexible coil conformation at higher temperatures, to an ordered, partly helical and more rod-like structure on cooling. At low temperatures, the network strands consist of both helical and coil-like portions. On heating, the gels 'melt' and the individual molecular chains revert to the coil state (thermoreversible gels), and this process can be repeated many times. Both systems are derived from natural polymers; as we have already noted, gelatin is extracted from collagen (protein), particularly originating from mammals or fish, whereas agarose is a polysaccharide extracted from marine algae. At one level, these gel networks consist of long, thin interconnected fibres. We analyse the mechanisms of network formation and the sol–gel transition. In both cases gelation is not driven by ionic interactions and so differs from the systems presented in Chapter 5. We focus in particular on the contribution of the helical assemblies to the rigidity of the networks for types of helical structures.

Chapter 8 Gelation through phase transformation in synthetic and natural polymers

Network junctions can be created by physical means, when gels are associated with a phase transformation (Keller, 1995), although the gels arising specifically from phase transformations usually do not correspond to a stable state for these systems. At its final stage, liquid–liquid phase separation generates two fluid layers; in a liquid-to-crystal phase transition the final state is a crystal with macroscopic dimensions. The liquid-to-glass transition in bulk polymers inhibits molecular diffusion and freezes the conformations.

In order to obtain a gel from such a polymer solution the final stages of phase transformation should not be attained; instead, the new structure needs to be arrested at a certain stage just where the network is formed. Every case reported in this chapter illustrates a different situation, although most of the examples examined involve synthetic polymers. In each case phase transformation leads to network formation under specific conditions. The systems described in this chapter include: poly(vinylchloride) (PVC) gels with various solvents, atactic poly(styrene) (PS) in *trans*-decalin, atactic PS in carbon disulphide, isotactic and syndiotactic poly(methylmethacrylate) (PMMA) gels and, finally, cryogels from poly(vinyl alcohol) (PVA) and from certain plant polysaccharides in water such as the galactomannan locust bean gum (LBG).

Chapter 9 Colloidal gels from proteins and peptides

This class of physical gels encompasses a range of systems including those formed from mildly denaturing globular proteins so as to generate partially folded but still essentially globular entities. These are formed when the protein in aqueous (or electrolyte) solution is altered by some physical (e.g. heating) or chemical treatment (e.g. changing pH). Under these apposite conditions, some of the hydrophobic groups, which normally remain buried in protein core, become exposed. This unstable condition leads to aggregation, giving rise either to coarse-stranded amorphous particulate structures or to more fine-stranded networks. The former are sometimes referred to as 'particulate' or colloidal gels. The latter, with a more uniform assembly of fine strands, are denoted 'fibrillar', or often now 'amyloid', gels because of the similarities between their structures and those found in disease conditions such as Alzheimer's. In this chapter we also discuss certain gels formed from milk (casein), the specific ordered fibrous assemblies produced, for example, by actin, as well as the very topical class of materials formed from the self-assembly of specially synthesized peptides.

Chapter 10 Mixed gels

Mixed gels consisting of two different polymers are classified into two groups: gels formed by 'simple' phase separation and those formed by specific (in the literature, commonly although not always accurately referred to as 'synergistic') molecular interactions, where the enthalpy of the specific interactions is sufficient to overcome the normal tendency for phase separation.

Phase separation tends to be produced by the so-called 'exclusion effect' between the two polymers, which then increases the effective concentration of each component within the mixture. On the other hand, gels formed by more positive interactions include coupled networks, which are formed by specific binding between two different polymers, and interpenetrating networks, sometimes now recognized as being formed by microphase separation. The mixtures of importance mainly involve certain polysaccharides or mixtures of polysaccharides and proteins. The microscopic structure, kinetic aspects and morphology of the phases and the overall rheological properties are illustrated for a large number of systems including gelatin with dextran, maltodextrin and agarose; protein–protein mixtures; and κ-carrageenan, gellan and xanthan mixtures, some of the latter showing synergistic effects.

Chapter 11 Innovative systems and applications

Our final chapter is intended to serve not only as a guide to novel systems but also to more recent and/or future potential applications. The novel systems include the cleverly synthesized 'sliding gels', polyelectrolyte complexes, gel micro- and nano-particles, recently developed multi-membrane hydrogels and hydrogels as ultra-sensitive cantilever materials.

As far as applications are concerned, the majority of developments, and arguably the most exciting ones, are in the pharmaceutical and biomedical areas, including the very fast growing area of scaffolds for tissue engineering. Here physical gels have been used as scaffold materials that act as supports and to allow *in vitro* cell growth to proceed unhindered throughout the gel sample (van Vlierberghe *et al.*, 2011). Pharmaceutical applications in drug release and in enabling efficient dispersion of dosage forms (for example, tablets) have also shown great progress. For example, fast-swelling materials are being used to encourage fast drug distribution (as so-called 'disintegrants').

That said, physical gels still have novel applications in the food and cosmetic areas. For example, in the former, there is a great interest in developing low-fat systems. Here a combination of fat replacement by biopolymer structurants and gel-stabilized aerated systems has been shown to be useful.

References

Almdal, M., Dyre, J., Hvidt, S., Kramer, O., 1993. *Polym. Gels Network* **1**, 5–17.

Burchard, W., Ross-Murphy, S. B. (eds), 1990. *Physical Networks: Polymers and Gels*. Elsevier Applied Science, Barking, UK.

Clark, A. H., Ross-Murphy, S. B., 1987. *Adv. Polym. Sci.* **83**, 57–192.

Devreux, F., Boilot, J. P., Chaput, F., Malier, L., Axelos, M. A. V., 1993. *Phys. Rev. E* **47**, 2689–2694.
Djabourov, M., 1988. *Contemp. Phys.* **29**, 273–297.
Ferry, J. D., 1980. *Viscoelastic Properties of Polymers*. Wiley Interscience, New York.
Flory, P. J., 1953. *Principles of Polymer Chemistry*. Cornell University Press, New York.
Flory, P. J., 1974. *Faraday Discuss.* **57**, 7–18.
Graham, T., 1861. *Philos. Trans. R. Soc. London* **151**, 183–224.
Haraguchi, K., Takehisa, T., 2002. *Adv. Mater.* **14**, 1120–1124.
Hermans, P. H., 1949. Gels. In Kruyt, H. R. (ed.), *Colloid Science*. Elsevier, Amsterdam, pp. 483–651.
Jordan Lloyd, D., 1926. The problem of gel structure. In *Colloid Chemistry*. Chemical Catalog Company, New York, pp. 767–782.
Keller, A., 1995. *Faraday Discuss.* **101**, 1–49.
Martin, J. E., Wilcoxon, J., Adolf, D., 1987. *Phys. Rev. A* **36**, 1803–1810.
Mülhaupt, R., 2004. *Angew. Chem., Int. Ed.* **43**, 1054–1063.
Nishinari, K., 2009. *Prog. Colloid Polym. Sci.* **136**, 87–94.
te Nijenhuis, K., 1997. *Adv. Polym. Sci.* **130**, 1–252.
Treloar, L. R. G., 1975. *The Physics of Rubber Elasticity*. Clarendon Press, Oxford.
van Vlierberghe, S., Dubruel, P., Schacht, E., 2011. *Biomacromolecules* **12**, 1387–1408.

Bibliography

Bohidar, H. B., Dubin, P., Osada, Y., 2003. *Polymer Gels: Fundamentals and Applications*. ACS Symposium Series 833. ACS, Washington DC.
Burchard, W., Ross-Murphy, S. B. (eds), 1990. *Physical Networks: Polymers and Gels*. Elsevier Applied Science, Barking, UK.
Cowie, J. M. G., Arrighi, V., 2007. *Polymers: Chemistry and Physics of Modern Materials*. CRC Press, Boca Raton, FL.
de Gennes, P.-G., 1979. *Scaling Concepts in Polymer Physics*. Cornell University Press, Ithaca, NY.
De Rossi, D., Kajiwara, K., Osada, Y., Yamauchi, A., 1991. *Polymer Gels: Fundamentals and Biomedical Applications*. Plenum, New York.
Doi, M., Edwards, S. F., 1986. *The Theory of Polymer Dynamics*. Clarendon Press, Oxford.
Erman, B., Mark, J. E., 1997. *Structure and Properties of Rubberlike Networks*. Oxford University Press, New York.
Ferry, J. D., 1980. *Viscoelastic Properties of Polymers*. Wiley Interscience, New York.
Guenet, J. M., 1992. *Thermoreversible Gelation of Polymers and Biopolymers*. Academic Press, London.
Lapasin, R., Pricl, S., 1995. *Rheology of Industrial Polysaccharides: Theory and Applications*. Blackie Academic and Professional, Glasgow.
Macosko, C. W., 1994. *Rheology: Principles, Measurements, and Applications*. Wiley-VCH, New York.
Osada, Y., Ross-Murphy, S. B., 1993. *Sci. Am.* **268**, 82–87.
Stauffer, D., 1985. *Introduction to Percolation Theory*. Taylor and Francis, London.
Stepto, R. F. T., 1998. *Polymer Networks: Principles of Their Formation, Structure and Properties*. Chapman and Hall, Glasgow.
Tanaka, T., 1981. *Sci. Am.* **244**, 124–138.
Young, R. J., Lovell, P. A., 2011. *Introduction to Polymers*. CRC Press, Boca Raton, FL.

2 Techniques for the characterization of physical gels

2.1 Introduction

Most of the techniques for physical gel characterization are well known in the physicochemistry of polymer networks, in colloid science or in both, and many standard texts describe these in more detail than we can hope to include. In this chapter, we have decided to adopt a somewhat eclectic choice, biased in some cases by the personal interests of the authors.

Techniques (van Holde *et al.*, 2006) such as nuclear magnetic resonance (NMR), particularly including broad-line NMR, 'classical' absorption spectroscopy and chiroptical (dichroic) methods (the latter of which have proved invaluable in the study of, for example, marine and microbial polymers) and mass spectrometry are not covered in this chapter, although reference to some of them appears in a number of places, in later, system-specific chapters. Instead, the techniques we concentrate upon include scattering, ultramicroscopy (cryomicroscopy and atomic force microscopy), differential scanning calorimetry (DSC), rheology and simulations. The order in which these are discussed follows, at least roughly, the so-called distance scale approach (Clark and Ross-Murphy, 1987), although most of the techniques are macromolecular or supramolecular (in modern parlance, nano- or micro-scale) in nature.

As a final remark, it is crucial to understand and appreciate that in physical gel systems several parallel investigative approaches are almost always needed. No single technique – for example, rheology or calorimetry – should be used in isolation because of the convolution of factors. These include (a) the several distance scales appropriate for determining the structure, (b) the influence that thermodynamic properties have on mechanisms of aggregation and (c) the fact that functional properties are often based on rheological properties. In general, appropriate molecular characteristics (such as molecular mass or composition of gel precursors) should be performed before investigating macroscopic gel properties.

2.2 Scattering techniques

Scattering techniques based on X-ray, neutron and light sources all play an important role in studying polymer gels and colloidal suspensions. Light scattering has been used for a long time for characterizing polymers in solution, whereas small-angle X-ray and neutron scattering have developed rapidly over the last two decades because of the availability and performance of large-scale facilities accessible to the whole scientific community. However, it is outside the scope of this book to treat the theoretical background of scattering techniques in detail, and the reader is referred to specialist monographs (Glatter and Kratky, 1982; Roe, 2000). Instead we wish to stress the type of information on the structure of physical gels that can be obtained by these techniques.

The junction zones in physical gels are somewhat analogous to a crystalline structure (for example, the multi-stranded helices in some biopolymer gels), so X-ray diffraction techniques (wide-angle X-ray scattering or WAXS) should be able to elucidate the periodicity of the crystals. However, the proportion of crystalline structure is very small and the scattering by the solvent, which is a major component, is superposed on the diffraction patterns of the crystals, and resolution is often poor.

When diffraction techniques are to be used, the gel can be conditioned as an oriented fibre, and allowed to dry slowly in order to eliminate the scattering due to the solvent. In turn, to observe a good diffraction pattern it is often necessary to create preferential chain orientation by withdrawing a liquid drop of a concentrated solution and stretching it between two tips, when the gel starts to form (for cold-set gels, for instance). The microcrystals in gels are normally randomly oriented, but when a gelling drop is pulled, the junctions (crystals) orient along the fibre axis. When the fibre is allowed to dry under tension, more crystals can be nucleated and, eventually, in the totally dried state the fibre has an oriented semi-crystalline structure. Such a stretching and drying procedure is not especially natural: chain conformations different from the junctions developed at rest may be adopted during the preparation step (Paul, 1967; Arnott *et al.*, 1974; Lemstra and Keller, 1978; Foord and Atkins, 1989), so the method must be used with caution.

2.2.1 Principles of scattering

Scattering techniques are based on the interaction of radiation with matter. X-rays and light (photons) are electromagnetic waves characterized by very different wavelengths, close to 0.1 nm for X-rays and about 600 nm for visible light. Neutrons are moving particles having a wavelength of several nanometres. X-rays interact with the electrons in atoms, while neutrons interact with the atomic nuclei. When light passes through a transparent medium, it creates a dipole and radiates an electromagnetic field proportional to the polarizability of the medium.

When X-rays strike an object, every electron becomes the source of a scattered wave and the scattered waves are coherent (incoherent scattering is neglected). Coherence means that the amplitudes scattered by different electrons are added. The amplitudes are of equal magnitudes and they differ only in their phase, ϕ, which depends on the position of the electron in space. The principle of a scattering experiment is to measure the spatial repartition of the scattered radiation. The secondary wave can be written in the complex form $e^{i\phi}$. The calculation of ϕ is illustrated in Figure 2.1.

Let us denote by $\boldsymbol{k}_\mathrm{i}$ ($|\boldsymbol{k}_\mathrm{i}| = 2\pi/\lambda$) the wave vector of the incident beam along $\boldsymbol{u}_\mathrm{i}$, and by $\boldsymbol{k}_\mathrm{s}$ the wave vector of the scattered wave in the direction $\boldsymbol{u}_\mathrm{s}$. (By convention, the angle between $\boldsymbol{u}_\mathrm{s}$ and $\boldsymbol{u}_\mathrm{i}$ is denoted θ in light scattering experiments, and 2θ in X-ray scattering.)

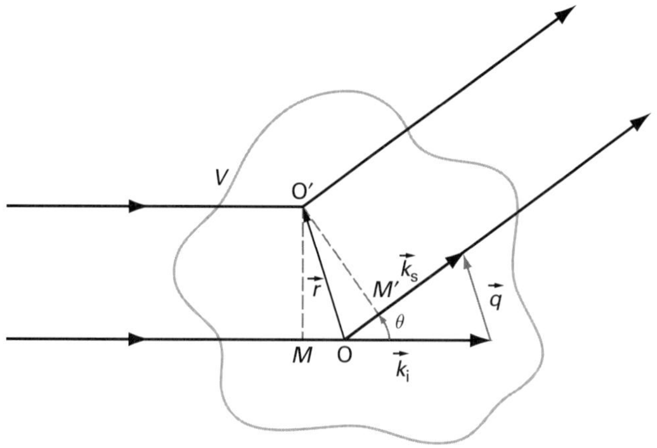

Figure 2.1 Definition of the scattering parameters in a light scattering experiment: scattering wave vector $\boldsymbol{q}$, scattering angle θ, scattering volume V.

Usually the wavelength of the incident beam is changed very little by the scattering process, so

$$|\boldsymbol{k}_{\mathrm{i}}| \cong |\boldsymbol{k}_{\mathrm{s}}|.$$

The scattering wave vector is determined by

$$\boldsymbol{q} = \boldsymbol{k}_s - \boldsymbol{k}_i. \tag{2.1}$$

In light scattering,

$$q = 2k_{\mathrm{i}} \sin\frac{\theta}{2} = \frac{4\pi}{\lambda_{\mathrm{i}}} \sin\frac{\theta}{2}. \tag{2.2}$$

In X-ray scattering, the scattering vector is generally defined by $\boldsymbol{s} = (\boldsymbol{k}_s - \boldsymbol{k}_u)/(2\pi)$, so that

$$s = \frac{2}{\lambda_i} \sin\theta,$$

or

$$s = \frac{q}{2\pi}. \tag{2.3}$$

The path difference of a point O′ is specified by the vector $\boldsymbol{r}$ against the origin O. Then the phase lag is

$$\phi = rq. \tag{2.4}$$

For X-rays the amplitude of the waves scattered by scattering volume dV is

$$dA(\boldsymbol{s}) = A_{\mathrm{e}}\rho(\boldsymbol{r})e^{-i2\pi \boldsymbol{s}\cdot \boldsymbol{r}}dV, \tag{2.5}$$

where A_e is the amplitude scattered by one electron, $\rho(\boldsymbol{r})$ is electronic density and the product $\rho(\boldsymbol{r})\, dV$ is the number of electrons inside the scattering volume dV.

In light scattering experiments, (2.5) becomes

$$dA(\boldsymbol{q}) = C\rho(\boldsymbol{r})e^{-i\boldsymbol{q}\cdot\boldsymbol{r}}dV, \tag{2.6}$$

where C is a coefficient which depends on detection conditions and $\rho(\boldsymbol{r})$ is the polarizability of an elemental volume dV.

Equations (2.5) and (2.6) show that the amplitude of the scattered wave is the Fourier transform of the functions $\rho(\boldsymbol{r})$. The total amplitude $A(\boldsymbol{q})$ scattered by volume V is obtained by integration of these equations. If the function $\rho(\boldsymbol{r})$ had a constant value in the whole volume, the integral would be a Dirac function $\delta(\boldsymbol{q})$: no wave would be scattered apart from the incident direction $\boldsymbol{q}=0$. Consequently, any light scattered in direction θ with scattering vector $\boldsymbol{q}$ is due to local fluctuations of $\rho(\boldsymbol{r})$.

The mean intensity $I(\boldsymbol{q})$ of scatter at scattering vector $\boldsymbol{q}$ is the average of the product, $\langle A(\boldsymbol{q})\,A^*(\boldsymbol{q})\rangle$. Here the brackets $\langle\ \rangle$ refer to the average over all orientations and all positions of the particles inside the scattering volume:

$$I(\boldsymbol{q}) \;=\; \langle A(\boldsymbol{q})\,A{*}(\boldsymbol{q})\rangle. \tag{2.7}$$

Particles (atoms, molecules, colloids) interact through intermolecular forces which, as a first approximation, depend on the relative distances between the centres of the particles (for example, steric repulsion, double layer repulsions in colloids and long-range van der Waals attractions). Assuming that the interaction between a pair of particles is isotropic in space (i.e. it depends only on the distance between particles and not on their relative orientations), a full statistical physics treatment allows the derivation of an expression for the intensity $I(\boldsymbol{q})$ of the radiation scattered by N particles in direction $\boldsymbol{u}_S$. The function $I(\boldsymbol{q})$ can be separated into two factors:

$$I(\boldsymbol{q}) \;=\; I(q) = N\,I_1(q)\,S(q), \tag{2.8}$$

where $I_1(q)$ is the intensity scattered by a single particle (such as a colloidal particle, polymer coil or micelle) or the 'form factor' measured in very dilute solution, and $S(q)$ is the so-called 'structure factor', which reflects the relative positions of the particles related to their interaction potentials. The structure factor is equal to 1 when the solution is extremely dilute and the scattering by N particles is just the addition of the intensities scattered by the individual particles. In more concentrated solutions, the structure factor becomes increasingly important and reflects the so-called 'short-range order' in the solution. Brownian motion does not allow ordering of the hard field repulsion core of spherical particles over large distances, beyond two to three times the centre-to-centre distance, but the short-range order is important and reflects the thermodynamic stability of the solution. The second virial coefficient can be derived from this measurement.

When scattering techniques are used to investigate the local structure of physical gels, these two factors are both very important.

2.2.2 Scattering by a single particle

Measuring the intensity scattered by a single particle $I_1(q)$ – in other words, the intensity measured in very dilute solutions – allows the identification of the structural unit or the block which builds the framework of the gel. The angular dependence of the scattered radiation by particles having simple shapes with well-defined geometries is well known, for instance for spheres, hollow spheres, ellipses, discs, cylinders, hollow cylinders, polymer random coils, semi-flexible polymers and micelles (Burchard, 1983, 1994). From the measurement of intensity versus scattering vector in dilute solutions we can derive not only the radius of gyration of the particles but also further information such as the thickness of a disc, the cross-section of a rod or the persistence length of a polymer coil. The limit of very small scattering vector is called the Guinier range; the scattered intensity then follows a simple law, whatever the shape of the particles:

$$I_1(q) \propto \exp\left(-\frac{q^2 R_g^2}{3}\right), \quad qR_g \ll 1. \tag{2.9}$$

In principle, in dilute solutions, when the logarithm of the intensity is plotted versus q^2, the initial slope gives $-R_g^2/3$, so the particle radius of gyration is derived from the negative slope (Guinier approximation). For the case of X-rays the radius of gyration is the mean square distance from the centre of gravity, where the role of mass is played by electrons. The overall intensities $I_1(q)$ scattered by single particles are monotonically decreasing functions of the scattering vector $\boldsymbol{q}$, as shown on a double logarithmic scale in Figure 2.2. This also shows the difference of the scattered intensity $I_1(q)$ for particles of

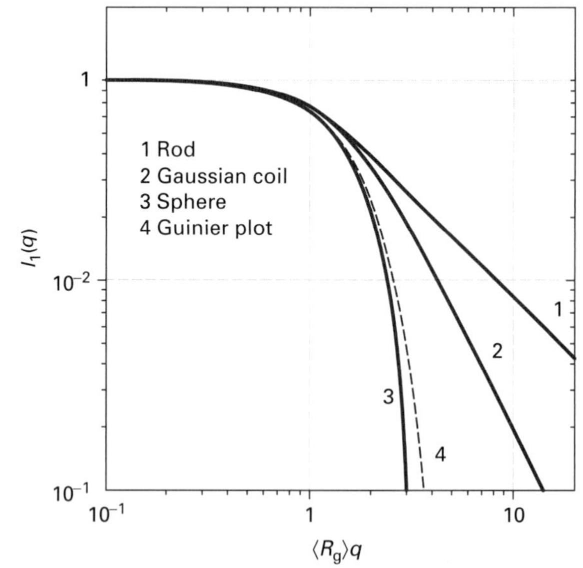

Figure 2.2 Scattering of single particles of various shapes: rod, Gaussian coil and sphere reduced to the same radius of gyration, in a double logarithmic plot. The Guinier plot (2.9) is shown as a dashed line ($\ln I(q) = 1 - \langle R_g^2 \rangle q^2/3)$). It is valid for any particle shape when $qR_g < 1$.

Table 2.1 Characteristics of scattered wavelengths.

Radiation	Incident wavelength (nm)	q range (nm^{-1})	Spatial resolution ($2\pi/q$) (nm)
Visible light	400–600	$5 \times 10^{-4} - 3 \times 10^{-2}$	200–10 000
X-rays	0.1–0.4	0.1–150	$5 \times 10^{-2} - 50$
Neutrons	0.1–3	0.01–150	$5 \times 10^{-2} - 500$

different shapes, which is the usual way of identifying these, in particular when scattering vectors such as $q > R_g^{-1}$ can be used.

In (2.9), the scattered intensity decays exponentially versus the square of the scattering vector. The characteristic length R_g such that $q \approx 1/R_g$ determines the limit of validity of the Guinier approximation. This means that the range of scattering vectors to be used for characterizing the size of the particles has to be selected according to the expected average size.

As an example, if $R_g = 10$ nm, the range of q vectors to be chosen is around $q \approx 0.1$ nm^{-1}. Considering Figure 2.2, when using classical light scattering equipment the usual range of θ is between 30° and 150°, so with a He-Ne laser, $\lambda = 633$ nm, this gives a q range of $7 \times 10^{-3} < q < 2.5 \times 10^{-2}$ nm^{-1}, which is obviously too small. In this case light scattering is not the best technique. Instead, using small-angle X-ray scattering (SAXS) with a CuK_α X-ray source, the incident wavelength is $\lambda = 0.154$ nm; with a specially designed small-angle collimation camera, the scattering angle $0.65 < 2\theta < 5 \times 10^{-3}$ rad, so the q range in this case is $0.02 < q < 0.2$ nm^{-1}, which is a much better choice.

Table 2.1 summarizes the ranges of scattering vectors for various techniques. It also indicates the spatial resolution, $2\pi/q$, meaning that any detail on the structure within a distance smaller than $2\pi/q$ cannot be observed with this technique.

2.2.3 Effect of particle concentration

Increasing the concentration of particles in solution increases the scattered intensity. The structure factor reflects the interactions between particles or the correlation between the relative positions of a pair of particles within a coordination shell. When dealing with spherical particles (such as colloidal particles), the function $S(q)$ exhibits a peak whose position is directly related to the interaction potential between the particles. An example is shown in Figure 2.3. The intensity scattered by a single spherical particle of radius R and the overall measured intensity are both shown, and the contribution of the factor $S(q)$ is well identified. These curves correspond to scattering spheres interacting with a hard core potential with a distance $2R$. The maximum in $S(q)$ should correspond approximately to $2\pi/q = 2R$ when the concentration is rather high. In this example, the distance between centres of a pair of spheres is larger than $2R$, corresponding to a volume fraction of spheres of $\phi = 12.5\%$.

The effect of concentration on the intensity scattered per particle in a suspension of spherical particles interacting with a hard core potential is shown in Figure 2.4. As can be

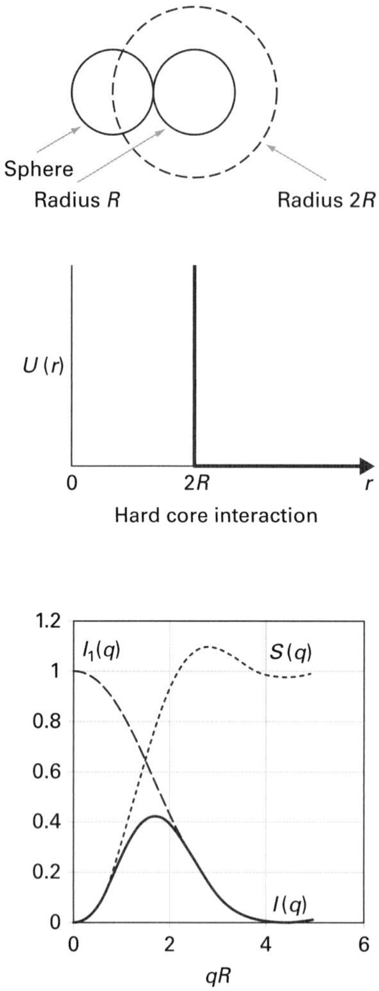

Figure 2.3 Scattered intensity for a suspension of hard spheres of radius R at a volume fraction $\phi = 12.5\%$. Two spheres interact through a hard core interaction potential. The scattered intensity $I(q)$ and the two contributions – the form factor $I_1(q)$ and the structure factor $S(q)$ – are shown separately.

seen, the scattered intensity normalized by concentration displays a distinct peak with progressively increasing particle volume fraction.

2.2.4 Polymer solutions

When dealing with polymer solutions one has to distinguish between dilute solutions and semi-dilute solutions or gels. In dilute solutions, inter-particle interference effects should be taken into account. Then the total intensity scattered is given by the well-known Zimm equation, valid for very dilute solutions ($c < 1\%$):

$$I(q) = KcM\left[I_1(q) - 2A_2cMI_1(q)^2\right], \tag{2.10}$$

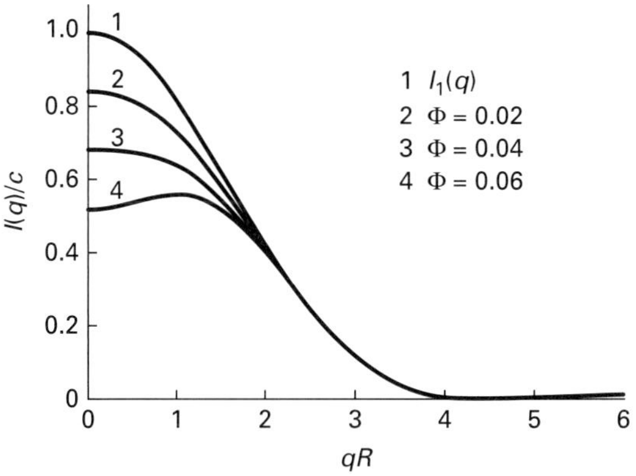

Figure 2.4 Total intensity normalized by the particle concentration for a dilute suspension of hard spheres at various volume fractions indicated in the figure. The volume fractions are smaller than in Figure 2.3. A peak in scattered intensity appears with increasing volume fraction.

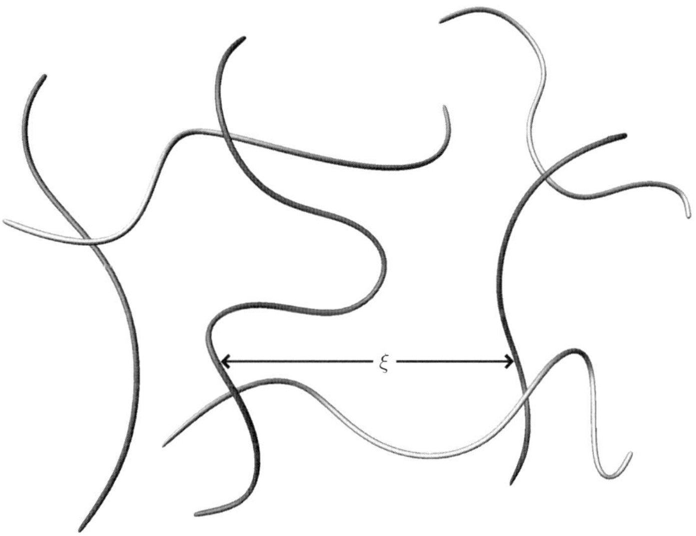

Figure 2.5 Definition of the correlation length ξ in a semi-dilute solution.

where K is a constant related to the scattering technique, c is the concentration of chains, M is the molar mass (or 'weight'), A_2 is the second virial coefficient and $I_1(q)$ is the normalized scattering function of a single chain. When the second virial coefficient is zero, the total intensity is proportional to concentration; when it is positive, the polymer is in a good solvent and the intensity divided by concentration decreases with increasing concentration, with the converse for poor solvents when $A_2 < 0$.

As the polymer concentration increases, a new parameter describes the structure: the correlation length ξ, shown schematically in Figure 2.5. The correlation length decreases with polymer concentration according to a scaling law:

$$\xi = R_g \left(\frac{c}{c^*} \right)^{-\frac{3}{4}} \quad \text{when} \quad c > c^*. \tag{2.11}$$

In this equation, c^* is the so-called overlap concentration, at which the radii of gyration of the individual polymers start to overlap; in semi-dilute solutions they begin to interpenetrate. Equation (2.11) is valid above c^* and the correlation length in semi-dilute solutions is smaller than the radius of gyration for the isolated macromolecule, because of the chains' interpenetration.)

The intensity scattered by semi-dilute solutions was calculated by de Gennes (1979) for polymers in good solvents, and has a Lorentzian form:

$$I(q) = \frac{I(0)}{1 + q^2 \xi^2}. \tag{2.12}$$

In principle, the scattering from polymer gels should be similar to that for an entangled solution, with an additional input from the detailed local structure of junctions. In particular, at high q vectors (local distances) this would include contributions from the cross-section of junctions or fibres. Besides, in many cases gelation is associated with some local heterogeneity, which does not seem to be associated with any particular superstructure but with defects in the colloidal size range (10–20 nm or even larger). In this case an additional contribution to the scattering intensity is observed at small scattering angles:

$$I(q) = \frac{I(0)}{1 + q^2 \xi^2} + \frac{B a_0^2}{1 + q^2 a_0^2}. \tag{2.13}$$

The second term of this equation is an excess scattering contribution, the Debye–Bueche term, in which B is a constant and a_0 is the typical heterogeneity size. Similar excess scattering was also observed for semi-dilute solutions in both good and Θ-solvents.

2.3 Calorimetric studies

Both chemical and physical gelation routes are usually accompanied by important enthalpy changes. Calorimetry measures the heat absorbed or generated in a material when it changes from an initial state to a final state and therefore is a useful method to follow structural changes during gelation. DSC and micro DSC (μDSC) can be used to follow a large number of gelation processes such as the conformational changes of biopolymers, micelle formation in solution, partial crystallization and chemical cross-linking of macromolecules – the latter both in solution and in melts. The enthalpy changes could occur under isothermal conditions or, for instance, for thermoversible gels through temperature differences during either heating or cooling stages.

Isothermal Titration Calorimetry (ITC) is another technique, albeit one less known in gel studies: it allows an independent investigation of the interactions responsible for the self-association of macromolecules or in complex formation between macromolecules and other small molecules or ionic species which underlie some gelation mechanisms.

We shall introduce the experimental methods in this section, and then briefly illustrate the various techniques by examples which help to illustrate the applicability of each approach to gelation studies. More details are given in the relevant systems chapters.

2.3.1 Basic concepts

DSC is a differential method in which the measured quantity is proportional to the heat flow rate. DSC can be operated in isothermal conditions, under a preset rate of temperature change (heating or cooling) or even in a modulated way (temperature-modulated DSC or TMDSC) where a modulation is superimposed on a constant rate of temperature change (Hatakeyama and Quinn, 1999; Brown, 2001; Höhne *et al.*, 2003; Matsuo, 2004).

2.3.2 Differential scanning calorimetry (DSC)

A typical DSC apparatus is shown in Figure 2.6. The device has a disc-like measuring system made of a good thermal conducting material and two sample containers positioned symmetrically to the centre of the disc. One of the containers is filled with the sample and the other is a reference – generally an empty pan, identical to the sample container. In gel studies the reference is often filled by approximately the same amount of solvent to balance the heat capacity of the sample and the reference, and achieve a good baseline.

The main heat flow passes from the furnace through the disc to the two containers. Temperature sensors (for example, thermocouples or resistance thermometers) are attached to the base of the sample and reference holders. A second set of thermocouples measures the temperature of the furnace and of the heat-sensitive plate. If the sample contains volatile components the pans must be sealed hermetically with a sealing press. Typical DSC pans contain approximately 10 mg of sample by weight.

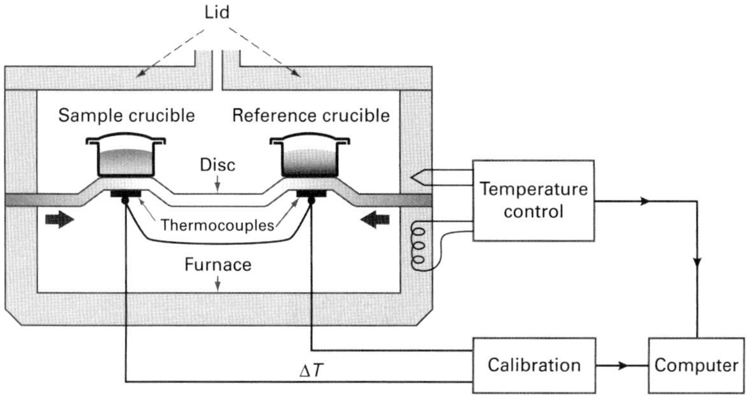

Figure 2.6 Schematic view of a differential scanning calorimetry (DSC) set-up with disc type measuring system.

When the furnace is heated, the same amount of heat should flow through the disc into both pans. Under steady-state conditions (when no reactions and no transitions occur), the differential signal ΔT – normally an electric potential difference – provides a baseline for heat flow due to differences in heat capacity between the sample and the reference. When a transition occurs, the steady-state equilibrium is disturbed and a differential signal is generated which is proportional to the difference in heat flow rates to the sample and to the reference.

Measured heat flow rates ϕ_m are proportional to temperature differences:

$$\phi_{\text{furnace}\to\text{sample}} - \phi_{\text{furnace}\to\text{reference}} \approx -\Delta T \tag{2.14}$$

$$\phi_m \approx -K_m \Delta T \tag{2.15a}$$

$$\phi_{\text{true}} = K_\phi \phi_m. \tag{2.15b}$$

Measured ϕ_m and true ϕ_{true} heat flow rates are obtained by the manufacturer by careful calibration with reference materials (to determine the parameters K_m and K_ϕ). The heat flux is expressed in μW or mW (power units).

The enthalpy difference ΔH in a phase transition, being the change in a state variable, is a well-defined parameter. This enthalpy is always a function of pressure p, temperature T and composition χ, which is, in turn, related to structural changes.

The heat capacity $C_{p,\chi}(T)$ at constant pressure and constant 'structural' composition of the sample is

$$C_{p,\chi}(T) = \left(\frac{\partial H}{\partial T}\right)_{p,\chi}. \tag{2.16}$$

The enthalpy change ΔH of the sample at constant temperature and pressure, due to a phase transition, a chemical reaction or a mixing effect associated with changes of composition χ, is given by

$$\Delta H = \left(\frac{\partial H}{\partial \chi}\right)_{T,p}. \tag{2.17}$$

DSC experiments measure the rate of change of the heat (or heat flux) through the sample during a temperature ramp, dQ_m/dt:

$$\phi_m = \frac{dQ_m}{dt}. \tag{2.18}$$

At constant pressure and in the absence of any external energy perturbation, the total heat flux is related to the enthalpy changes arising from the two contributions shown in (2.17) and (2.18). Differentiating these equations with respect to time,

$$\phi_m(T,p,\chi) = C_{p,\chi}(T)\frac{dT}{dt} + \left(\frac{\partial H}{\partial \chi}\right)_{p,T}\frac{d\chi}{dt}, \tag{2.19}$$

where dT/dt is the rate of change of the temperature (in general taken as constant). The heat capacity $C_{p,\chi}(T)$ is proportional to the sample mass m, while $c_{p,\chi}(T) = C_{p,\chi}(T)/m$ is the heat capacity per unit mass.

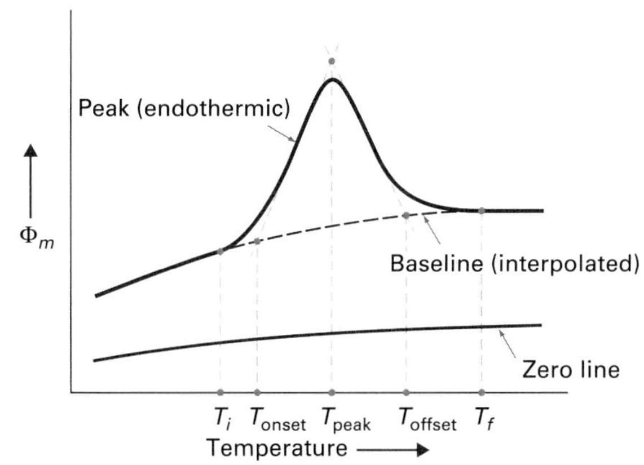

Figure 2.7 Heat flux versus temperature in a DSC experiment with a sample undergoing a phase transition.

The first term in (2.19) is related to the baseline. To get the true sample heat flow and so to derive its heat capacity, the zero line has to be subtracted from the baseline. The zero line is the curve measured with an empty sample pan and conditions identical to the experiment itself. The second term in (2.19) is associated with the enthalpy change due to a reaction or a phase transition. In this case, a peak appears during the temperature ramp, as can be seen in Figure 2.7, where heat flux is plotted versus temperature.

By international convention, when heat is added to a system (as in melting transitions) the endothermic peak is plotted upwards (or in a positive direction) and when heat is released from a system, as in crystallization, the exothermic peak is plotted in the opposite direction (this convention is not always applied in the literature).

This peak can be characterized by several points (Figure 2.7): T_i, the initial deviation from the baseline; T_{peak}, the maximum/minimum position; T_{onset}, the extrapolated onset temperature obtained from the intersection of the linear ascending peak slope with the baseline; and T_f, the temperature when the measured heat flux rate again reaches the baseline.

In general, the melting of pure substances gives a very steep rise in the heat flux, and the melting temperature is determined at the onset temperature. The onset temperature, peak position and peak shape depend on the rate of heating. In particular, some gels contain very small crystalline domains and the peaks are rather wide. This requires that any characteristic temperature chosen to describe the thermal evolution of system should mention the rate of change of the temperature.

Another important parameter derived from DSC measurements is the area under the peak, which provides the total enthalpy of the transition. The enthalpy change between the limiting temperatures T_i and T_f can be calculated by integrating the heat flux, after the baseline has been subtracted. When dealing with the melting of pure substances, the heat capacities of the solid and the liquid states differ in absolute value and in their temperature dependence, so the baseline is shifted after the transition. To derive the exact melting enthalpy, the baseline correction must be performed very carefully. By contrast, for gels the solvent phase is the major component and no changes are expected for the baseline during gelation or melting.

However, in thermodynamics the melting or crystallization enthalpies are defined for infinite phases, where surface energies are neglected. In practice the heat of melting of fine powders differs from large crystals and so the melting temperature changes as well. The crystalline junctions in physical gels are necessarily small and their thermal behaviour normally depends on the whole thermal history of the sample.

Chemical reactions may be carried out isothermally or non-isothermally, in a scanning mode or following a special temperature programme, and can be analysed by DSC. In (2.19), the reaction enthalpy $(\partial H/\partial \chi)_{T,p} = \Delta H_r$ is generally replaced by the average reaction enthalpy $\langle \Delta H_r \rangle$, independent of the instantaneous composition of the system. To carry out chemical reactions in the isothermal mode, the sample has to be brought extremely rapidly from the initial inert state to the desired reaction temperature. Then (2.19) can be written in a simple form:

$$\phi_m(T, p, \chi) = \langle \Delta H_r \rangle \frac{d\chi}{dt}. \tag{2.20}$$

When the reaction progresses, the total heat exchanged is obtained by integration of the heat flux over time, between 0 and time t:

$$Q_m(t) = \int_0^t \phi_m(t')\, dt' = \langle \Delta H_r \rangle \int_0^\chi d\chi'. \tag{2.21}$$

When the reaction is completed, $\chi = 1$, $Q_m = \langle \Delta H_r \rangle$.

In the scanning mode, the heat of reaction is generally obtained by integrating the total area of the heat flux below the baseline.

2.3.3 Microcalorimetry: μDSC

Microcalorimetry is an alternative to heat flux DSC measurement, and operates with a cylindrical type of measuring system. The microcalorimeter has a sensitivity of the order of 1 μW. The block type furnace is provided with two cylindrical cavities which contain identical cells. In the μDSC III instrument (Setaram, Caluire, France), the cells are surrounded by high-sensitivity Peltier elements ensuring thermal contact with the calorimetric block. Closed batch cells and mixing cells of the special thermal conducting alloy Hastelloy® are available with a volume of 850 μL, as shown in Figure 2.8. The mixing cells contain two containers (Figure 2.8b): the upper one is filled with the reactant and the lower one with the polymer solution. By pushing on the piston separating the two compartments and stirring rapidly, the two liquids are brought into contact and rapidly mixed and the chemical reaction starts. Such microcalorimetric measurements are well suited both to studies of thermal gelation using batch cells and to isothermal chemical cross-linking using mixing cells.

Such a microcalorimeter has the advantages of relatively large sample size, very high sensitivity (resolution 0.02 μW) and operation in either isothermal or scanning mode. Scan rates are necessarily low, at 10^{-3}–1°C min^{-1} (typically 0.1°C min^{-1}), to allow a uniform temperature inside the sample. The basic principles of operation are the same as in DSC.

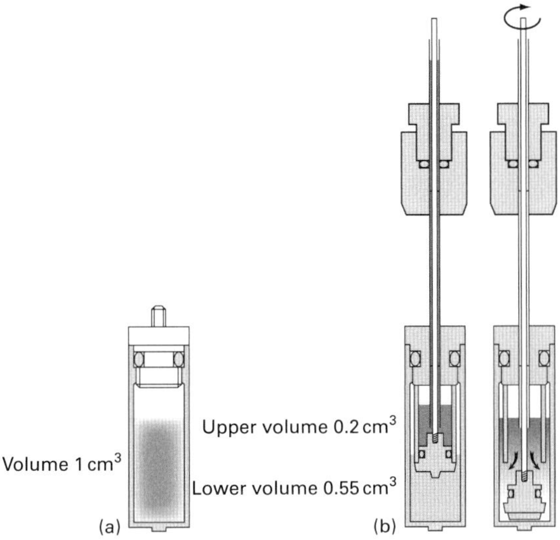

Figure 2.8 Schematic views of the two types of cells in μDSC III from Setaram (Caluire, France): (a) closed batch cell; (b) mixing batch cell before and after piston displacement. Reproduced courtesy of Setaram.

2.3.4 Isothermal titration calorimetry (ITC)

Up until now, few results are known in which gelation has been studied using isothermal titration calorimetry, even though ITC provides a complete thermodynamic profile of molecular interactions, and will increase in importance. Thermodynamic data reveals the energy contributions which drive molecular interactions and molecular complex formation: conformational changes, hydrogen bonding, hydrophobic interactions and electrostatic interactions, and how such interactions can be understood at a molecular level.

Specific binding and molecular recognition between biological molecules are essential in biological processes (Blandamer *et al.*, 1998; O'Brien *et al.*, 2001). Many proteins have specific binding sites for small molecules, including sugars, drugs, metals and surfactants, referred to as ligands. Knowledge of these interactions is important in understanding the role of proteins in biological activity. The method is well known to biochemists, but nowadays it is used for other types of systems which are relevant to the physical-chemical domain, such as association of amphiphilic polyelectrolytes (Raju *et al.*, 2001), dissociation of copolymer micelles upon dilution below the critical micellar concentration (Paula *et al.*, 1995; Garidel and Hildebrand, 2005) and binding of multivalent ions to a gelling polysaccharide (Fang *et al.*, 2007).

An ITC instrument is composed, like DSC and μDSC, of two identical cells made of a good conducting material and surrounded by an adiabatic jacket (Figure 2.9). Very sensitive thermocouples are used to detect temperature differences between the reference cell, filled with the solvent (for example, water or buffer), and the titration cell, filled with the macromolecule in solution. Solutions with well-known concentrations of small molecules

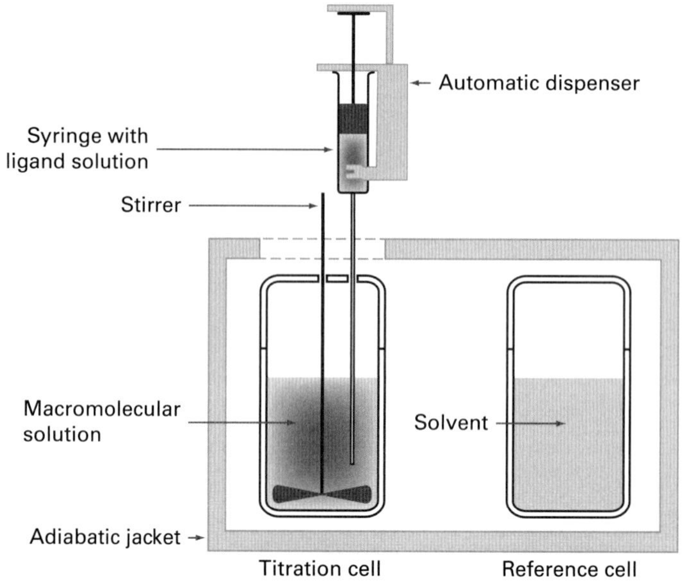

Figure 2.9 Schematic view of an ITC calorimeter.

are placed in an automated syringe and injected into the titration cell, which is maintained at a constant temperature. Consecutive injections of small volumes (5–10 μL) are performed and the heat exchange during the dilution is monitored. Experiments must be conducted with very dilute solutions.

The system works with a power compensation method, accomplished by continuously regulating the amount of heat applied to the titration cell, so as to drive the temperature difference between the two cells towards the baseline, the steady-state value. The thermal power required to return to the steady-state temperature differential as a function of time is the quantity determined, and this is directly proportional to the heat of the reaction (Freire *et al.*, 1990). In the most sensitive instruments, very low heat exchanges, *c.*40 nW, can be detected.

The basic equations for ITC are known from standard thermodynamics. The equilibrium constant K_{eq} which characterizes a molecular complex formation is related to the Gibbs free energy change ΔG:

$$K_{eq} = e^{\frac{-\Delta G}{RT}}, \tag{2.22}$$

where R is the gas constant and T the temperature in Kelvins (K). The equilibrium constant for chemical reactions in solutions containing different species i is expressed as a function of the activity a_i of the solutes:

$$K_{eq} = \prod_i a_i^{\nu_i}, \tag{2.23}$$

where ν_i are the stoichiometric coefficients.

When K_{eq} is expressed in terms of concentrations instead of activities, it is called the affinity constant, K_a.

For a bimolecular interaction between a macromolecule M and a ligand L, the affinity constant K_a is the ratio of the molar concentration of the complex [ML] to the product of the molar concentrations of the species [M] and [L]:

For the reaction $M + L \Leftrightarrow ML$

$$K_a = \frac{[ML]}{[M]\,[L]}. \tag{2.24}$$

Knowledge of the affinity constant does not, in itself, establish the association mechanism; such interactions are better characterized by separating the enthalpic and entropic contributions:

$$\Delta G = \Delta H - T\Delta S. \tag{2.25}$$

The enthalpy term ΔH is related to the difference in energy required to create and to break the bonds between L, M and the solvent during the complex formation. The entropic term characterizes the number of accessible states or microscopic configurations which, in aqueous solution, are generally dominated by rearrangements of the water molecules. ΔG, being a combination of these two, may make either a net additive or subtractive contribution.

Isothermal titration calorimetry measures the binding affinity K_a, enthalpy change ΔH, binding stoichiometry (n) and the entropy change ΔS simultaneously from a single experiment involving a macromolecule and a ligand. When substances bind, heat is either generated or absorbed. An example of a titration curve is shown in Figure 2.10, which also shows, albeit under ideal conditions, the experimental procedure.

ITC allows the direct measurement of the temperature dependence of ΔH and the variation of the heat capacity of the solution:

$$\frac{\delta(\Delta H)}{\delta T} = \Delta C_p. \tag{2.26}$$

The latter is related, for instance, to changes of the polar surfaces accessible to the solvent during complex formation.

2.4 Microscopy of gel networks

Traditionally, studies of physical gels have used optical or electron microscopy. The various techniques of transmission electron microscopy (TEM) have been quite widely employed. Scanning electron microscopy (SEM) has also been used, but images tend not to be very revealing. More recently, atomic force microscopy has been employed. Many of the published images have proved informative but, as with TEM, there are always caveats, since the gel sample under study is no longer in its native state.

2.4.1 Transmission electron microscopy (TEM)

The assumption that polymers in the gel state are interconnected throughout the liquid phase is one of the major arguments for the origin of elasticity in gels. The ability of gels

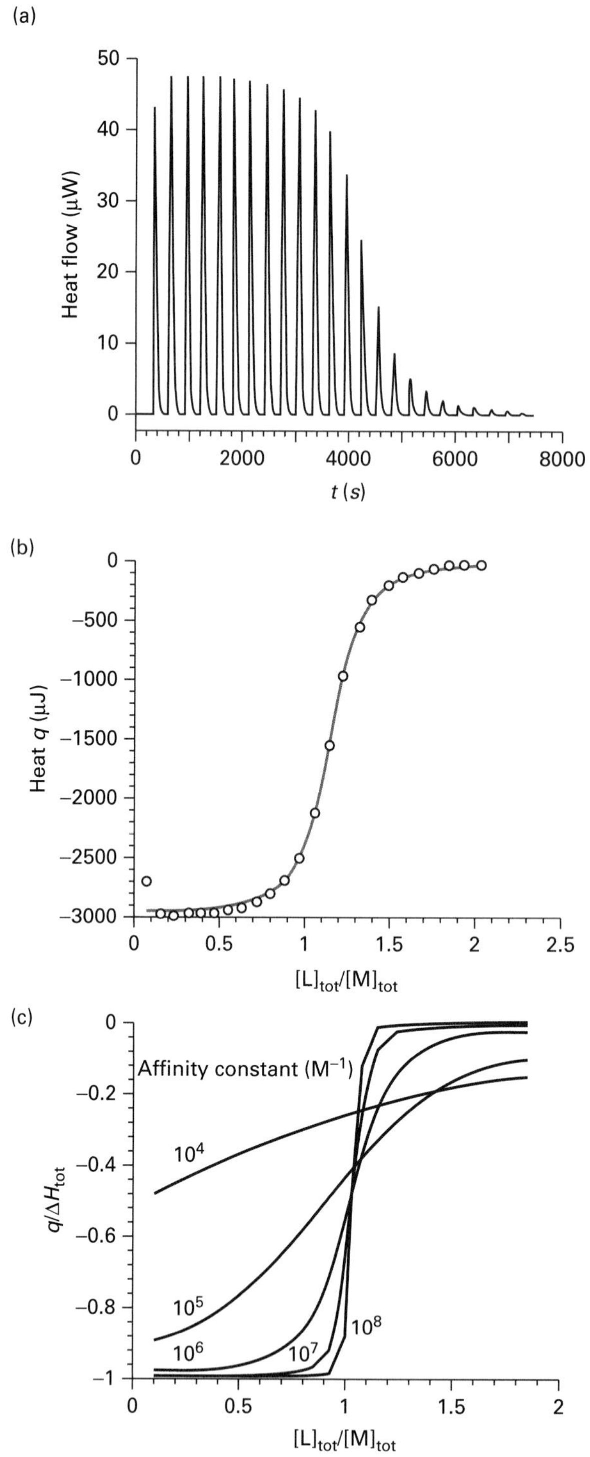

Figure 2.10 (a) Power dissipation in an ITC experiment after successive injections of 10 μL solutions containing the ligand into a macromolecular solution; (b) total heat exchanged in terms of the ratio of molar concentrations of ligand to macromolecule, showing the saturation of the complex formation; (c) simulation of titration curves of a 10^{-4} M solution for various values of the affinity constant K_a. The vertical axis shows the heat released, normalized by the total heat if all the ligand injected interacts. The optimum choice of affinity constant for such experiments is on the order of 10^6 M^{-1}, where the product $c = K_a[M]_0$ of the affinity constant and the molar concentration of the receptor in the cell is approximately 100. Adapted with permission from Bruylants and Bartik (2005).

to hold a large amount of solvent (99%) is strongly related to their microstructure. The size of the pores, or 'mesh' size, has a direct influence on the capillary forces which are especially strong for the smallest pores, of size 1–100 nm (from the Laplace law). Direct observation of the structure is of primary importance. However, images of gel networks are difficult to obtain. In this chapter we try to explain the reasons for this, and to briefly describe the techniques based on TEM imaging used so far.

The difficulties in obtaining good-quality electron micrographs, in particular of aqueous physical gels, are similar to those encountered for most biological specimens (Hermansson and Langton, 1994). First is the lack of contrast: many such polymers consist of elements with low atomic numbers (C, H, O, N), so do not absorb the incident beam, a basic requirement for observing amorphous structures. However, labelling techniques can be used which make such specimens visible by a selective 'staining' with solutions containing heavy atoms (such as osmium, tungsten or uranium). The staining methods usually include chemical fixing of the specimen and embedding with epoxy resins. Another difficulty arises from the high degree of hydration: while the electron beam in the microscope is under vacuum, water must not be in the liquid state. In such gels, the high degree of hydration and the weakness of the bonds holding the network make all manipulations very delicate and so likely to disrupt the structure. As an alternative to chemical fixing, which is traditionally used for biological specimens, cryotechniques, based on a physical fixing method, can be used. The physical network is 'locked' by very fast freezing of the sample. Although the idea – freezing water – seems very natural and easy to perform, a certain number of precautions must be followed (Ayache *et al.*, 2010).

When a liquid is cooled below its freezing temperature, crystallization involves first a nucleation step, then a nuclei growth step which allows the crystals to extend throughout the liquid. Liquid water has a particular structure such that water molecules can establish hydrogen bonds between each other, and this structure is labile, because of thermal agitation. In ice, below 0°C and at atmospheric pressure, the crystal structure is hexagonal, allowing the maximum number of hydrogen bonds to form (ice has an open structure, with a lower density than water at 0°C). The development of this hexagonal ice can cause very severe damage to the local structure of the gel network, which destroys the original architecture. Crystal size is a function of the cooling rate and of the freezing temperature, so the lower the temperature, the slower the diffusion of liquid molecules towards the crystal surfaces and the smaller the crystals. The number of crystal nuclei is also important, but when they are small enough that they do not show any diffraction pattern under the electron beam in TEM experiments, water is assumed to be in a vitreous state. This state can be reached at temperatures below −196°C (liquid nitrogen temperature). In pure water, the vitreous state is reached at cooling rates of around 10^9 °C min^{-1}. In aqueous solutions containing species such as ions, macromolecules or small organic molecules, the presence of these solutes stabilizes the system and reduces the diffusion of water molecules during freezing, so the vitreous state can be accessed with (still very high) cooling rates of 10^6 °C min^{-1}, and at higher temperatures, say around −148°C for a biological system with a solute concentration of 300 mM.

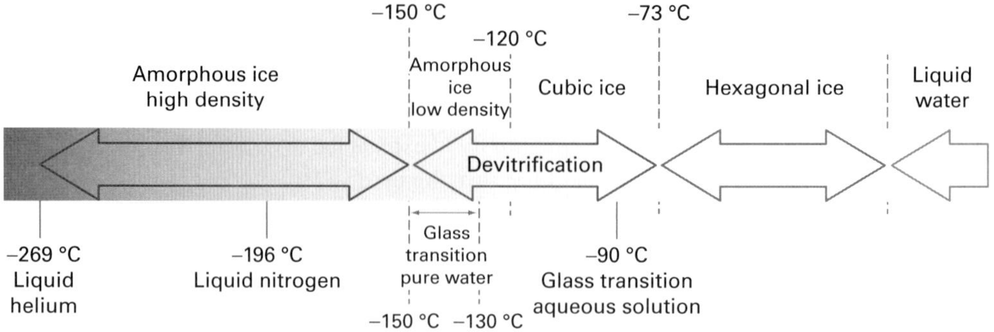

Figure 2.11 Freezing diagram for pure water and for an aqueous solution with 300 mM organic solute.

Figure 2.11 summarizes the various states of water as a function of temperature for pure water and for a biological solution. The glass transition temperature of pure water is between −150°C and −130°C, whereas for a biological aqueous solution it is around −90°C. In general, between −150°C and −73°C the crystals can grow very slowly, but still fast enough that amorphous ice will be converted into cubic ice when the temperature approaches −73°C. Therefore, in all circumstances where frozen water must remain in the vitreous state, any increase in temperature above −150°C for pure water or above −90°C for a biological system is likely to induce recrystallization (devitrification) of the amorphous ice, and should be avoided.

The phase diagram of water also depends on pressure: at high pressures (2100 bar), formation of vitreous water becomes easier and the rate of cooling can be lowered. At high pressures the viscosity of liquid water is also higher (1500 times higher than normal pressure) and the diffusion of molecules is slowed. At this pressure, the amorphous ice is in a 'high-density' amorphous state (density $d = 1.17$), while at atmospheric pressure it is in a 'low-density' state ($d = 0.94$).

Cryotechniques (Benedetti and Favard, 1973; Ayache *et al.*, 2010) can be used to investigate gel networks, and ultra-rapid freezing can be achieved with various devices. These devices allow the freezing of small samples containing colloidal particles or proteins in suspension by placing them on grids and immersing them in liquid nitrogen or slush nitrogen (close to its solid state at −210°C, at a low pressure), and jet freezing. The latter involves freezing under high pressure, and is the most suitable method for bulk samples (200 μm thickness) containing large quantities of water (cell cultures or cell suspensions). The sample is placed between two copper or gold cups filled with a fluid (hexadecane) giving good thermal conductivity. The sandwich is placed under high pressure and cooled under a jet of ethanol at liquid nitrogen temperature. However, such a device is complex and the operation procedures are delicate.

Examples of the difficulties in quick-freezing hydrated samples such as gelatin gels are provided below. The technique of freezing is here based on an electropneumatic system which pushes the piece of gel against a copper block, previously cooled with liquid helium, down to −269°C (4 K) and maintained under vacuum. After freezing, the sample is transferred to a vacuum evaporator at −130°C under very low pressure. In order to

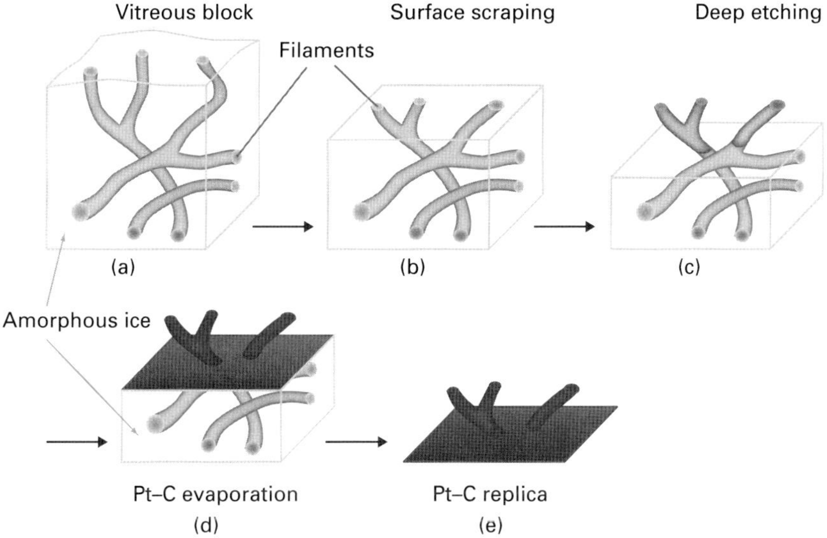

Figure 2.12 (a) A gel sample with vitrified ice; (b) scraping of the surface; (c) surface after deep etching (*c*.0.5 μm), exposing the network; (d) rotary shadowing with heavy metals (Pt, Au, W), followed by carbon evaporation under normal incidence; (e) observation of the gel replica by EM, after chemical dissolution of the underlying organic material.

display the structure of the network, a knife is used to scrape the surface that was in contact with the copper block, first at a small angle (see below), then at various depths below the free surface. After scraping, the water is allowed to sublimate at −90°C. This stage, called 'deep etching', is followed by platinum evaporation (shadowing under small-angle incidence) to create the relief contrast of the structure, and then by carbon evaporation normal to the surface, to create a continuous film and fix the structure. The underlying organic material is then removed by chemical dissolution. This is the method used to prepare a 'replica' of the network. Figure 2.12 summarizes the various preparation steps.

The scraping of the gel surface prior to replica preparation is shown schematically in Figure 2.13. It is seen that additional labelling was used to mark the surface of the gel, in order to reveal the part which was frozen at the highest rate.

In Figure 2.14, the progressive distortion of the gel network with distance from the surface can be observed. The scale is the same (0.5 μm) in all panels but the distance from the surface varies from 0–4 μm to 24–28 μm from top to bottom. The development of ice crystals of more than 1 μm in size can be seen in the bottom image. Only the upper part of the figure represents the exact structure of the gel network.

2.4.2 Atomic force microscopy (AFM)

Atomic force microscopy has rapidly emerged as a versatile imaging tool for investigations in various areas of chemistry, biology and physics (Eaton and West, 2010), and it has been used by a number of workers, particularly on pre-gel systems, and in

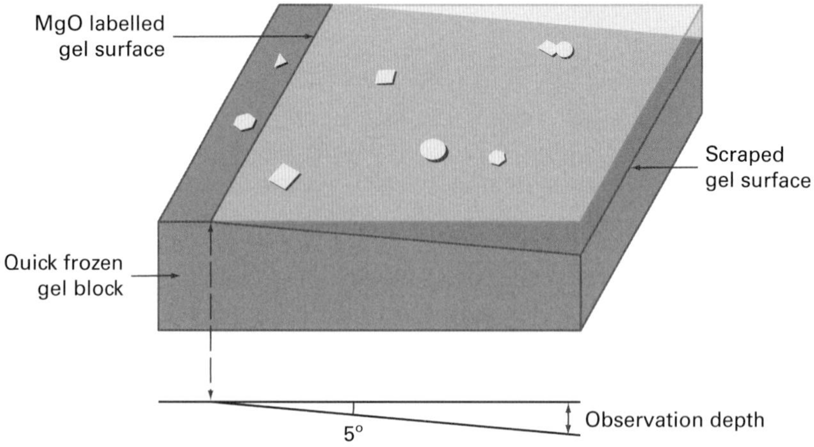

Figure 2.13 Surface scraping of a quick frozen gelatin block surface labelled with MgO crystals. Scraping occurs obliquely with an inclination angle of 5°. Originally published by Favard *et al.* (1989) © Portland Press.

investigations of their structure. The instrument is based on the scanning tunnelling microscope, which was developed in the 1980s and provided the first atomic resolution images of surfaces. Since then, various developments have led to a group of related microscopes, collectively known as the scanning probe microscope family. The fundamentals behind these microscopes are essentially the same.

A fine tip, usually *c.*20 nm and most commonly made from silicon nitride, is attached to a spring-like device called a cantilever. The cantilever is also made from silicon and is typically around 150 nm in length and 1 nm in thickness. When an image is sampled, the probe or tip is lowered on to the sample surface, generating a force between the tip and the sample which depends on the physical and chemical properties of the sample. As the probe, controlled by a piezoelectric tube, moves along the surface of the sample, the cantilever is deflected.

Measurement of this cantilever deflection is the key to obtaining topological information. In early work this deflection was measured by the so-called 'tunnelling effect' of electrons just above the cantilever, although soon afterward optical methods employing a laser were developed. Using this method, deflections as low as 0.1 nm can be detected with the laser and an appropriate optical detector. The detector is used to feed back information to the equipment, which can use the data to control the vertical movement of the sample.

Two different techniques are currently used to obtain image information. In the 'contact mode', the distance between the sample and the tip is adjusted to keep the cantilever at constant deflection. The voltage supplied to the piezoelectric tube to maintain constant deflection is used to provide an image. However, this technique has proved to be unsuitable for imaging soft materials since it involves relatively high lateral shear forces that can easily disrupt the sample.

Alternatively, and more widely, the so-called 'tapping' or 'non-contact mode' is used to image fragile – including biological – materials. Here, the cantilever is oscillated in a

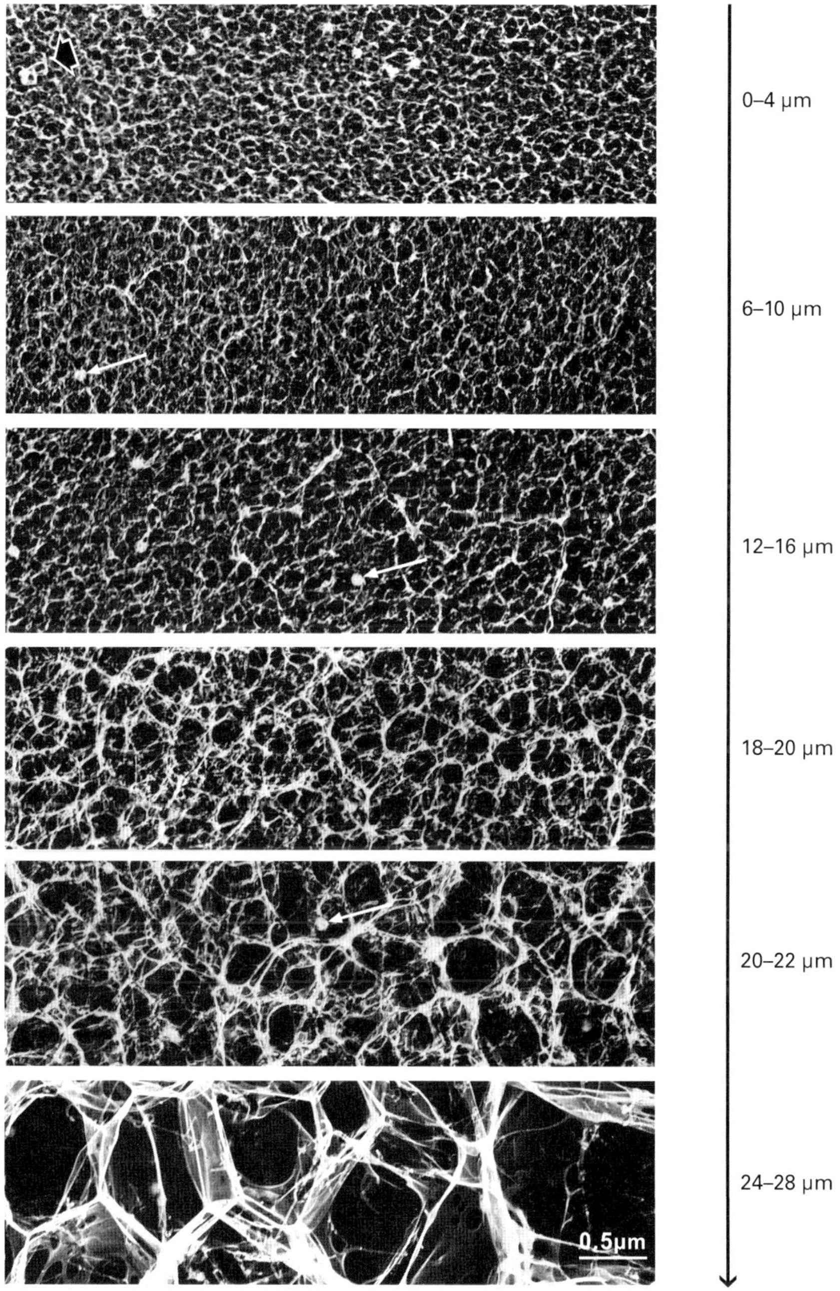

Figure 2.14 Electron micrographs showing the effects of ice crystal growth in a gelatin gel ($c = 2\%$) according to their depth. MgO crystals used to label the surface are seen on the top image (thick arrow) and some calibrated latex spheres (0.09 μm) included in the gel (thin arrow) on the other pictures. The network appears more and more disorganized by the growth of ice crystals. Only images corresponding to the superficial areas, around 5 μm from the surface, are interpreted as being the true gel structure. Originally published in Favard *et al.* (1989) © Portland Press.

sinusoidal manner at its resonance frequency. The amplitude of this oscillation is dampened when the tip approaches the sample, and information is relayed to the piezoelectric tube, which keeps the vibration at constant amplitude by adjusting the vertical position of the sample. This virtually eliminates lateral forces and allows imaging of weakly absorbed samples. Further technological advancements have enabled the use of this technique under pseudo-physiological conditions. In the latter case, samples can be imaged under physiological buffer, with the result that the sample, although immobilized, is kept 'alive'. This gives it a fundamental advantage over traditional electron microscopy, where the sample is static and is usually partially destroyed during the process of imaging. Thus, a portion of the flat surface can be constantly scanned for time-evolving changes, from which some time information can be derived.

The resolution obtained with AFM very much depends on the size of the tip that is used. Most commercial silicon nitride tips have a size of 10–20 nm, although carbon nanotube technology has been used to enhance the resolution that is obtained. Here, the tips can have a nominal size as low as 5 nm. In order to immobilize the sample, mica (negatively charged) is frequently employed, although glass, gold and hydrophobically surfaced highly orientated pyrolytic graphite (HOPG) have also all been used. Such materials should have the 'atomically smooth' surface necessary for such imaging. Often, when a thin water film at the sample surface forms a capillary meniscus, capillary forces reduce the resolution obtained with biological samples scanned in air. A number of interesting applications have been partially developed including so-called 'nanotools', using the tip to cut or unfold and manipulate samples, and chemical contrast imaging, which is adding specific ligands or chemically active groups to the tip, which can help to map chemical interactions.

Applications of the technique will be discussed in the relevant chapters, but work has been published on both polysaccharide and protein gel precursors (Ikeda *et al.*, 2001; Gosal *et al.*, 2002; Round *et al.*, 2010). Although these studies have concentrated on imaging sub-gelling concentrations of aggregates, particularly for ionic polysaccharide gels (Chapter 5), progress has now been made in imaging aqueous gels, by forming a thin hydrated film on the mica substrate (Morris, 2009). The imaging process compresses the polymer network, but without obvious structural damage. This approach has been particularly successful for gellan gels (Chapter 5). At the moment, however, there appear to be few such compressive gel studies for other systems such as protein gels.

2.5 Rheological characterization

The characterization of gels by rheological measurements has been carried out for many years, since the viscous and elastic properties of gels are among the most significant in practical applications. This is not the place for a rigorous introduction to rheological techniques, since there are many sources (Ferry, 1980), but it is important to consider a few basics before discussing, in more detail, the characterization techniques employed.

As we noted in Chapter 1, gels (and again, arguably all real systems) are viscoelastic – that is to say, under various conditions of time, temperature or other factors they can be regarded as having both elastic (solid) and viscous (liquid) properties. For that reason the majority of measurements are made using conventional commercial small-strain oscillatory instruments. However, it is very important to point out here that such small-deformation experiments may have little to do with applications, which often involve large-deformation fracture or rupture.

2.5.1 Small-deformation measurements

There is no absolute measure of 'small' and 'large'; the (static) elastic modulus of any solid is defined as the ratio of stress to strain. In some experimental set-ups it is the stress that is varied, and in others the strain, but there is a clear implication in the above definition that stress and strain are proportional. This is always true in the limit of small enough stresses and strains, but the absolute value of this linear viscoelastic strain limit can be as high as 25% or as low as 0.01%, depending upon the nature of the gel system.

This does not mean that measurements outside these limits are pointless, just that they have a different purpose. They are concerned, say, with the absolute stress or strain 'to failure' – fracture or rupture – and the area measured under the stress/strain curve under various conditions (for example, rate of deformation or temperature). Traditional methods of gel characterization include those based on falling or oscillating microspheres, and gels oscillating in a U-tube manometer (Ross-Murphy, 1994). Although cheap to construct, they are now mainly of historic interest, but some recent gel work still employs falling-ball and tilting-tube methods (Chapters 8 and 9).

2.5.1.1 Small-deformation oscillatory shear methods

Nowadays the vast majority of physical measurements on gels are made using oscillatory shear rheometry (Ferry, 1980; Kavanagh and Ross-Murphy, 1998). The essential feature of a typical rheometer for studying gels is a vertically mounted motor (which can either drive steadily in one direction or oscillate). In a controlled-stress machine, this is usually attached to the upper fixture. A stress is produced, for example by applying a computer-generated voltage to a DC motor, and the strain induced in the sample is measured using an optical encoder or radial position transducers attached to the driven member. In a controlled-strain instrument, a position-controlled motor, which can be driven from above or below, is attached to one fixture, and opposed to this is a transducer housing with torque, and in some cases normal force, transducers. Figure 2.15 represents a typical controlled-stress instrument. The sample geometry can be changed from, for example, Couette to cone/plate or disc/plate, and the sample temperature controlled.

In a typical experiment, a sinusoidal oscillation of maximum strain γ_M and oscillatory frequency ω is applied to the sample. If the material is perfectly elastic, then the resultant stress wave is exactly in phase with the strain wave, and does not depend upon the oscillatory frequency, so we can define the (Hookean) equilibrium elastic modulus G as the ratio of stress to strain. If the material is a purely viscous fluid, since the rate of change of the sinusoidal oscillation is a maximum when the strain is zero, the resultant stress wave will be exactly 90° out of phase with the imposed deformation.

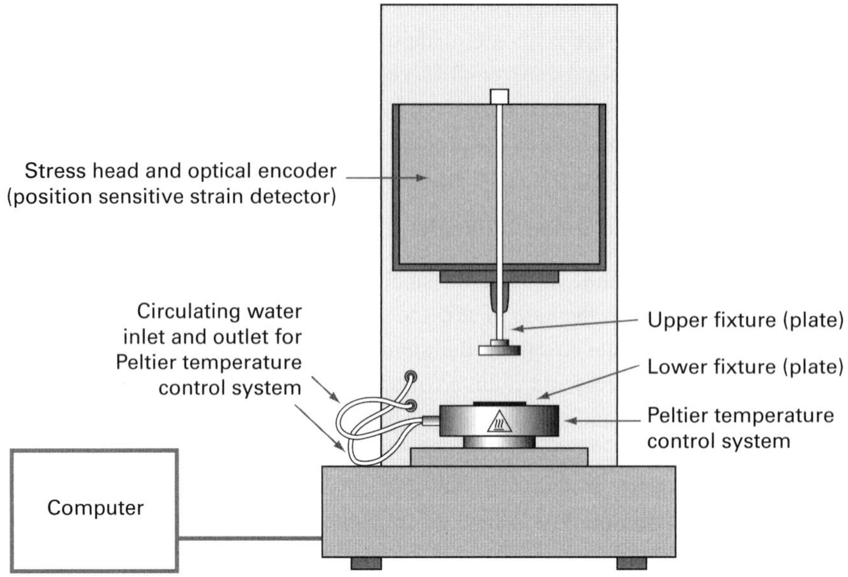

Figure 2.15 A typical controlled-stress instrument.

More generally the stress wave will have a frequency-dependent phase difference δ ($0 < \delta < 90°$) so that δ, or more usually $\tan\delta$, is a measure of the viscous/elastic ratio for the material at the given frequency ω. The elastic (in-phase) and viscous (out-of-phase) components of the stress wave are separable, and they define the shear storage modulus G' as the ratio of in-phase stress to strain and the shear loss modulus G'' as out-of-phase stress to strain. Clearly the values of both G' and G'' will in turn depend upon ω, the oscillatory shear (radial) frequency – with ω equal to 2π times the frequency f (in Hz).

Having defined G' and G'', we can evaluate a number of other commonly used rheological parameters, since all are interrelated. For example, G^*, the amplitude of the complex modulus, is given by

$$|G^*| = \sqrt{(G')^2 + (G'')^2}, \tag{2.27}$$

and

$$\frac{G''}{G'} = \tan(\delta). \tag{2.28}$$

In the early days of oscillatory rheometry, the phase angle δ, rather than its tangent, was an experimentally observed parameter. Finally, the amplitude of the complex viscosity η^* is given by

$$|\eta^*| = \frac{|G^*|}{\omega}. \tag{2.29}$$

This helps to define and subsequently measure the so-called *mechanical spectrum* – the trace of (log) G' and (log) G'' versus (log) ω – and to establish whether or not a system is a gel, or simply a very viscous fluid. Finally, oscillatory measurements can also be made in tension/compression, leading to alternative parameters, such as the storage (E') and loss (E'') Young's moduli, with all other corresponding relationships still holding. For gels and networks, these measurements are relatively uncommon, but they do have the major advantage that this geometry helps to reduce slip, often a real problem with low-concentration ionic gels of polysaccharides (Chapter 5). Although shear mode oscillation is more frequently used, it sometimes suffers from slippage. The longitudinal vibration method is free from slippage. It can be used only for self-supporting gels for which the strain from its own weight is almost negligible. When the gel is surrounded by oil, the temperature dependence of the complex Young's modulus can be easily determined. We do not discuss these further here, although a number of such measurements are discussed in Chapter 5 and other chapters.

2.5.1.2 Controlled strain versus controlled stress

Nowadays the majority of modern instruments are of the controlled-stress type. However, they usually still generate results in controlled-strain form, that is, as the modulus components G' and G''. Strictly speaking, since stress is applied and the strain is measured, then results should be reported as the components of the complex compliance J' and J''. However, most instruments circumvent this by applying a stress and measuring the strain, but in a servo or feedback mode, so that it appears that they are indeed controlling the strain. For many applications and systems this is acceptable, but for systems very close to gelation, it is certainly not ideal. This is because there is no sure way of controlling the feedback when the system just changes from solution (sol) to gel, and yet at the same time guaranteeing that the strain remains very low. For such systems there is a further advantage in a genuine controlled-strain technique, in that the mechanical driving head and the measurement transducer are completely separate assemblies – the only link between them is the test sample and geometry.

At this point it is important to point out the potential (and real) problems that can be caused by slippage of the sample, particularly for those where syneresis ('weeping' of the gel, i.e. loss of solvent at the gel surface) occurs. Much of the data collected for such materials is unfortunately flawed by slippage at the geometry surface. For example, certain carrageenan gels (Chapter 5) are particularly susceptible, as was shown convincingly by Richardson and Goycoolea (1994). They assembled a special punctured cylindrical fixture designed to eliminate slippage, and compared measurements made with this and with a standard concentric cylinder. Results for gelatin (not susceptible to slip) were identical; for the carrageenan samples very different results were obtained. The use of roughened or serrated surfaces for the standard geometries is an alternative solution for avoiding slippage. The oscillating tension/compression experiment mentioned above also tends to reduce this problem. Non-standard geometries, such as rotating vane geometries, have been widely used in complex fluid rheology such as food formulations (Barnes and Nguyen, 2001), in particular under shearing conditions, allowing the determination of a yield stress.

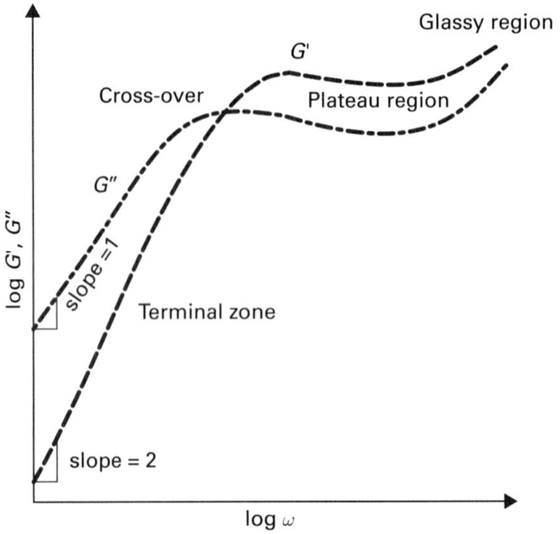

Figure 2.16 Mechanical spectrum of a polymer solution – viscoelastic liquid.

2.5.1.3 Frequency and strain dependence

Polymer solutions

The mechanical spectrum of a polymer solution (a viscoelastic liquid) has the general form illustrated in Figure 2.16 (Ross-Murphy, 1994; Kavanagh and Ross-Murphy, 1998). At low frequencies (note the double log scale) G'' is greater than G', but as the oscillatory frequency increase, G' increases more rapidly than G'' (with a slope ~2 in the log–log representation, compared to a slope of 1 for G''), and at some frequency there is a cross-over. After this, both G' and G'' become much less frequency-dependent – the so-called 'rubbery plateau' region.

Whether the cross-over region is reached in the frequency window of conventional oscillatory measurements depends upon polymer concentration, relative molecular mass (M) and chain flexibility. At the same time, the mechanical spectrum measured will be essentially independent of the amount of shear strain, out to, say, 100% in 'strain units' (i.e. a strain, in terms of the geometry of deformation, of unity). Rheologists may express this by saying that the linear viscoelastic (LV) strain extends out to *c.*100%.

Polymer gels

The mechanical spectrum of a viscoelastic solid will have a finite G', with a value usually well above (say 5–50 times) that of G'' at all frequencies, as in Figure 2.17 (Clark and Ross-Murphy, 1987; te Nijenhuis, 1997). In this respect it shows some similarities with the plateau region of the solution mentioned above; such a plateau is often referred to, somewhat imprecisely, as gel-like, for exactly this reason.

The strain-dependent behaviour, for polymer gels is more difficult to generalize, although the LV strain is rarely as great as 100% (some gelatin gels may be the exception here), and may be extremely low – say 0.1% or less. At values just greater than the LV

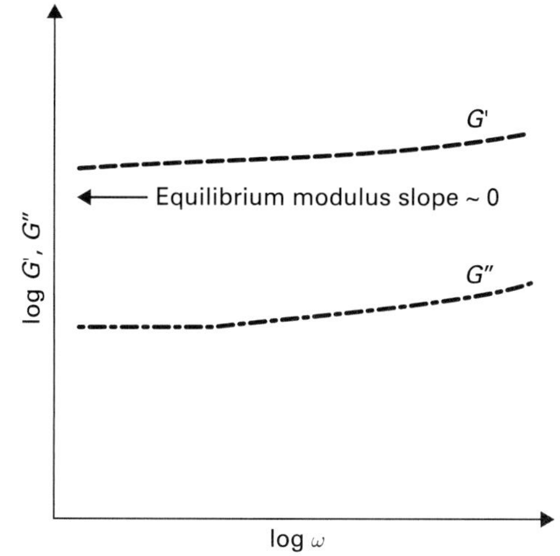

Figure 2.17 Mechanical spectrum of a viscoelastic solid.

strain, G' and G'' may show an apparent increase with strain. This is, of course, largely an artefact of the experiment, since G' and G'' are only *defined* within the LV region. This is then followed by a dramatic decrease, caused by failure, either by rupture or fracture, sometimes macroscopic – but more often failure occurs at the geometry interface.

2.5.1.4 Creep and stress relaxation

Although there are limitations to using a controlled-stress instrument in pseudo-controlled-strain mode, such controlled-stress instruments are ideal for time domain experiments, in which a small fixed stress is applied to a gelled sample and the strain ('creep') is monitored over time. In creep, the static compliance J is defined as the ratio of strain to stress, and so is simply the reciprocal of the ideal (or Hookean) elastic shear modulus G. For a viscoelastic sample, however, the value of time-dependent compliance $J(t)$, at the time given by $t = 1/\omega$, is not simply equal to the reciprocal of dynamic modulus $1/|G^*(\omega)|$, although the two can be related by appropriate transforms. However, it is just as useful to use the time domain expression, and to evaluate J as a function of time as the sample 'creeps'.

The long-time behaviour of $J(t)$ is particularly valuable, since it can help us to understand the behaviour of physically cross-linked gels. As a rule of thumb, for times shorter than, say, 100 s (corresponding radial frequency $\omega > 2\pi/100 \approx 6 \times 10^{-2}$ rad s^{-1}) the oscillatory strain experiment is the best, whereas for longer times (and for time stationary systems) the creep experiment has advantages. The disadvantage in a long-time stress-relaxation experiment is that the detectable stress signal decays to zero with time (rather than increasing monotonically as does the strain in creep), and tends to disappear into the zero-signal background of the rheometer.

2.5.1.5 Temperature dependence

Many polymer gels show so-called 'glassy' behaviour at high enough frequencies or low enough temperatures, and the study of gels under these conditions, perhaps induced by measuring in highly viscous low molar mass solvents such as saturated sucrose, is an active area of interest.

What the above does suggest is the well-known time–temperature superposition (TTS) effect in polymer materials science, namely that high frequencies and low temperatures may be regarded as equivalent (Ferry, 1980). Very often this works well, but caution should always be applied. The glass transition itself is related to polymer free volume, and temperature discontinuities in said free volume should make the approach invalid. If we are to follow the principles outlined by Ferry – one of the co-devisers of the method, and its strongest protagonist – then TTS should never be applied within 50°C of a phase transition within the system. This would eliminate all TTS approaches for water-based (hydro-)gels from −50°C to +150°C, i.e. more than the whole regime of potential interest. This is an extreme view and, provided no deep interpretation is given, the approach may still be useful. It has been employed by several workers.

Here it is important to point out that many gelation experiments in the literature have been performed using temperature 'ramps', i.e. heating, say, from 25°C to 85°C at a rate of +1°C min^{-1}. This applies whether the temperature ramp is positive, as here, in order to form heat-set gels or melt out cold-set thermoreversible systems, or negative, i.e. a cooling ramp applied to 'set up' cold-set gels such as gelatin. From the theoretical and interpretation viewpoint this is not ideal, since the rate of heating is conflated with the temperature rise, so analysing the real kinetics of gelation becomes impossible. For this, it is far better to perform the experiment as an isothermal temperature jump, although this approach is experimentally more testing because it obviously requires very fast and efficient heating, cooling and re-equilibration.

In early instruments, where a hot-air heating system was used and cooling involved boiling off liquid nitrogen, ±20°C min^{-1} was easily achieved. Later, fluid circulation systems, which are intrinsically slow to change and to re-equilibrate, were used. Then the only way to obtain fast rates was to have a swappable liquid bath system, with one bath at a low temperature and the other at a high temperature, and a tap turned to change from one to the other. Nowadays, Peltier systems are all but ubiquitous even for controlled-strain instruments, and so there is less of a problem. That said, some of the data discussed in other chapters has been obtained using the temperature-ramp approach, and the 'ramp' approach is still widely used, not least because it tends to mirror the DSC technique.

2.5.1.6 Time-dependent systems

The kinetic gelation experiment

Clearly if we are by some physical method – say, heating – converting a solution to a gel, we will change the initial sol mechanical spectrum (Figure 2.16) to the gel spectrum (Figure 2.17) (Kavanagh and Ross-Murphy, 1998). In a typical experiment, following the progress of gelation using mechanical spectroscopy, the oscillatory frequency is kept constant at *c.*1 Hz (6.28 rad s^{-1}) – for convenience many workers use a frequency of 10 rad s^{-1} – and the strain is maintained constant and low, say typically 10% or less. The choice of frequency is always a compromise – we need a high enough value that a single frequency measurement does not take too long, so we can collect enough data, but not so high that instrumental artefacts begin to appear. In our experience these can be seen quite

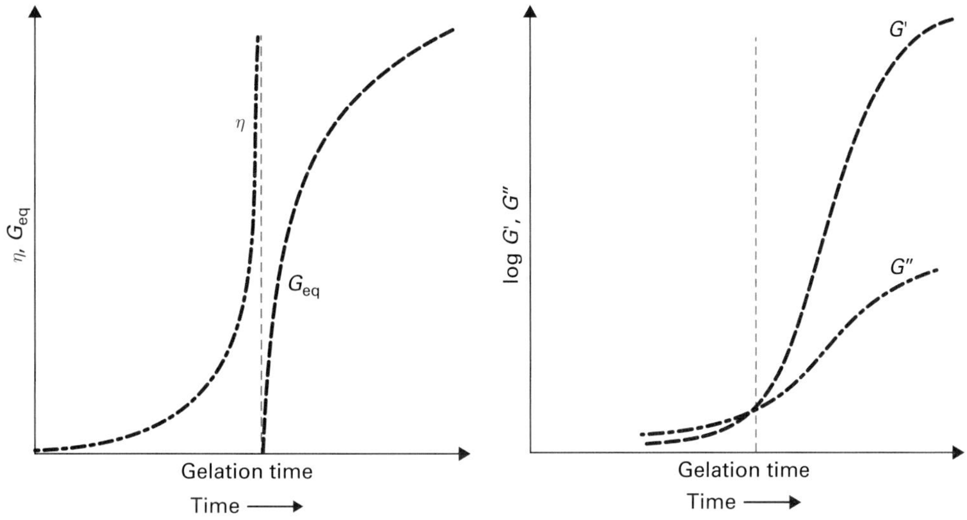

Figure 2.18 (a) Time growth of equilibrium gel modulus. G_{eq} rises very rapidly, even on a log scale, at or just after the gelation time, before reaching a final asymptotic level. (b) Because of the finite frequency effect and the contribution of non-ideal network assembly contributions, both G' and G'' will tend to rise before the true gelation point.

commonly for frequencies greater than, say, 30 rad s^{-1}. The measuring strain may also have to be a compromise so sufficient signal can be detected pre-gel, but with sensitive modern instruments this is less of a problem.

Gelation time measurement

There are a number of approaches to the determination of, say, gelation time, but first we consider the expected self-assembly time profile. For the equilibrium gel modulus, the ideal profile is shown in Figure 2.18a. Initially there is no response, but then G_{eq} rises very rapidly, even on a log scale, at or just after the gelation time, before reaching a final asymptotic level. Such behaviour is a simple consequence of the positive-order kinetics of self-assembly (cross-linking) and the requirement of a minimum number of cross-links per 'chain' at the gel point. We note that some phenomenological models have neglected the pre-gel behaviour and simply fitted the G_{eq} (>0) versus t behaviour to an n-order kinetic model. From the data-fitting viewpoint this is quite acceptable, provided it is appreciated that the underlying physics of self-assembly has been perverted.

The above scenario is, of course, complicated by the consideration that what is being evaluated by the instrument is not G_{eq} but G' and G''. Both of these are finite even for a solution, although the respective moduli values may be very low. However, because of the finite frequency effect and the contribution of non-ideal network assembly contributions, both G' and G'' will tend to rise before the true gelation point, and something akin to Figure 2.18b is usually seen. The flattening off of G'' is not something predicted from theory – indeed some would expect a pronounced maximum in G'' after gelation – but this is rarely observed for physical gels, except for some low-concentration gelatin systems. This asymptotic level G'' behaviour has been associated with the 'stiffness' of the network strands. The gelation time can then be estimated from the cross-over of G' and G'' (if seen, although this is not always the case), the time where there is a sudden increase in G' or by back-extrapolation of the G' versus time curve. In practice the results are not very different, as discussed elsewhere (Tobitani and Ross-Murphy, 1997).

2.5.1.7 Range of viscoelastic linearity

Yet another aspect of gel time measurement, and arguably one of greater significance, is the effect of finite strain on the tenuous mechanical system close to gelation (Ross-Murphy, 2005). In performing the kinetic gelation experiment it is usual practice to employ the smallest strain consistent with obtaining reliable data. In principle this can be checked to be within the linear viscoelastic region both before and after gelation by stopping the experiment and performing a so-called strain sweep. However, as we discussed above, many experiments are performed using controlled-stress instruments in their pseudo-controlled-strain mode, and such instruments do have more problems measuring a gelling system when the properties are changing quite rapidly within the oscillatory cycle, especially when using the 'controlled' strain mode. This is because one might expect that the linear viscoelastic strain of the gelling system, rather than being constant, would tend to change during the gelation process, and would be a minimum just at the gel point (Rodd *et al.*, 2001).

2.5.1.8 Failure of the Cox–Merz rule

In the introductory chapter a *rheological* definition of a gel as a viscoelastic solid was given (Ferry, 1980). However, as this chapter makes clear, systems such as surfactant based 'shower gels' fail the first definition. Many of these are what rheologists refer to as 'structured liquids', although sometimes – misleadingly – these have been called 'weak gels'.

For many polymer solutions and almost all melts, the shear-rate dependence of viscosity shows the expected 'pseudoplastic' behaviour, in which the viscosity becomes increasingly shear-rate dependent at high deformation rates but at low enough rates is constant (so-called Newtonian behaviour). For the same class of materials, the frequency dependence of η^* and the shear-rate dependence of η are observed to be closely superimposable, when the same numerical values of ω and $\dot{\gamma}$ are compared. This empirical correlation, often called the Cox–Merz rule (Ferry, 1980), has been observed for many solutions, but within the perspective of the present volume there are a number of exceptions. In these cases, useful information may be extracted from the comparison of $(\eta, \dot{\gamma})$ and (η^*, ω) profiles.

A number of these materials, particularly at low strains, have a gel-like frequency spectrum but with very pronounced strain dependence. In this case, if subjected to a steady (shear) deformation, they will apparently flow rather than fracture. More significantly, such systems do *not* obey the Cox–Merz rule; in fact $\eta^*(\omega)$, *measured in the small-strain limit*, tends to lie above $\eta(\dot{\gamma})$, except at high frequencies when the curves may appear to converge (Richardson *et al.*, 1987; Ross-Murphy *et al.*, 1993). Such behaviour is usually associated with a tendency to form aggregated structures or dispersions, which are then broken down under the applied strain.

One such system is solutions of xanthan, a microbial exopolysaccharide, discussed in Chapter 5. In steady shear measurements, apparent viscosity shear rate power-law exponents are high (≈0.9), and the overall viscosity versus shear rate profile is more like that of a so-called Bingham or viscoplastic material (Goodwin and Hughes, 2000) rather than a pseudoplastic fluid. For the pseudoplastic material, as the shear rate $\rightarrow 0$, the viscosity levels off to give the constant Newtonian viscosity, while for viscoplastic systems the apparent viscosity $\eta \rightarrow \infty$ (Richardson and Ross-Murphy, 1987).

Failure of the Cox–Merz rule, and the observation of power-law (log–log) viscosity versus shear rate curves have, nevertheless, been reported in a number of other systems, especially including those having long-range ordering and structure. The converse behaviour, in which η lies above η^*, is seen in certain hydrophobically modified polymer solutions and also in a number of micellar 'gels'. Such behaviour is predicted by the Tanaka–Edwards theory (Tanaka and Edwards, 1992a, 1992b, 1992c) and is presented in Chapter 6 in terms of the short and reversible network lifetimes; no current theory seems able to predict the $\eta < \eta^*$ behaviour, although it is intuitively easier to understand.

2.5.2 Large-deformation measurements

Many published large-deformation measurements take a very simplistic approach in that they make cylindrical samples of gels and simply crush them between two cylindrical plates. The force and amount of compression are measured, and various signature traces can be obtained. This is the basis of the technique of 'texture analysis' used in quality-control laboratories. In such compression measurements, the usual configuration is a vertical movement motor and fixed normal force transducer, and this is used in many instrument designs (Ross-Murphy, 1994). Although originally these were quite bulky, a number of miniature versions are available and are probably more useful for gel systems.

However, the problem with this approach is that in compression the point of failure of the sample can be quite difficult to establish simply from the trace obtained, and much of the compression involves further maceration of an already ruptured sample. Interpretation of such results in terms of gel macromolecular properties is almost impossible. The surface properties of the gel and the interaction with the (metal) interface of the compression rig are also crucial, and results may be quite different if the gel surface is lubricated.

A much cleaner and more easily interpretable approach is to measure in tension, rather than in compression, but there are problems there as well. One approach is to 'superglue' the top and bottom surface of the gel to the instrument and then set the instrument to 'pull' on the gel. This is more reliable that using the usual (dumbbell-shaped) solid samples. In either case an external extensometer needs to be employed, since the samples tend to 'draw' in the narrower central region. Even then, computation of stress and strain is not trivial. Some success has been achieved by casting ring samples and 'pulling' these over rods mounted top and bottom (McEvoy *et al.*, 1985). However, for very friable or 'weak' samples, failure may occur simply in removing the sample from the mould or in fixing it to the instrument. An alternative approach uses a 'shear sandwich' geometry.

Experiments can also be performed using the same apparatus as for small-deformation oscillatory or steady shear measurements, provided the sample can be made to adhere so firmly, e.g. again using 'superglue', that deformation occurs in the bulk of the sample, rather than by de-adherence at the interface. Alternatively the sample surface can be well lubricated; a combination of the two experiments can be used to eliminate errors due to friction.

Large-deformation and failure measurements must be carried out over a wide range of strain rates (including very fast strain rates, where the experiment time is less than 1 s). If the instrument is calibrated, and extensometers are fitted, it is easy to convert force and amount of deformation into stress and strain. In this way the initial linear region can be seen, and then the sample may either strain 'harden' or 'soften' (show upward or downward curvature) before a clear failure point is seen.

A number of approaches can then be employed, but one which we have found useful, although it has so far been employed sparingly in this area, is to construct the so-called 'failure envelope', originally suggested in the 1960s. By measuring the stress and strain at failure (in tension) for a range of gels, each measured as a series of replicates and over a wide range of extensional strain rates (say $\sim 10^{-5}$–10^{-1} s^{-1}), and plotting failure stress versus failure strain at various deformation rates, a graph – the failure envelope (Smith, 1963) – can be constructed.

Under certain circumstances the compression of cylindrical gel samples, in practice by far the simplest geometry and one still widely employed in industry, can still provide useful information. For example, the work by Nakamura and co-workers (Nakamura *et al.*, 2001) has used this approach to compress samples of gellan gel (Chapter 5) in a range of compression rates covering more than five orders of magnitude. At the slowest rates, a compressive de-swelling takes place; in other words, solvent is squeezed out of the sample, but this can be recovered perfectly reversibly (and as expected) by allowing the gel to re-swell in excess solvent. At higher compression rates, the gel fractures (ruptures) instead. Unfortunately, analysis in terms of the more fundamental stress and strain parameters is limited.

2.5.3 Particle tracking microrheology

Conventional rheological techniques as described above are widely used, but they do have some limitations. In particular it is difficult to measure extremely tenuous gel networks at the very small deformations needed to ensure that the reacting system is within the linear viscoelastic limit. For this reason other 'non-perturbing' techniques have a major potential advantage, and options employed over the last decade or so fall under the general heading of particle tracking microrheology (PTM).

There are a considerable number of techniques in operation but, so far as we are aware, no commercial equipment is available. The basic principle goes back many years, and simple variations were apparently in use in the 1920s. Essentially a set of small micron-size probes are placed in a (pre-gelled) solution and their position and velocity under Brownian motion are followed, for example by laser scattering techniques. The experimentally relevant measure is that of the set of (one-dimensional) mean square displacements of the particles, $\langle x^2 \rangle$. If the embedding fluid is Newtonian, then $\langle x^2 \rangle$ will be linear with time t, following classical diffusion, and can be written

$$\langle x^2 \rangle = 2Dt.$$

In three dimensions this gives

$$\langle r^2 \rangle = 6Dt. \tag{2.30}$$

Here D is the Fickian diffusion coefficient, and the Stokes–Einstein equation gives

$$D = \frac{k_B T}{6\pi\eta r_p}. \tag{2.31}$$

Since k_B is Boltzmann's constant and r_p is the hydrodynamic or Stokes radius of the probe particle, the Newtonian viscosity η is straightforwardly obtained.

Conversely, if the embedding system is purely elastic, then particles will move as a result of thermal motion but never actually diffuse. The amount of movement depends upon the elastic modulus of the medium; in this case the movement can be equated to that of a spring with constant K_e, so that, again in one dimension,

$$K_e = \langle x^2 \rangle k_B t. \tag{2.32}$$

For viscoelastic materials, more sophisticated methods need to be employed. For example, Mason and Weitz (1995) showed how the Stokes–Einstein (SE) equation can be generalized to a new relation – the generalized Stokes–Einstein relation (GSER) – which can be used for deriving the frequency-dependent viscosity, or as shown in the review by Waigh (2005), the creep compliance. A limitation of this simple approach is that it is less successful if sets of single particles are tracked, since these may simply probe sample inhomogeneities. In order to overcome this limitation, two-particle correlations can be followed.

The idea, as the description suggests, is that a set of the particles is tracked using, for example, a laser and digital correlator set-up. A variety of experimental devices have been reported, and among these are those which make use of dynamic light scattering. The resultant intensity fluctuation correlation function can be inverted to give the mean square displacement. However, there is no need to go to such trouble: results can be obtained relatively directly, although smart software needs to be written and implemented, using the technique of video particle tracking (VPT).

This, as the name suggests, images the probe particles using a video camera (Corrigan and Donald, 2009a, 2009b). Digital image sets (frames) of the sample are captured, and the positions of embedded particles calculated frame by frame by the software, to form particle trajectories and mean square displacements. This approach allows large numbers of particles to be tracked simultaneously, while retaining information on their individual positions. Consequently, such experiments are relatively fast to perform and the local microstructure of the sample can be investigated.

It is appropriate to consider both advantages and disadvantages of the techniques. Advantages are that the technique, as already pointed out, is potentially non-perturbing and that the apparatus can be assembled from commercial scattering kit and so is relatively cheap to construct. This is particularly true for the video particle tracking method. Also, the sample volume can be extremely small, as little as 20 μL, and for valuable samples that is a major benefit; potentially, measurements can be made over a whole range of length scales so, for example, analysis of a gel close to the sample edge can be made. There is also a good deal of work on both whole-cell assemblies and cellular components (e.g. actin; see Chapter 9) where the technique has been able to probe, and where classical rheology cannot be used.

The disadvantages of the technique are not to be discounted. For a gelling system, it cannot be assumed that the local strain close to the bead is small, and at higher modulus the bead size may not be negligible relative to the mesh size of the network. These two factors can even lead to the tunnelling effect seen, for example, in earlier magnetic microsphere experiments. Also, the nominal frequency is often quite high, in the kHz region for some scattering methods, a thus a long way from the required zero-frequency limit. Overall, the technique is still developing, and remains in the hands of the developers rather than being available for general applications.

Attempts have been made to compare these methods and classical rheology, but so far the results are not clear. Some papers show results in agreement (Dasgupta and Weitz, 2005), while others do not (Valentine *et al.*, 2004). Perhaps most significant of all is that no study of which we are aware has made both classical rheological and PTM measurements on a physical gel during and through the sol–gel transition.

2.6 Role of numerical simulations

The relationship between the phase diagram and the physical properties of colloidal solutions is an area where experimental findings can be addressed by numerical simulations. Simulations may provide strong support for experimentalists in understanding the origins of the effects they observe, and may also help to predict what should happen when experimental conditions are changed. For example, for particulate gels, one fundamental approach to understanding the gel formation mechanism is provided by the analogy between a colloidal suspension and a molecular liquid such as argon. Typically, colloidal particles are arranged in a gas-like state at low concentrations and in a liquid-like, glassy or crystal-like state at much higher concentrations. Interactions between colloidal particles can often be treated as the sum of pairwise additive potentials, as in a molecular liquid. When the attraction between colloidal particles is strong enough and has a long enough range, this attraction leads to a vapour–liquid phase transition as predicted in the well-known van der Waals equation of state, familiar in thermodynamics.

Numerical simulations on colloidal suspensions are often performed using molecular dynamics (Allen and Tildesley, 1989) and more recently using the Brownian dynamics (BD) model invented in 1975 by Ermak (1975). In this model the particles are treated as hydrodynamically isolated, interacting with solvent via a simple friction coefficient and Brownian forces only. Hydrodynamic forces between particles are neglected, an assumption that should be valid at least in the absence of external field (Ansell and Dickinson, 1987). This simplified description permits relatively large systems to be followed for long enough times to observe any structural changes.

The interaction between two model colloidal particles can be represented by the generalized Lennard–Jones (LJ) *m-n* potential:

$$U(r) = A(m,n)\,\varepsilon\left[\left(\frac{\sigma}{r}\right)^m - \left(\frac{\sigma}{r}\right)^n\right], \tag{2.33}$$

where r is the inter-particle centre-to-centre separation, σ is the colloidal particle diameter and $A(m, n)$ is a normalization factor which ensures that ε is the potential well depth.

Varying the exponents m and n generates various interaction ranges. The minimum of the potential occurs at a separation $r = 2^{1/n}\sigma$, displaced to smaller separations as n or m increases. Lodge and Heyes (1997, 1998, 1999a) performed simulations using 12:6, 24:12 and 36:18 potentials ($m = 2n$) and $A = 4$. Along this series, the attractive part of the potential becomes shorter-range. The thermodynamic properties are given in units of σ for length and ε for energy. The reduced temperature T^* is given by

$$T^* = \frac{k_B T}{\varepsilon}. \tag{2.34}$$

The units of time are a^2/D_0, where $a = \sigma/2$ is the particle radius, D_0 is the self-diffusion coefficient of the colloidal particle in very dilute suspensions and ζ_0 is the infinite dilution friction coefficient:

$$D_0 = \frac{k_B T}{\zeta_0} \tag{2.35}$$

In reduced units, $\zeta_0^* = 1$ and then $D_0^* = T^*$.

A typical phase diagram for a colloidal system of attractive particles is given in Figure 2.19, plotted as reduced temperature versus density ρ for the vapour–liquid two-phase system (or versus particle volume fraction in the case of a liquid–liquid phase separation in a suspension). The system can be quenched into various domains within the two-phase region (phase separated region) with a binodal boundary line. Inside the binodal line there is a spinodal line marking the transition from metastable to unstable regions.

Inside the spinodal boundary, it is possible to form a rigid but mechanically unstable network which has a finite yield stress (see below): a ‘gel line’ is drawn on Figure 2.19, a line which marks the boundary of states, to the right of which they display a gel-like structure. These systems are referred to as ‘transient gels’ because the colloidal particles do not have permanent bonds and therefore the structure tends to collapse.

Experimentally, phase separation can be induced in a number of ways, for example by adding non-absorbing polymers (depletion flocculation), changing the pH or salt concentration for aqueous systems or changing the temperature. In real experiments, adding polymers or salts is widely used since the quench is essentially instantaneous, so there is minimal change within the system during the quench. In computer simulations, changing the temperature is often the most convenient way of initiating phase separation. Starting from an equilibrium configuration, above the critical point, the system is quenched in a single time step. The change of temperature enters the model through the Brownian force term. Experimental and simulation procedures are equivalent, as T^*, proportional to the

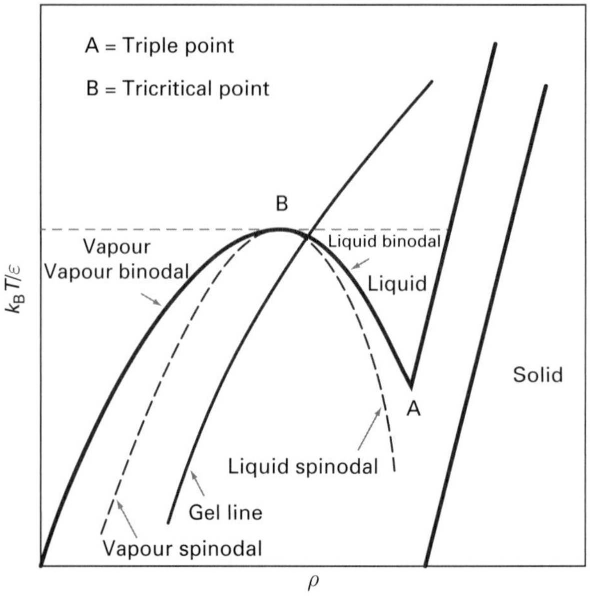

Figure 2.19 Typical phase diagram for a colloidal system of attractive particles. The position of the transient gel line is somewhat imprecise because there is a lifetime associated with persistence of the equilibrium modulus G_{eq}. Therefore, the exact location of the gel line will depend on the value given to this lifetime (Lodge and Heyes, 1999b). Reproduced by permission of the PCCP Owner Societies.

ratio of temperature T to interaction potential well depth ε, is the key parameter that determines phase separation.

Shortly after the temperature quench, the development of local structure can be seen in instantaneous (snapshot) configurations (Lodge and Heyes, 1997). The snapshots shown in Figure 2.20 represent colloidal suspensions after quenching at a time $t = 82a^2/D_0$, a long elapsed time in the simulation, for each of three interaction potentials, at the same reduced temperature $T^* = 0.3$ and the same volume fraction of particles $\phi = 0.1$:

$$\phi = \frac{\pi}{6} \frac{N\sigma^3}{V}. \tag{2.36}$$

Here N is the total number of particles and V is the volume of the cubic simulation box. In Figure 2.20 the spheres represent 1σ and are drawn to scale relative to the simulation box (Lodge and Heyes, 1999a). At this low temperature, all of the systems displayed phase separation, eventually forming dense aggregates that spanned the simulation box.

With the LJ 12:6 potential, the particles start clustering soon after the quench and, at a later stage as shown in Figure 2.20a, they have collapsed into dense structures. As the attractive part of the potential becomes more short-range (24:12 potential, Figure 2.20b) the aggregates adopt a more diffuse morphology. The particles still cluster, but they no longer collapse into a dense mass. For the system with the 36:18 potential (Figure 2.20c)

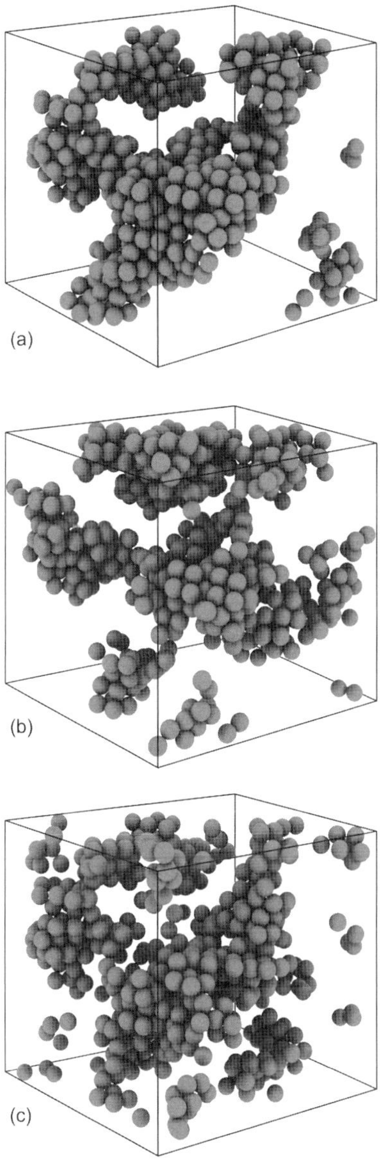

Figure 2.20 Snapshots showing the structure in a system of 864 particles at $\phi = 0.1$, $T^* = 0.3$ for the interaction potentials (a) 12:6, (b) 24:12 and (c) 36:18. Snapshots are taken at the final simulation step at time $t = 82a^2/D_0$. The spheres represent 1σ and are drawn to scale in relation to the simulation box. Adapted from Lodge and Heyes (1999a). Reproduced with permission of the American Institute of Physics for The Society of Rheology.

the particles again cluster, but the clusters are very diffuse and there is little order beyond the range 2σ, as expected with a short attraction range between particles.

Quantitative analysis of the radial distribution function reveals that for the LJ 12:6 potential the structure remains amorphous, in a glassy state (liquid-like structure) for a long period of time, while the particles interacting over a shorter range (24:12 or 36:18 potential) show the local development of a crystalline FCC structure, but at a larger scale they form a tenuous network. In other words, for the same reduced temperature the LJ 12:6 particles are 'kinetically trapped' in an amorphous glassy structure.

At lower volume fractions, $\phi < 0.1$, the particles assemble into clusters with convex surfaces which form strand-like structures through the box. At higher volume fractions, $\phi > 0.1$, suspensions evolve into structures with many more concave surfaces. In systems quenched deep into the two-phase region, phase separation proceeds via a spinodal decomposition mechanism: a peak appears in the structure factor at low wave vectors that displays an immediate increase in intensity and a decrease in position, indicating the growing importance of longer length scales in the system.

It is clear that the structure of the aggregated phase is sensitive to small variations in parameters such as interaction potentials, quench depth and volume fraction. In the LJ system, cluster growth is relatively simple. On the time scale of the simulation, any rearrangement towards a crystalline ordered phase is retarded for the 12:6 potential. For short-range interaction potentials, particles adopt a crystalline structure more readily, an observation which could be relevant in protein crystallization (Bonneté *et al.*, 1996; Wolde and Frenkel, 1997).

2.6.1 Fractal dimensions

In irreversible aggregation of spherical particles, three spatial scales of structure are expected (Dickinson, 1987): (i) short-range order from packing and excluded-volume effects, (ii) medium-range disorder associated with the ramified structure of the aggregating clusters, and (iii) long-range uniformity for a homogeneous material. The pair distribution function $g(r)$ measures the probability of finding a particle at a distance r from another particle. In the first region (shortest distances), strongly damped liquid-like oscillations of $g(r)$ are expected out to a few particle diameters ($r < r_0$), while in a second region a fractal scaling regime is found with

$$g(r) \approx \left(\frac{r}{\xi}\right)^{d_f - 3} \qquad r_0 \leq r \leq \xi. \tag{2.37}$$

A non-fractal, uniform structure with ($g(r) = 1$) appears beyond some characteristic correlation length ξ. When a fractal structure appears, a power-law region in $g(r)$ is observed for distances over which the fractal structure holds, with d_f the fractal dimension in (2.37).

For the cases investigated by Lodge and Heyes (1999a, 1999b), the fractal dimension of the clusters was almost always found close to $d_f \approx 3$ ($g(r) = 1$). For the systems investigated, and on a relatively short time scale after the quench, other authors have found lower fractal dimensions, down to $d_f \approx 1.4$. Their fractal dimensions nevertheless increased with time, persisting only for very slowly separating systems, i.e. at high temperatures. It is clear that fractal dimensions are mostly associated with particles forming permanent bonds, and then the fractal dimension depends on sticking probability: for high sticking probability, in diffusion limited cluster aggregation (DLCA) the fractal dimension of the clusters is $d_f \approx 1.7$ (Weitz and Oliveria, 1984), while for low bonding probability, in reaction limited cluster aggregation (RLCA) the fractal dimension is higher, $d_f \approx 2.3$ (Weitz *et al.*, 1985). For weak reversible interactions, there is a substantial restructuring and condensation from the earlier stages of the phase separation, and therefore the structure approaches a uniform structure with increasing simulation times.

The effect of the strength of inter-particle interactions on the microstructure of particle gels generated by Brownian dynamics simulation was also examined by Bijsterbosch and Bos (1995). The authors simulated various potentials other than LJ, for example where particles can form flexible linkages when they approach within a certain bonding distance $r < d_{bond}$ and where once the bond is formed it is permanent. In addition, pairs of particle were assumed to interact with distance-dependent forces which could be either repulsive ($r < d_{max}$) or attractive ($r > d_{max}$). The authors observed that, under conditions of thermodynamic phase separation, with an LJ potential, the clusters have time-dependent structures, with a process of short-range densification with formation of coarser blob-like structures and large voids, as previously described by Lodge and Heyes. On the other hand, with the second, flexible linkage interaction potential, they clearly observed a fractal structure with a low dimensionality ($d_f = 1.9$) of clusters within an intermediate range of r, typically less than $r < 10a$, with a the particle radius.

However, for such systems the value of d_f by itself does not give a full description of the microstructure of the aggregated phase. It is interesting to observe how different the two snapshots simulated by Bijsterbosch and Bos (1995) appear (Figures 2.21 and 2.22); these exhibit the same fractal dimension in the same intermediate range but were produced with different interaction potentials. The very different appearance of the two network structures in terms of porosity and strand thickness indicates the danger of adopting fractal dimensionality uncritically as an appropriate single parameter to describe the structure of 'simulated particle gels'.

The fractal dimension d_f of the intermediate regime is only one aspect of the gel structure. Also important is a scaling prefactor n_0 that Bijsterbosch and Bos introduced explicitly for the average number $n(r)$ of particles within a range r of another particle:

$$n(r) = n_0 \left(\frac{r}{r_0}\right)^{d_f}. \tag{2.38}$$

In this formulation, r_0 may be an arbitrary length. If r_0 is taken as the primary particle radius, then $n_0 = 1$. The value of n_0 is a measure of the average number of particles in the primary clusters from which the fractal scaling regime is built: a large value of n_0 indicates a coarse microstructure; low values of n_0 (possibly even less than unity) imply open structures. The rate of restructuring can be measured by the rate of increase of n_0 and depends on the attractive interactions. For instance in Figure 2.21, $n_0 = 0.5$ and in Figure 2.22, $n_0 = 2$.

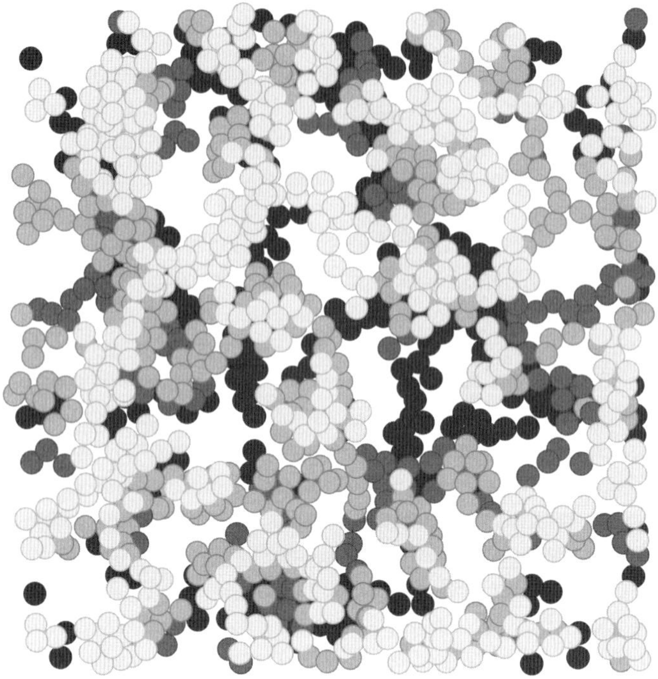

Figure 2.21 Snapshot of the three-dimensional network structure produced from simulated aggregation of particles with $\phi = 0.05$, no interactions. The darkest particles lie furthest away from the observer. Bijsterbosch and Bos (1995) introduced a scaling factor n_0 for the average number $n(r)$ of particles within a range r of another particle. Here $n_0 = 0.5$. Particles can form irreversible flexible linkages with other particles when they approach to within a certain bonding distance. Assuming a bond probability of 1, at low volume fractions the simulation produces a fine-stranded network structure having a fine-pore distribution. Reproduced with permission of the Royal Society of Chemistry.

Ultimately, the overall microstructure is likely to be dependent on the volume fraction of particles and on the detailed nature of the particle–particle interactions during and after phase separation. In such simulations, when the volume fraction of systems reaches $\phi = 0.3$ or above, no fractal character can be found.

2.6.2 Gelling or non-gelling systems?

The method developed by Lodge and Heyes allows the computation of linear rheological properties without any assumption about the microstructure and without applying a shear strain to the system but, instead, attributing a purely thermodynamic origin to each instantaneous configuration (Lodge and Heyes, 1999a). In this model, rheological properties, purely thermodynamic in origin, are computed from the stress tensor elements. The stress relaxation function, which contains a complete description of the linear response viscoelasticity ($G'(\omega)$ and $G''(\omega)$), relaxes to zero in the stable single phase. In the phase separated system, it does not relax to zero but reaches an apparent plateau within a certain time interval, of the order of several tens of a^2/D_0, where the structure does not change appreciably, so suggesting the presence of an elastic network. The residual stress can be

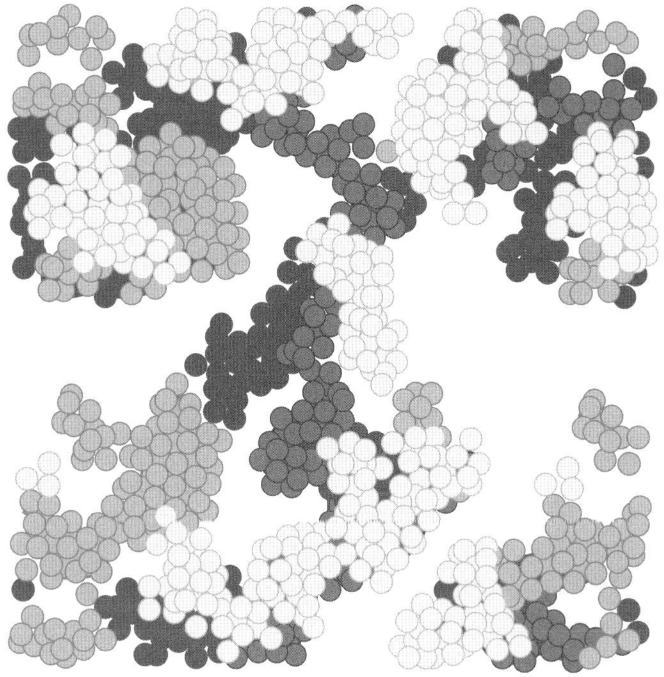

Figure 2.22 Snapshot of the three-dimensional network structure produced from simulated aggregation of particles with $\phi = 0.11$, $\varepsilon = 4k_BT$, $n_0 = 2$. The pair interaction potential has a shorter finite attractive range than the LJ potential. A coarse, phase separated structure is observed. Adapted from Bijsterbosch and Bos (1995) with permission of the Royal Society of Chemistry.

attributed to a percolating network. This plateau is referred as an 'equilibrium modulus' G_{eq}, on intermediate time scales, even though there are no formal links.

G_{eq} increases in time after the quench, and depends on the interaction potential. As seen in Figure 2.23, with $\phi = 0.2$, a transition appears between temperatures $T^* = 0.5$ and 0.7, which suggests the formation of a network and determines the gel line in Figure 2.19.

Obviously, the definition of the gel line from these simulations is somewhat imprecise because there is a lifetime associated with the persistence of the equilibrium modulus. Gel-like behaviour was observed in the two-phase region of LJ potentials, with comparatively large G_{eq} values measured during relatively short lifetimes (of the order of a^2/D_0). With the other potentials, the network is more diffuse (see Figure 2.20c). For the shortest-ranged (36:18) interaction, any potential phase separation was largely arrested on long length scales for a significant part of the simulation times. This system displayed the weakest of all rheological gel-like features.

A finite equilibrium modulus G_{eq} is associated with a gelation threshold, the formation of a percolating network (Chapter 3) with distances between the particles close to the minimum in the pair potential. In this situation the 'structural' and 'mechanical' transitions coincide. The value of G_{eq} decreases with increasing temperature for the 12:6 potential. If temperature is maintained constant, G_{eq} also increases with volume fraction.

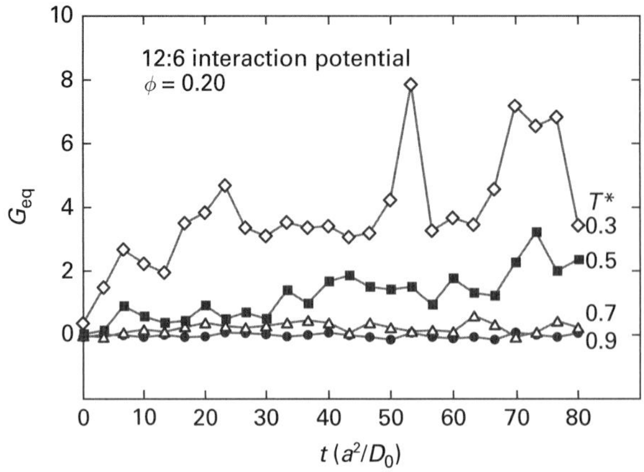

Figure 2.23 Time dependence of G_{eq} for the 12:6 interaction potential with $\phi = 0.2$ and for a series of temperatures. A transition appears between temperatures $T^* = 0.5$ and 0.7, which suggests the formation of a network, and determines the gel line in Figure 2.19. From Lodge and Heyes (1999a), reproduced with permission of the American Institute of Physics for The Society of Rheology.

For the higher volume fractions, which form fairly large dense clusters, the magnitude of G_{eq} eventually increases to a similar value for all volume fractions.

2.6.3 Improvements of the interaction potentials

In more recent work Del Gado and Kob (2005, 2007, 2008) investigated the role of directionally effective interactions in the formation of colloidal gels at very low volume fractions. These interactions could account for the low coordination numbers and the local rigidity reported in experimental studies. For the first time, the use of more complicated interactions allowed formation of open – i.e. ramified and fractal – structures that are also stable at relatively high temperatures and low volume fractions; this is definite progress compared with previous models. In particular, the case of a relatively simple network structure, with chains connected by bridging points, was studied, allowing analysis of the role of the various parts of the structure in the dynamics.

Using molecular dynamics simulations, Del Gado and Kob (2008) introduced an effective potential that is the sum of two- and three-body terms. The two-body potential is itself the sum of a generalized LJ 18:16 potential and a term that depends on the relative orientation of the particles. This new approach gives a more realistic view of the interaction potentials for colloids that are seldom uniform and smooth. Then, when they come into contact, their 'roughness' could produce rigid links between particles. Accordingly, each particle was decorated with 12 points in such a way that interactions became more favourable when the centre of the particles was aligned preferentially in the direction of the points. In addition, an explicit short-range, three-body interaction was included. Such a potential introduces a mechanism that favours the formation of an open network relative to phase separation induced by purely radial interactions. As a consequence, there is an effective functionality for the particles, which is a result of the competition between entropic and internal energy contributions in the network formation.

Overall, then, it appears that colloidal suspensions are expected to exhibit phase diagrams that are much more complex than those of molecular liquids. Forces between colloids in solution have all sorts of forms that can be tuned (see Frenkel, 2002) by choosing an appropriate combination of solvent, solutes, additives etc. Experimental and simulation approaches exhibit the same effects.

2.7 Conclusions

In this chapter a range of experimental tools has been introduced to study the structure and rheological properties of physical polymer gels. The 'toolbox' also contains new and powerful techniques based on numerical simulations, which bring very useful insights to fundamental aspects of network formation, starting from molecular interaction potentials well known in colloidal science. It is important to realize that the rich behaviours found in experiments on polymer gels, and which we report in this book, can be tuned with respect to the range, strength, coordination etc. of effective interactions, both in numerical simulations and, practically, with a large choice of available synthetic and biological polymers and solvent compositions.

References

Allen, M. P., Tildesley, D. J., 1989. *Computer Simulation of Liquids*. Oxford University Press, New York.

Ansell, G. C., Dickinson, E., 1987. *Faraday Discuss.* **83**, 167–177.

Arnott, S., Fulmer, A., Scott, W. E. *et al.*, 1974. *J. Mol. Biol.* **90**, 269–284.

Ayache, J., Beaunier, L., Boumendil, J., Ehret, G., Laub, D., 2010. *Sample Preparation Handbook for Transmission Electron Microscopy*. Springer, New York.

Barnes, H. A., Nguyen, Q. D., 2001. *J. Non-Newtonian Fluid Mech.* **98**, 1–14.

Benedetti, E. L., Favard, P. (eds), 1973. *Freeze Etching: Techniques and Applications*. Société Française de Microscopie Electronique, Paris.

Bijsterbosch, B. H., Bos, M. T. A., 1995. *Faraday Discuss.* **101**, 51–64.

Blandamer, M. J., Cullis, P. M., Engberts, J. B. F. N., 1998. *J. Chem. Soc., Faraday Trans.* **94**, 2261–2267.

Bonneté, F., Vidal, O., Robert, M. C., Tardieu, A., 1996. *J. Cryst. Growth* **168**, 185–191.

Brown, M. E., 2001. *Introduction to Thermal Analysis*. Kluwer Academic Publishers, Dordrecht.

Bruylants, G., Bartik, K., 2005. *Chim. Nouv.* **23**, 9–14.

Burchard, W., 1983. *Adv. Polym. Sci.* **48**, 1–119.

Burchard, W., 1994. Light scattering techniques. In Ross-Murphy, S. B. (ed), *Physical Techniques for the Study of Food Biopolymers*. Blackie Academic and Professional, Glasgow, pp. 151–213.

Clark, A. H., Ross-Murphy, S. B., 1987. *Adv. Polym. Sci.* **83**, 57–192.

Corrigan, A. M., Donald, A. M., 2009a. *Eur. Phys. J. E* **28**, 457–462.

Corrigan, A. M., Donald, A. M., 2009b. *Langmuir* **25**, 8599–8605.

Dasgupta, B. R., Weitz, D. A., 2005. *Phys. Rev. E* **71**, 0215041–9.

de Gennes, P.-G., 1979. *Scaling Concepts in Polymer Physics*. Cornell University Press, Ithaca, NY.

Del Gado, E. D., Kob, W., 2005. *Europhys. Lett.* **72**, 1032–1038.

Del Gado, E. D., Kob, W., 2007. *Phys. Rev. Lett.* **98**, 028303.

Del Gado, E. D., Kob, W., 2008. *J. Non-Newtonian Fluid Mech.* **149**, 28–33.

Dickinson, E., 1987. *J. Colloid Interface Sci.* **118**, 286–289.

Eaton, P., West, P., 2010. *Atomic Force Microscopy*. Oxford University Press, Oxford.

Ermak, D. L., 1975. *J. Chem. Phys.* **62**, 4189.

Fang, Y., Al-Assaf, S., Phillips, G. O. *et al.*, 2007. *J. Phys. Chem.* **B111**, 2456–2462.

Favard, P., Lechaire, J. P., Maillard, M. *et al.*, 1989. *Biol. Cell* **67**, 301–207.

Ferry, J. D., 1980. *Viscoelastic Properties of Polymers*. Wiley Interscience, New York.

Foord, S. A., Atkins, E. D. T., 1989. *Biopolymers* **28**, 1345.

Freire, E., Mayorga, O. L., Straume, M., 1990. *Anal. Chem.* **62**, 950A–959A.

Frenkel, D., 2002. *Science* **296**, 65–66.

Garidel, P., Hildebrand, A., 2005. *J. Therm. Anal. Calorim.* **82**, 483–489.

Glatter, O., Kratky, O., 1982. *Small Angle X-ray Scattering*. Academic Press, London.

Goodwin, J. W., Hughes, R. W., 2000. *Rheology for Chemists*. Royal Society of Chemistry, Cambridge, UK.

Gosal, W. S., Clark, A. H., Pudney, P. D. A., Ross-Murphy, S. B., 2002. *Langmuir* **18**, 7174–7181.

Hatakeyama, T., Quinn, F. X., 1999. *Thermal Analysis: Fundamentals and Applications to Polymer Science*, 2nd edn. John Wiley & Sons, Chichester.

Hermansson, A. M., Langton, M., 1994. Electron microscopy. In Ross-Murphy, S. B. (ed), *Physical Techniques for the Study of Food Biopolymers*. Blackie Academic and Professional, Glasgow, pp. 277–341.

Höhne, G. W. H., Hemminger, W. F., Flammersheim, H.-J., 2003. *Differential Scanning Calorimetry*, 2nd edn. Springer, London.

Ikeda, S., Morris, V. J., Nishinari, K., 2001. *Biomacromolecules* **2**, 1331–1337.

Kavanagh, G. M., Ross-Murphy, S. B., 1998. *Prog. Polym. Sci.* **23**, 533–562.

Lemstra, P. J., Keller, A., 1978. *J. Polym. Sci., Polym. Phys. Ed.* **16**, 1507.

Lodge, J. F. M., Heyes, D. M., 1997. *J. Chem. Soc, Faraday Trans.* **93**, 437–448.

Lodge, J. F. M., Heyes, D. M., 1998. *J. Chem. Phys.* **109**, 7567.

Lodge, J. F. M., Heyes, D. M., 1999a. *J. Rheol.* **43**, 219–244.

Lodge, J. F. M., Heyes, D. M., 1999b. *Phys. Chem. Chem. Phys.* **1**, 2119–2130.

Mason, T. G., Weitz, D. A., 1995. *Phys. Rev. Lett.* **74**, 1250–1253.

Matsuo, T., 2004. *Comprehensive Handbook of Calorimetry and Thermal Analysis*. John Wiley & Sons, Chichester.

McEvoy, H., Ross-Murphy, S. B., Clark, A. H., 1985. *Polymer* **26**, 1483–1492.

Morris, V. J., 2009. Single molecule techniques: Atomic force microscopy and optical tweezers. In *Modern Biopolymer Science*. Elsevier, San Diego, pp. 365–398.

Nakamura, K., Shinoda, E., Tokita, M., 2001. *Food Hydrocolloids* **15**, 247–252.

O'Brien, R., Ladbury, J. E., Chowdry, B. Z., 2001. Isothermal titration calorimetry. In Harding, S. E., Chowdhry, B. Z. (eds), *Protein-Ligand Interactions: Hydrodynamics and Calorimetry*. Oxford University Press, Oxford, pp. 263–286.

Paul, D. R., 1967. *J. Appl. Polym. Sci.* **11**, 439–455.

Paula, S., Sues, W., Tuchtenhagen, J., Blume, A., 1995. *J. Phys. Chem.* **99**, 11742–11751.

Raju, B. B., Winnik, F. M., Morishima, Y., 2001. *Langmuir* **17**, 4416–4421.

Richardson, R. K., Goycoolea, F. M., 1994. *Carbohydr. Polym.* **24**, 223–225.

Richardson, R. K., Ross-Murphy, S. B., 1987. *Int. J. Biol. Macromol.* **9**, 257–264.

Rodd, A. B., Dunstan, D. E., Ross-Murphy, S. B., Boger, D. V., 2001. *Rheol. Acta* **40**, 23–29.
Roe, R.-J., 2000. *Methods of X-ray and Neutron Scattering in Polymer Science*. Oxford University Press, Oxford.
Ross-Murphy, S. B., 1994. Rheological methods. In Ross-Murphy, S. B. (ed), *Physical Techniques for the Study of Food Biopolymers*. Blackie Academic and Professional, Glasgow, pp. 342–392.
Ross-Murphy, S. B., 2005. *J. Macromol. Sci., PartB: Phys.* **44**, 1007–1019.
Ross-Murphy, S. B., Shatwell, K. P., 1993. *Biorheology* **30**, 217–227.
Round, A. N., Rigby, N. M., MacDougall, A. J., Morris, V. J., 2010. *Carbohydr. Res.* **345**, 487–497.
Smith, T. L., 1963. *J. Polym. Sci., Part A: Gen. Pap.* **1**, 3597–3615.
Tanaka, F., Edwards, S. F., 1992a. *J. Non-Newtonian Fluid Mech.* **43**, 247–271.
Tanaka, F., Edwards, S. F., 1992b. *J. Non-Newtonian Fluid Mech.* **43**, 273–288.
Tanaka, F., Edwards, S. F., 1992c. *J. Non-Newtonian Fluid Mech.* **43**, 289–309.
te Nijenhuis, K., 1997. *Adv. Polym. Sci.* **130**, 1–252.
Tobitani, A., Ross-Murphy, S. B., 1997. *Macromolecules* **30**, 4845–4854.
Valentine, M. T., Perlman, Z. E., Gardel, M. L. *et al.*, 2004. *Biophys. J.* 86, 4004–4014.
van Holde, K. E., Johnson, W. C., Shing Ho, P., 2006. *Principles of Physical Biochemistry.* Pearson, Upper Saddle River, NJ.
Waigh, T. A., 2005. *Rep. Prog. Phys.* **68**, 685–742.
Weitz, D. A., Huang, J. S., Lin, M. Y., Sung, J., 1985. *Phys. Rev. Lett.* **54**, 1416–1419.
Weitz, D. A., Oliveria, M., 1984. *Phys. Rev. Lett.* **52**, 1433–1436.
Wolde, P. R., Frenkel, D., 1997. *Science* **277**, 1975–1978.

3 The sol–gel transition

Physical gels exhibit a wide range of properties which determine their use in many applications. New materials are synthesized for particular industrial requirements and/or for biomedical applications. Gel characterization requires a large palette of instruments, techniques and methods, as presented in Chapter 2. Because of the large number of systems already known and because new ones are invented every day, there is a need to understand and predict the behaviour of these systems. It would be ideal if we were able to understand and map the relations between the (macro)molecular composition of the initial system and the functional properties (mechanical, swelling, drug delivery and so on) in the final gelled state.

However, such an objective is very ambitious and, for the moment, not realizable. Consequently, we shall concentrate on a particular aspect that still occupies many chemists and physicists. This is the very special limit when the solution switches from a liquid state, where polymers are free to diffuse, to a solid-like medium, where the polymers entrap the solvent. The transition from one state to the other may happen quite suddenly. Furthermore, despite the large number of theoretical and experimental publications dealing with the sol–gel transition, there is still a debate even on the best way to determine the 'gel-point'.

In other words, can there really be a unique definition of the transition from sol to gel? Do we have a single method for all systems or could everyone choose the most convenient way to describe their own particular system (Li and Aoki, 1997)? In this chapter we shall try to present the various ways of defining the gel point from theoretical and experimental viewpoints, and put particular stress on both the problems and the potential resolutions.

3.1 Flory–Stockmayer ('classical') theory

It was Paul Flory, in work begun in the Carothers group at DuPont in the late 1930s, who first tried to develop a quantitative theory for the linear end-linking reaction of small molecular mass species, a typical exemplar being the reaction of adipic acid and hexamethylenediamine to produce Nylon 66. From this he was able to derive equations relating number and weight average molecular mass to the 'degree of conversion'.

For the linear reaction of a single species, for example the self-condensation of a diol, the weight average degree of polymerization DP_w is given by

$$DP_w = \frac{1+p}{1-p}. \tag{3.1}$$

Here p is the proportion of chemically reacted – in this case diol – groups.

The derivation of this equation explained how and why it was extremely difficult to obtain very high molecular mass product for such a reaction, since even when $p = 0.999$ – an arguably unfeasible amount – the degree of polymerization is only *c.*2000, so the weight average molecular mass M_w is typically of order 100 000 g mol^{-1}.

The expansion of this approach to non-linear (branched and cross-linked) systems has much more relevance to the present volume. Work by Flory (1941, 1942) and extensions by Stockmayer (1943) identified that p has still more significance for these systems. Unlike the linear systems, for non-linear reactions, when there are more than two potential reactive groups, very high molecular masses are quite easily achieved, and for a simple symmetric triol system, when p becomes greater than 0.5, the Flory–Stockmayer (FS) theory predicts that DP_w will diverge to infinity.

The non-linear analogue of (3.1) is

$$DP_w = \frac{1+p}{1-p(f-1)}. \tag{3.2}$$

In this case, f is the functionality – in this case simply the number, 3, of reactant groups on the triol. The denominator of this equation becomes zero when the product $p(f-1) = 1$, and this marks the so-called Flory–Stockmayer gel point, so that p_c, the critical degree of conversion, is given by

$$p_c = \frac{1}{(f-1)}. \tag{3.3}$$

Of course substituting $f = 2$ in (3.2) simply returns (3.1).

The applicability of this approach to vulcanization systems is also of interest. Flory deduced that if a preformed chain had n_0 possible cross-linking sites distributed along the polymer chain backbone, then trivially $f \approx n_0$. A consequence of this approach is that a fully cross-linked elastomer can be formed if, on average, each chain is linked to at least one other. Since n is typically quite large – 1000 or more – in the absence of so-called wastage reactions, only 1 in 1000 units in the chain needs to have been cross-linked to pass the gel point.

Early efforts to test this theory used f-functional ('polycondensation') end-linking of esters and alcohols, for example the reaction of a multifunctional f_m-ol with, say, a multi-functional f_n-acid, e.g. pentaerythritol ($f_m = 4$) and the dibasic adipic acid ($f_n = 2$), as used in a series of experiments by Stockmayer and Weil (Flory, 1953).

The prime purpose of this work was to establish whether or not the degree of conversion p_c at the gel point was as given by (3.3). These and many other systems were used, and in most cases it was found that measured values of p_c were greater than the value predicted, but it was soon realized that the main reason for this was that, as well as intermolecular bonds being formed, a small proportion of intramolecular reaction took place, forming cycles. Consequently later work concentrated on other systems, including those from

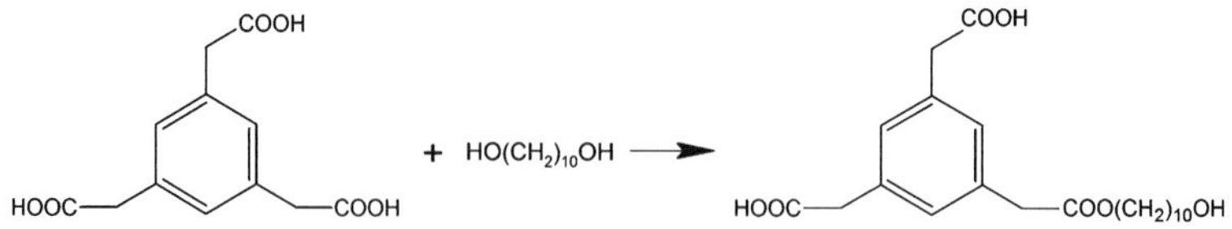

Figure 3.1 Step-addition (polycondensation) of BTA–DMG.

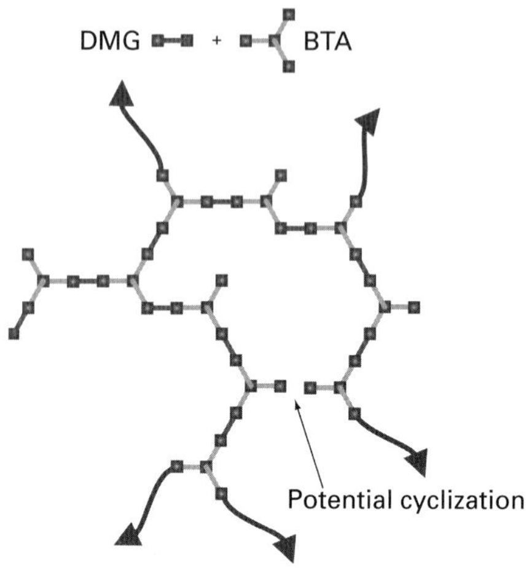

Figure 3.2 Graph-like representation of BTA–DMG step addition leading to gelation, showing a potential intramolecular linking site.

benzene triacetic acid ($f_n = 3$) and decamethylene glycol ($f_m = 2$) (BTA–DMG), which were designed specifically as a model polyester system by separating the acid groups with a benzene ring and the diol reactants by an extended chain of 10 $–CH_2–$ units. This system has been studied by a number of groups.

Figure 3.1 shows the structure and chemistry of the BTA–DMG system, and Figure 3.2 shows the same with the chemical detail reduced to ball and stick form. In this case, where $f_m = 2$ and $f_n = 3$, according to Flory's gel criterion, gelation should occur when $p_c = (2)^{-1/2} = 0.707$ (Stepto, 1998). To quote Flory himself, this system 'appears to be a particularly felicitous choice in this respect, the vitiation of the gel point being less than one percent' (Flory, 1974), so that measurements typically give $p_c \approx 0.71–0.72$ (Gordon *et al.*, 1975; Ross-Murphy, 1975). The increase is due, in this case, to the presence of *c*.1% of intramolecular polyester linkages – cycles – most of which can be eliminated by following the procedures introduced by Stockmayer and co-workers (Jacobson *et al.*, 1950). They measured p_c in the presence of various concentrations c of diluent, and then extrapolated to $1/c = 0$. In this way the value of p_c found for the BTA–DMG system was essentially the classical Flory value of 0.707.

3.2 Percolation model

At the end of the 1970s, physicists specializing in second-order phase transitions in the vicinity of the critical point of liquid–vapour mixtures or of magnetic transitions came up with an encouraging proposal that the sol–gel transition should obey universal 'laws' similar to those established for these other critical points, at least within the narrow range of cross-linking parameters governing network formation close to the transition. In thermodynamics, this parameter is temperature. In the case of gelation one has to determine the microscopic parameter which enables the network formation for any system.

One suggestion was to classify gelation as a fifth-order transition in the Ehrenfest sense (Gordon and Ross-Murphy, 1975). This hypothesis is, of course, purely theoretical, as it seems very unlikely that any experimental evidence could ever establish its truth or otherwise. In any case, subtle effects such as the non-equilibrium states observed in many physical gels, and reported throughout this book, make the real situation still more complicated.

In any second-order phase transition, the Landau theory defines an 'order parameter' which determines this approach to the critical temperature, and this parameter characterizes the onset of order at the phase transition. It is a measure of the degree of order in a system, with its extreme values being zero for total disorder, usually above the critical point, and non-zero and approaching unity for complete order. In such a second-order phase transition, the order parameter decreases continuously, following 'universal laws' when the measurements are performed very close to the critical temperature. In magnetic transitions, the spontaneous magnetization M is the 'order parameter'. At the critical temperature, called the Curie temperature T_C, when temperature decreases a spontaneous magnetization appears and increases:

$$\begin{aligned} &M \neq 0 \quad \text{for } T < T_C, \\ &M = 0 \quad \text{for } T > T_C, \\ &M \to 0 \quad \text{when } T \to T_C \text{ with } T < T_C. \end{aligned}$$

At the critical point of the vapour–liquid phase transition, the 'order parameter' is the difference between liquid and vapour densities, $\rho_{liq} - \rho_{vap}$, a difference which decreases smoothly towards zero, when temperature approaches the critical point. The universal laws are in general power laws such as, for instance, for spontaneous magnetization:

$$M \sim (T_C - T)^{\beta} \quad \text{for } T < T_C. \tag{3.4}$$

The exponent β is the same for spontaneous magnetization near the Curie point and near the vapour–liquid critical point, $\beta \approx 0.32$, as has been determined experimentally for such diverse fluids as helium, xenon and water (Stauffer *et al.*, 1982). This same exponent, when derived from van der Waals equations, however, gives a different value, close to $\beta \approx 0.50$.

The difference between these two values defines a different 'universality class'. The latter is that from the classical theory, where fluctuations are neglected near the phase transition, while the former is deduced from, for example, renormalization group theory, which takes into account the local fluctuations of parameters, such as magnetization, which actually diverge towards an infinite value at the approach of the critical point. The theory of critical phenomena (Stanley, 1998) states that the thermodynamic properties of

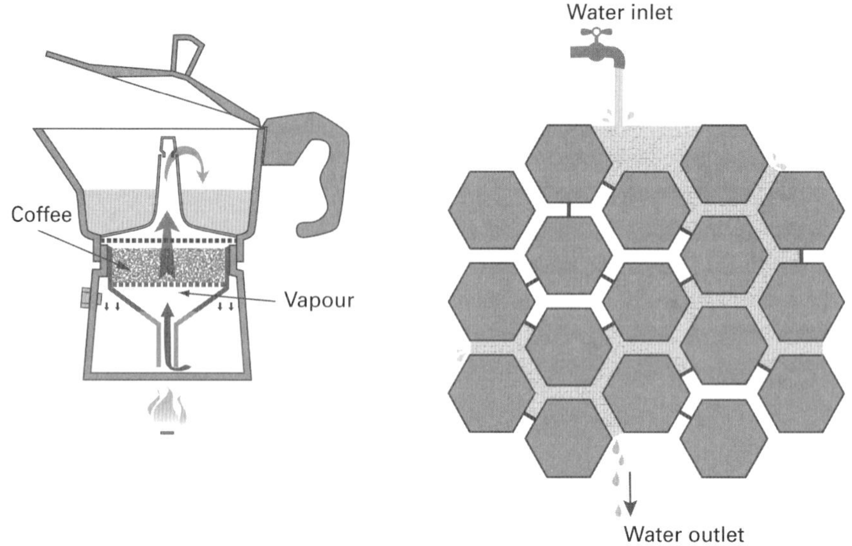

Figure 3.3 Percolation transition in a porous medium, some passages being blocked. The connectivity between the two sides of the porous medium is achieved by a single path via which water can flow.

a system near a phase transition depend only on a small number of features, such as dimensionality and symmetry, and are insensitive to the underlying microscopic properties of the system. This presentation may seem obscure for non-specialists in phase transitions, but we shall try to show that this is the origin of a new and unified description of gelation in terms of what is now known as percolation.

History tells us that the term 'percolation' was given by the mathematician J. M. Hammersley (1957) to a statistical geometric model which reminded him of the passage of a fluid through a network of channels in which of the channels, randomly distributed, were blocked. A sketch of the percolation transitions through a porous medium, in this case using a traditional Italian espresso pot, is shown in Figure 3.3.

The percolation path allows the liquid to flow through the medium. Another analogy is provided by the electric circuit represented by a square lattice of interconnections, where electric current circulates between two electrodes connected to a voltage source; see Figure 3.4. The interconnections are cut randomly. The current decreases until a critical fraction of bonds is cut and then the current vanishes. Of course, when the current stops circulating not all bonds have been cut, but there is no connected path between the two electrodes below a critical fraction of bonds. This critical fraction depends on the type of network. The transition from conducting to non-conducting network can be extremely sharp when the lattice is very large, i.e. when the L, the distance between the electrodes, is very large compared to a segment a; the mathematical limit is $L/a >> 1$.

In the limit of an infinitely large system, the critical value of the fraction of bonds is also sharply defined. For small lattices, of 'finite size', the threshold is distributed around the critical value. When the analogy between percolation and a physical phenomenon is

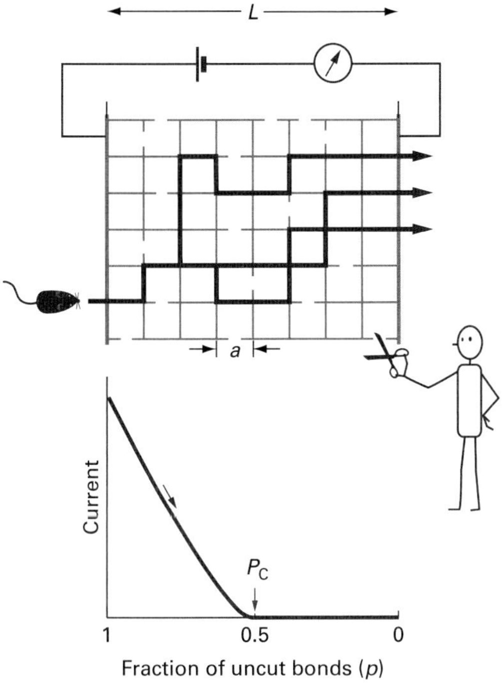

Figure 3.4 Electrical network randomly cut and the electric current corresponding to the fraction p of uncut bonds.

used, the thermodynamic limit is reached when $L/a \to \infty$, where L is the macroscopic size of the sample and a is an atomic or molecular scale.

Percolation deals with dramatic changes in the connectivity of systems of infinite size around this so-called percolation threshold. The electrical analogy in Figure 3.4 starts in a connected state and the process describes a progressive decrease in the number of bonds. A convenient way of comparing percolation and gelation is to imagine the process in the other direction, i.e. by increasing the number of connections beginning with a totally unconnected state. In Figure 3.5 one can see an example of the site percolation process on a square lattice.

The three frames show the lattice at three stages of the filling process. The sites of the lattice are initially empty, and they are occupied randomly by the heavy dots, which represent the filled sites. Nearest-neighbour filled sites are connected by heavy lines and belong to the same cluster. In the first frame, Figure 3.5a, one can see clusters containing between 1 and 4 filled sites.

The important variable is p, in this case the proportion of filled sites; s is the number of sites belonging to a cluster. In Figure 3.5a, $p=0.25$; in Figure 3.5b, $p=0.50$ and, while the fraction p has increased, the size of the clusters has also increased.

Let $n(s)$ be the normalized number of clusters of size s:

$$n(s) = \frac{\text{number of clusters of size } s}{\text{total number of sites}}. \tag{3.5}$$

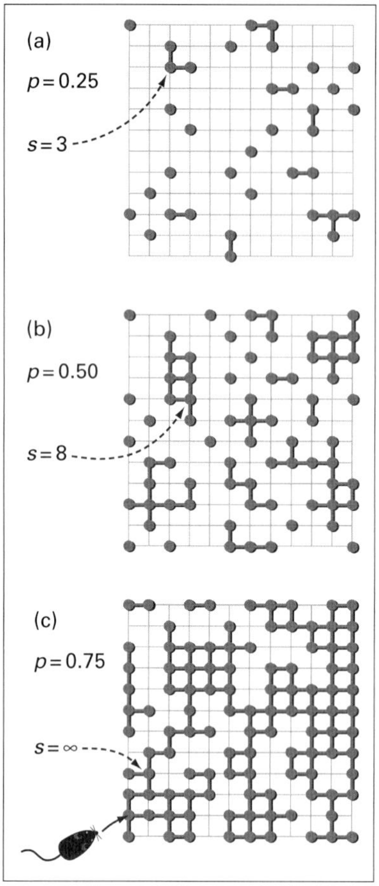

Figure 3.5 Site percolation on a square lattice for three values of the fraction p of filled sites.

Then, the number of sites in clusters of size s is proportional to $sn(s)$, and the site-weighted average cluster size s_{av} for each value of the parameter p is given by

$$s_{av} = \frac{\sum_{s=1}^{\infty} s^2\, n(s)}{\sum_{s=1}^{\infty} s\, n(s)}. \tag{3.6}$$

The sum in the denominator is equal to p. s_{av} starts at 1 for $p=0$ and increases as p increases. In Figure 3.5b, where $p=0.50$, there are fewer singlet sites and more clusters of larger size than in Figure 3.5a. In Figure 3.5c, $p=0.75$. Here there is a major difference compared to the two previous frames: there is now a very large cluster which connects the four edges of the frame. As the lattice grows infinitely large, $s_{av} \to \infty$.

Note that there are still independent clusters which do not belong to the infinite cluster, and also empty sites (at least when $p<1$). The infinite cluster is called the percolation

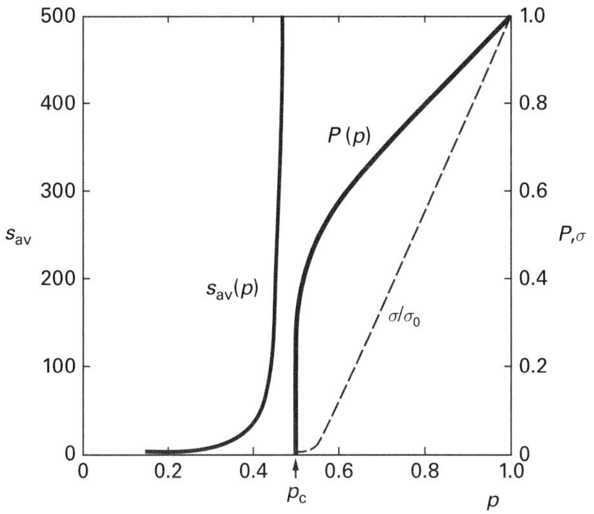

Figure 3.6 Behaviour of various quantities as a function of the fraction p of filled bonds, in a bond percolation simulation on a two-dimensional lattice.

cluster or the percolation path. The percolation path for an infinitely large cluster, as observed in Monte Carlo simulations, appears at the percolation threshold p_c. The percolation threshold for site percolation on a square lattice is found at $p_c = 0.59$. Below the threshold there is no percolating path; above the threshold there is one path. When the simulation on a square lattice takes into account the number of *bonds* created instead of the number of *sites* occupied, the so-called bond percolation on a square lattice has a threshold of $p_c = 0.5$.

Another important parameter is the probability that a site belongs to the infinite cluster when $p > p_c$. Referring to the *bond* percolation simulation on a square lattice, in Figure 3.6 one can see the probability $P(p)$ of a bond belonging to the infinite cluster. This probability is exactly zero below the threshold, and then increases very steeply at the threshold. As p approaches 1, $P(p) \approx p$. Other important parameters are seen in Figure 3.6: in particular, the increase in s_{av} as p approaches p_c from below. This simulation also represents the increase of the conductance of a random resistor network, σ/σ_o versus p, after the threshold, as seen in Figure 3.6. There is a small but increasing curvature at and just after p_c, and then a monotonic but essentially linear increase up to a value of 1 at full occupancy ($p = 1$).

It is important, at this stage, to clarify that the FS and later theories are not qualitatively different from those based on percolation – in fact the FS theory corresponds exactly to percolation on a so-called Bethe lattice or Cayley tree, a cycle-free lattice in which there is no overall connectivity until $p > p_c$ and then the increase is monotonic. The difference is simply that this lattice is totally ‘floppy’, i.e. it is not embedded in any dimensional space. Instead, the difference comes from quantitative predictions of the various critical exponents.

3.3 Percolation and phase transitions

The shape of the curve $P(p)$ is similar to one known in critical phase transitions for the 'order parameter', as explained above. At the threshold, the slope of $P(p)$ is infinite and the behaviour in the close vicinity of the threshold, when $|p - p_c| \rightarrow 0$, is controlled by a set of so-called 'critical exponents'. The reason why the probability $P(p)$ increases so sharply is related to the size distribution of the clusters near threshold, since it only needs, in some cases, just a few extra bonds (or the filling of a few sites) to convert several large clusters into one of infinite size. However, conductivity does not increase so sharply, nor at the same time, because the addition of new bonds from a large cluster increases the number of loops in the network but not necessarily the connectivity of the whole lattice. Many dead ends (or what we call 'network defects' in Chapter 4) are present near the threshold in the percolation path, and only a small fraction of the bonds, called the 'backbone', contributes to the conductivity. This is why the increase in conductivity is very small near the threshold. When the limit $p \rightarrow 1$ is reached, all the bonds will contribute to the conductivity.

Referring to the analogy between percolation and thermodynamic phase transitions, we can examine Table 3.1.

We have already discussed the analogy between temperature and the percolation threshold. Spontaneous magnetization M, density $\rho_{liq}-\rho_{vap}$ and percolation probability $P(p)$ are the respective order parameters for these transitions. All of them are zero in the disordered state (unconnected bonds or high-temperature limit) and they increase very sharply as the percolation threshold or the critical temperature is approached. The magnetic susceptibility, χ defined by $\chi = \partial M/\partial H$, describes the ability of a ferromagnetic material to increase its magnetization under the effect of a magnetic field H. When the Curie temperature is approached from above, $T > T_C$, in the paramagnetic state (no spontaneous magnetization) the susceptibility diverges. It can be shown that the analogue of the magnetic susceptibility is formally the isothermal compressibility in a fluid or the mean cluster size in the percolation model. These parameters diverge when the phase transition is approached from the high-temperature side or, in percolation, from below the threshold.

The correlation length in thermal phase transitions is the mean cluster size below the threshold. This mean cluster size l_{av} or mean spanning length is the average of the

Table 3.1 Analogy between percolation and thermodynamic phase transitions.

Percolation	Ferromagnet	Liquid–vapour critical point
Percolation threshold p_c	Critical temperature T_c	Critical temperature T_c
Percolation probability	Spontaneous magnetization M	$\rho_{liq} - \rho_{vap}$
Mean cluster size s_{av}	Magnetic susceptibility χ	Isothermal compressibility κ
Mean cluster spanning length l_{av}	Correlation length ξ	Correlation length ξ

spanning length l of the clusters. l is defined as the maximum separation of two sites or two bonds in the simulated network:

$$l \equiv \max\{|r_i - r_j|\}_{i,\, j\,\text{in cluster}}. \tag{3.7}$$

The divergence of the correlation length or average spanning length is an essential property, as we will see below. Such continuous phase transitions can be characterized by power laws containing the critical exponents, and perhaps the most important of these is the exponent describing the divergence of the correlation length when the transition is approached.

In Figure 3.6, three curves are drawn: s_{av} is related to the geometry of clusters of finite size below the threshold, which diverge as p_c is approached, whereas above p_c the cluster size is infinite. The other two curves, $P(p)$ and σ/σ_0, vanish below p_c, whereas above p_c they have finite values. The critical region is located near the threshold, with the condition that $|p - p_c| << 1$. In this region the important parameters describing the percolation or thermal phase transitions follow various power-law behaviours versus the distance to the threshold. For instance, for

$$S_{av} \approx \frac{1}{(p_c - p)^{\gamma}} \text{ with } (p_c - p) \to 0, \tag{3.8}$$

the exponent is γ. For the average cluster size

$$l_{av} \approx \frac{1}{(p_c - p)^{v}} \text{ with } (p_c - p) \to 0, \tag{3.9}$$

the exponent is v.

For the other parameters, $P(p)$ and $\sigma(p)$, the power laws are given by

$$P \approx (p - p_c)^{\beta} \text{ and } \sigma \approx (p - p_c)^{t} \text{ as } (p - p_c) \to 0. \tag{3.10}$$

The exponents γ, v, β and t are positive numbers that are not integers. For all simulations they appear to be independent of the lattice geometry, but strongly dependent on the space dimensionality d. The values of the exponents for various parameters deduced from simulations are gathered in Table 3.2.

One can see from Table 3.2 that the exponents are different for percolation and for thermal phase transitions, meaning that they do not belong to the same 'universality class'. However, the general structure of the theory is the same; the exponents are not independent of each other but are related by scaling laws (Stauffer *et al.*, 1982). The exponent $\beta < 1$, while $t > 1$, which explains the difference in the shape of the curves just beyond the threshold: the slope of $P(p)$ is infinite, while the slope of σ is null. For large clusters, at $p = p_c$ there is a relation between the size l and the cluster size s:

$$s \approx l^{f_d}, \quad s \to \infty, \tag{3.11}$$

where f_d is the fractal dimension of the cluster at the percolation threshold. It is found in simulations that $f_d = 1.9$ if $d = 2$, and $f_d = 2.6$ when $d = 3$.

Table 3.2 Comparison between percolation and thermal phase transitions.

Function near the threshold	Exponent	Space dimensionality		
		$d=2$	$d=3$	classical ($d>3$)
Percolation				
$P \approx (p - p_c)^{\beta}$	β	0.14	0.40	1
$\sigma \approx (p - p_c)^{t}$	t	1.1	1.65	3
$s_{av} \approx (p_c - p)^{-\gamma}$	γ	2.4	1.7	1
$l_{av} \approx (p_c - p)^{-\nu}$	ν	1.35	0.85	1/2
Magnetism				
$M(T) \approx (T_c - T)^{\beta}$	β	0.25	0.32	1/2
$\partial M/\partial H \approx (T - T_c)^{-\gamma}$	γ	1.75	1.24	1
$\xi(T) \approx (T - T_c)^{-\nu}$	ν	1.0	0.63	1/2
Liquid–gas critical point				
$(\rho_{liq} - \rho_{vap}) \approx (T_c - T)^{\beta}$	β	0.25	0.32	1/2
$\kappa = -1/V(\partial V/\partial P)_T \approx (T - T_c)^{\gamma}$	γ	1.75	1.24	1
$\xi(T) \approx (T - T_c)^{-\nu}$	ν	1.0	0.63	1/2

The fractal dimensionality of the cluster at the percolation threshold and the critical exponents are related. The fractal dimension reflects the ramified and open structure of the clusters right at the threshold. Away from the critical region ($p - p_c \sim 0.05$), the fractal dimension $f_d = d$, where d is the space dimension.

It is also stated that there are only two independent scaling exponents, the following relations being established:

$$f_d = d - (\beta/\nu), \tag{3.12}$$

$$f_d = \frac{d + (\gamma/\nu)}{2}, \tag{3.13}$$

$$2\beta + \gamma = \nu d. \tag{3.14}$$

These equations as written are valid for percolation as well as for magnetism and critical fluids. This property is an argument in favour of 'universality' in the description of percolation and phase transitions.

3.3.1 Extent of the critical domain

The theoretical analysis presented so far is valid for percolation in the 'critical domain'. We need to define the limits of the critical domain. These limits are given by the Ginzburg criterion (Ginzburg, 1960): when fluctuations in the number of sites inside the volume are large compared to the average value, the behaviour of the various parameters listed in Table 3.2 are given by critical exponents that depend on the space dimensionality. The critical domain is the range of values of $|p-p_c|/p_c$ where the critical exponents can be observed.

Away from the threshold, fluctuations decrease in amplitude, and in this case the expected exponents are independent of the space dimensionality and are characterized by the so-called mean field values. These values are listed in the last column of Table 3.2. It was shown by de Gennes (1977) and Daoud (1979) that the exponents expected in the random cross-linking of high molecular mass polymers in the molten state are mean field exponents, and this case is known as the vulcanization limit, by analogy with the cross-linking of a rubber in a dense medium. In semi-dilute solutions, the width of the critical domain is a function of the ratio c/c^*, where c^* is the overlap concentration; for instance, for $c = 10\ c^*$, the width could be nearly 0.4.

3.4 Percolation and gelation

The connection between percolation and gelation was first suggested by Frisch and Hammersley (1963) and was analysed in greater depth in independent papers by de Gennes (1976) and Stauffer (1976) in 1976. The best analogy recognized is between permanent cross-link formation in a chemical reaction and bond formation in a numerical simulation. The extent of the reaction is the probability of bond formation; clusters are branched macromolecules, while f is the functionality in a step growth process, as discussed in Section 3.1.

Table 3.3 gives the equivalence between the simulated process and a real cross-linking process.

The equivalence between electrical conductivity and shear modulus appears at the bottom of the table. Kirchhoff's current law states that the sum of the electric currents at each node of a network is zero:

$$\sum_j g_{ij}(v_i - v_j) = 0, \tag{3.15}$$

Table 3.3 Analogy between percolation and gelation.

Percolation	Chemical gelation
Percolation threshold p_c	Gel point
Connected bond	Reacted functional group
Fraction of connected bonds p	Extent of reaction p
Lattice for $p<p_c$	Sol state
Lattice for $p>p_c$	Gel state
Mean cluster size $s_{av}(p)$, $p<p_c$	Average molecular mass $M_w(p)$ ($p<p_c$)
Infinite cluster	Gel network
Percolation probability $P(p)$, $p>p_c$	Gel fraction
Coordination number z	Functionality f
Network conductivity σ	Equilibrium elastic modulus

where g_{ij} is the bond conductance and v_i is the equilibrium voltage at each node (or site) i. The formal analogy between Kirchhoff's current law and the equilibrium between forces acting upon a monomer appears in (3.15). A monomer i connected to bonds j with spring constants k_{ij} is in equilibrium when the sum of the forces exerted upon it by the j neighbours is zero:

$$\sum_j k_{ij}(u_i - u_j) = 0, \tag{3.16}$$

where u_i is the displacement of monomer i.

In the same way that the conductivity of a random network of resistors depends on the fraction of conducting bonds, the shear modulus should depend on the fraction of cross-links. The shear modulus is the measure of the elastic stiffness which opposes deformation, when volume is conserved. In the sol state, the shear modulus is zero. It is expected by analogy with conductivity that the shear modulus should vary with the fraction of bonds with the same exponent t (Table 3.2). The critical exponent of the conductivity for a resistor–insulator mixture is $t = 1.6$.

Before proceeding further, it is important to estimate the extent of the Ginzburg critical domain for the gelation case. In general, this is very difficult to do; evaluation of the extent of the critical domain remains the realm of the theorist, but the best estimate we have, due to Stauffer, Coniglio and Adam (Stauffer *et al.*, 1982), and a useful and practical guide, is that the upper limit is $|p/p_c - 1| \leq 10^{-1}$. In practice this can be very difficult to achieve, particularly for physical gels and networks. Indeed, it suggests that, in order to test theory and experiment, really good data needs to be collected when $p/p_c > 0.99$ or better: a very testing requirement.

The increase of the cluster size near the threshold in the sol state induces significant changes in the viscosity of the solution as the system approaches p_c, corresponding to the formation of branched species of larger size. Percolation on a lattice does not include the presence of solvent. How can we predict the changes of the sol viscosity from numerical simulations on lattices? The first attempt (Stauffer *et al.*, 1982) was to express the viscosity contribution of each cluster size in terms of the cluster radius and to refer to a monodisperse suspension of solid spheres in solution, occupying a volume fraction ϕ. The Einstein equation for dilute suspensions at very small volume fractions states that

$$\eta = \eta_0 \left(1 + \frac{5}{2}\phi + \ldots \right), \tag{3.17}$$

where η_0 is the solvent viscosity. In the case of gelation, when the distribution of polymer sizes in solution is highly polydisperse, the volume fraction of the clusters of size s is $\phi_s \approx n_s R_s^{\,3}$. If one neglects the cluster–cluster interactions and considers the clusters as solid spheres, Einstein's approximation leads to

$$\eta = \eta_0 \left(1 + const. \sum_s n_s R_s^3 + \ldots. \right). \tag{3.18}$$

In percolation theory, the sum diverges logarithmically at the gel point:

$$\eta \approx \log(p - p_c), \tag{3.19}$$

while in the original Einstein theory it remains constant, so using this equation for describing gelation is unrealistic. It is necessary to include other terms with quadratic contributions from the volume fractions of virial expansion type:

$$\eta = \eta_0 \left[1 + \frac{5}{2} \sum_s \phi_s + const. \left(\sum_s \phi_s \right)^2 + \ldots \right]. \tag{3.20}$$

At the gel point, the interaction between the clusters becomes too strong and this approximation also becomes unreliable: the third term in (3.20) becomes larger than the second term and the Taylor expansion in ϕ_s is no longer valid.

A completely different approach was proposed by de Gennes (1979) based on the analogy between viscosity and elasticity. The viscosity below p_c is the counterpart of the elasticity above p_c. To predict the behaviour of the viscosity in the sol state, he suggested simulating a random mixture of superconductors with fraction p and normal conductors with fraction $1-p$. According to this analogy, the viscosity's critical exponent equals the critical conductivity exponent which, for $d=3$, is given by $k=0.7$:

$$\eta \approx (p_c - p)^{-k}, \quad p < p_c. \tag{3.21}$$

Results based on molecular dynamics and Monte Carlo simulations show good agreement with this prediction (Farago and Kantor, 2000; Vernon *et al.*, 2001).

Another approach is the so-called Rouse approximation (Stauffer *et al.*, 1982), in which the contribution of a cluster s to the viscosity is proportional to sR_s^2: near the threshold, it states that the viscosity varies as

$$\eta = \eta_0 \left(1 + const. \sum_s n_s \, s \, R_s^2 + \ldots \right), \quad p < p_c. \tag{3.22}$$

The exponent k for viscosity is related to the exponents β and ν by $k=2\nu-\beta$, which gives $k=1.3$ for $d=3$. This approximation neglects hydrodynamic interactions and the excluded-volume interactions between monomers, and so may not be valid near the gel point. Subsequently, however, the Bound Fluctuation method (Carmesin and Kremer, 1988), based on an algorithm which takes into account excluded-volume interactions, supported the prediction that, for $d=3$, $k=1.3$.

Despite these attempts, there is no definite answer as to the value of the exponent k for the divergence of the viscosity near the threshold. A review by de Arcangelis (2003) suggests that there could be two distinct universality classes for the viscoelastic critical behaviour, with a crossover between different dynamic regimes. As we have already mentioned, the problem is very complex, since the distribution of cluster sizes is very (arguably tending to infinitely) polydisperse as the gel point is approached, and the relationship between size and M_w is complicated even for linear polymers.

The elasticity of the gel near the threshold can also be predicted by the scaling relation $t = d\nu$, which in three dimensions gives $t = 2.6$ (Daoud and Coniglio, 1981). The model assumes that elasticity is dominated by the entropic term. When there is a bond-bending contribution to the elastic energy of the network, as expected for particulate gels, the predictions for the exponent of elasticity are higher: $t = 3.7$ for $d = 3$ (Feng and Sen, 1984; Feng *et al.*, 1984; Kantor and Webman, 1984). More recently, a direct investigation of viscoelastic properties was performed by introducing bond fluctuation dynamics into the percolation model, taking into account the conformational changes of the polymer and the excluded-volume interactions. Simulations have shown a critical exponent $t = 2.5$ for $d = 3$ (Del Gado *et al.*, 2002), in agreement with the prediction $t = d\nu$ for entropic networks. Here again the numerical work suggests the possibility that there are two distinct universality classes, one characterized by an exponent $t = d\nu \approx 2.64$ and the other, based on the electrical analogy, by $t \approx 1.6$–2.

Before examining experimental results on gelation, we present another approach widely used to determine the 'gel point', known as the Winter–Chambon criteria.

3.4.1 Winter–Chambon criteria

Winter and Chambon (1986) investigated the end-linking reaction of polydimethylsiloxane (PDMS) chains, to provide a good model system. Divinyl terminated PDMS was linked with a tetrasilane ($f = 4$, functional) moiety, and the molecular mass of the prepolymer was below the entanglement limit. Using the time–temperature superposition procedure, they were able to reconstruct mechanical spectra over a very wide range of frequencies, far from the glass transition temperature. They were also able to stop the chemical reaction at intermediate times of conversion. They observed that the spectra of the shear storage modulus G' and loss modulus G'' of their system exhibit, at some stage of the conversion, a power-law behaviour in the entire measurable radial frequency domain.

They proposed to define the gel point by the following properties:

$$G'(\omega) \approx \omega^n \qquad 0 < \omega < \infty, \tag{3.23}$$

$$G''(\omega) \approx \omega^m \qquad 0 < \omega < \infty. \tag{3.24}$$

The Kramers–Kronig relation requires the two exponents n and m to be equal, so the complex moduli are then given by

$$G'(\omega) = \frac{G''(\omega)}{\tan \frac{n\pi}{2}} = \frac{G''(\omega)}{\tan \delta(\omega)} \qquad n < 1, \tag{3.25}$$

and, the relation being independent of the frequency, the loss angle $\delta(\omega)$ is independent of the frequency.

Three cases can be envisaged:

$$n < \tfrac{1}{2}, \quad G'' < G'$$
$$n > \tfrac{1}{2}, \quad G'' > G'$$
$$n = \tfrac{1}{2}, \quad G'' = G'.$$

In the particular case where $n = \tfrac{1}{2}$, $G' = G''$, the gel point is easily located by measurements at a single frequency. This was the case when they investigated samples with appropriately balanced stoichiometry.

For $n = m$, the relaxation modulus $G(t_{relax})$ of the solution exhibits time dependence, with a power law of slope $-n$:

$$G(t) \approx t_{relax}^{-n}, \tag{3.26}$$

where t_{relax} is the elapsed time in the relaxation spectrum. The authors exclude the high-frequency limit, where the material behaves as a glassy system, from their analysis. The steady state viscosity η_0 can be calculated from the time dependence of the relaxation modulus $G(t_{relax})$: it can be shown that, for any exponent $n < 1$, the viscosity diverges to infinity and the equilibrium modulus at infinite time decays to zero for $n > 0$. The solution is very viscous, but not yet elastic at the point defined by Equations (3.23) and (3.24). The authors commented that a system which exhibits power-law behaviour exhibits the 'classical attributes' of the gel point.

These two theoretical approaches to the definition of the gel point are therefore based on two different approaches. Most percolation models are essentially static theories. They do not take into account the mobility of the molecules. They predict the relations between the structure of the solution schematically represented by lattices and the fraction of bonds created at random. During gelation, larger and larger clusters modifying the viscosity of the solution are formed, until an infinite cluster is created and the solution starts to exhibit a finite elasticity at zero frequency (or infinite time of observation). The predictions for the average molecular masses of the clusters or the gel fraction are derived straightforwardly from simulations. The predictions for viscosity or elasticity are still debated and depend on the particular context of the analysis (Monte Carlo simulations, analogy with conductance of static network, particular model for network connectivity etc.). There is still an important debate about the universality classes for these parameters, where identical critical exponents can be predicted with various hypotheses.

The Winter–Chambon criteria are based first of all on the experimental observations these authors made by following an end-linking reaction of primary chains in the absence of any physical association (crystallization, phase separation etc.), where the precursor polymers were not entangled. At one particular moment in the reaction, both viscoelastic moduli G' and G'' measured in the linear regime exhibit power-law dependences on the frequency. This indicates that an infinite viscosity at zero frequency is also expected from the percolation model. The authors suggest that both approaches – percolation and mechanical spectrum – lead to the same determination. They also verified that, at the gel point, the large cross-linked cluster became insoluble.

3.5 Experimental investigations of gelation transitions

3.5.1 Percolation exponents

A large number of experimental investigations have been devoted solely to exploration of the vicinity of the gel point, in order to examine if either the classical or the percolation model is valid. As we have already hinted, experimental difficulties and the potential differences between behaviours of various chemical systems make the exercise delicate.

For example, Patton and co-workers (Patton *et al.*, 1989) investigated the bulk condensation polymerization of a neopentyl glycol-dimethyl terephthalate based polyester to probe the applicability of the scaling approach. They applied an effective quenching of the reactions which allowed the time evolution of the molecular mass distribution and other parameters below the gel point to be determined during the chemical reaction. The samples were diluted and characterized by dynamic and static light scattering and intrinsic viscosity measurements. The authors were able to evaluate several critical exponents, through various combinations of their experimental data. They first established the correlation between M_w and the extent of reaction p and found (Tables 3.2 and 3.3) the exponent $\gamma = 1.8 \pm 0.3$, in good agreement with the $d = 3$ percolation prediction. The relation between the radius of gyration R_g and molecular mass M_w of the branched molecules, or between intrinsic viscosity η and M_w, are predicted to follow power laws and this is indeed what the authors found, the experimentally determined exponents being much closer to critical percolation than to classical theory. Interestingly, the scaling relationships also applied experimentally over a broader range of extent of the reaction than was expected.

Colby and co-workers (Colby *et al.*, 1992) continued the analysis beyond the gel point, for the same reaction. Samples were separated into a sol and a gel fraction, the sol being constituted by the clusters which are not yet connected to the gel network. The models predict a scaling law between the gel fraction and the average molecular mass M_w of the independent clusters. The swelling ratio of the gel is also expected to vary with the gel fraction: near the threshold $p > p_c$, swelling is very important, but it decreases rapidly when the reaction proceeds towards a totally cross-linked network ($p = 1$). These very careful investigations established that the measured exponents were intermediate between percolation and classical theories, but also suggest that the distance to the gel point was not close enough and measurements exhibited a crossover between the two limiting theories. Indeed measurements do not appear to have been made within the Ginzburg region, as delineated above. Moreover, the chemical system employed in the two studies can be criticized since, compared for example with the BTA–DMG system (Section 3.1), it is potentially much more susceptible to cyclization, 'wastage' and other non-idealities (Stepto, 1998).

Following this work, the same group (Lusignan *et al.*, 1995) re-examined this point and designed a system with a very low degree of polymerization between cross-links, in an attempt to approach the ideal percolation case, and the Ginzburg limit. They performed similar experiments in an attempt to determine the relation between scaling exponents in agreement with, for example, percolation predictions. They also measured the viscosity of the sol and found the exponent $k = 1.36 \pm 0.09$ and the exponent for elasticity $t = 2.71 \pm 0.30$. At first sight, the results seem convincing, but the authors themselves point out, in connection with analyses of others' work, that 'without an independent measure of the critical point . . . the determination of critical exponents . . . is extremely imprecise and subjective'.

Other experiments by the same authors (Lusignan *et al.*, 1999) were designed to explore the vulcanization limit for the cross-linking of a melt of long linear chains of N monomers, where mean field (classical) exponents are predicted (Ginzburg, 1960; Daoud, 1979). Here it is expected that the critical (percolation) domain is limited to the range $|p - p_c|/p_c < N^{-1/3}$, which is very much reduced when the chains are long enough, so that classical exponents are expected to be observed experimentally. The analysis turned out to be more complex: the exponent for viscosity s was found to increase with the ratio N/N_e, where N_e is the entanglement length. The values for s are constant, $s = 1.33$ while $N < N_e$, and rose to $s = 6$ or 7 for $N = 30\ N_e$. The authors concluded that, for long chains, the exponents change systematically with the number of entanglements N/N_e per chain. The exponent for elasticity was found to be $t = 3.2 \pm 0.2$.

Other investigations on epoxy resins (Martin *et al.*, 1988) and silica particles (Gauthier-Manuel *et al.*, 1987) tried to derive the critical exponent of viscosity by assuming that the relative distance to the gel point $\varepsilon = |p - p_c|/p_c$ could well be replaced by the distance in reaction time t_{reac}, $\varepsilon = (t_{reac} - t_c)/t_c$. However, for any reaction the extent p is not linear with reaction time and, even if one explores a narrow domain of the cross-linking reaction, there is no fixed proportionality between these parameters (Ross-Murphy, 2005). Therefore such approximations should not be considered as corroborating critical exponent determinations.

In summary, in this experimental work it appears that only the melt polymerization of branched molecules (polyesters) with $N = 2$ monomers between junction points gives a good illustration of the percolation critical exponents. For longer chains the scaling behaviour seems to depend strongly on topological constraints, and no universal behaviour can be observed. The direct proof of the percolation models is best established by direct measurement of the cluster characteristics, where there is a clear prediction for exponents such as β, γ or ν. It appears, as Stauffer himself pointed out some years ago (Stauffer, 1998), that such experiments are very testing to carry out, and he even goes so far as to suggest a 'failure of cooperation between physics and chemistry'!

When it comes to rheological measurements, there is no consensus either from the theoretical or experimental viewpoint about what is expected for viscosity or elasticity. Percolation requires evaluating the extent of reaction using an independent method (such as spectroscopy or calorimetry). Very careful experiments are needed for an experimental determination of the critical exponents and of the gelation threshold. When dealing with kinetics of gelation that cannot be stopped at any moment, rheological measurements are limited by the time evolution of the system, and the limit of $t_{relax} \to \infty$ or $\omega \to 0$ is not accessible to experiments. In dynamic oscillatory measurements, the Newtonian behaviour of viscosity is defined by

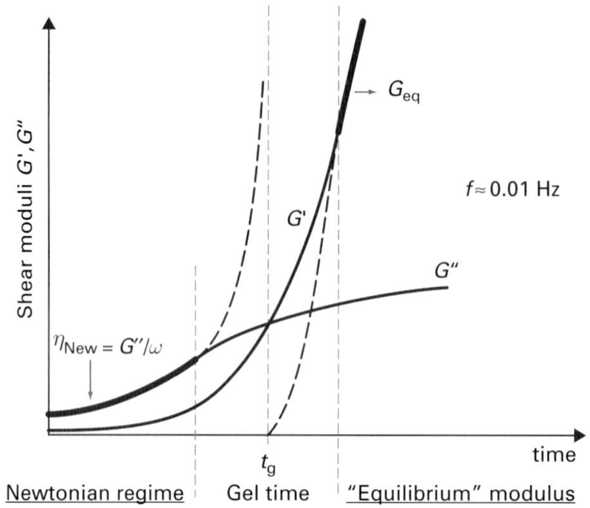

Figure 3.7 Determination of the gel point and of critical exponents for kinetic gelation processes. Adapted from Djabourov *et al.* (1988) with permission of EDP Sciences.

$$\eta_{New} = \lim_{\omega \to 0} \frac{G''}{\omega} \quad \text{below the gel point,} \tag{3.27}$$

and the equilibrium shear modulus G_{eq} is given by:

$$G_{eq} = \lim_{\omega \to 0} G'(\omega) \quad \text{beyond the gel point.} \tag{3.28}$$

In the near vicinity of the gel point, the experimental determination of these limits is incompatible with kinetics, since this does not allow the use of very low frequencies, on the order of 10^{-3} Hz. This limitation is seen schematically in Figure 3.7. It is generally suggested that any experiment be carried out at the lowest accessible frequency, the data analysis carried out in the reaction time domain to minimize errors, and the intersection $G' = G''$ be used as a guide to identify the approximate location of the gel time. The determination of critical exponents can be carried out once the parameter p is identified and its time dependence measured independently. The gel point then appears by a fitting procedure, assuming, for instance, a power-law dependence of viscosity or elasticity versus the distance to the gel point (i.e. $p/p_c - 1$). An approximate gel point can be located by this procedure, although this is an intrinsically weak approach, as pointed out by Lusignan *et al.* (1995) and by others much earlier (Gordon and Torkington, 1981).

Figure 3.7 illustrates that, in the close vicinity of the gel point and for any finite frequency, the equilibrium modulus is smaller than $G'(\omega)$, while the static viscosity is larger that $G''(\omega)/\omega$.

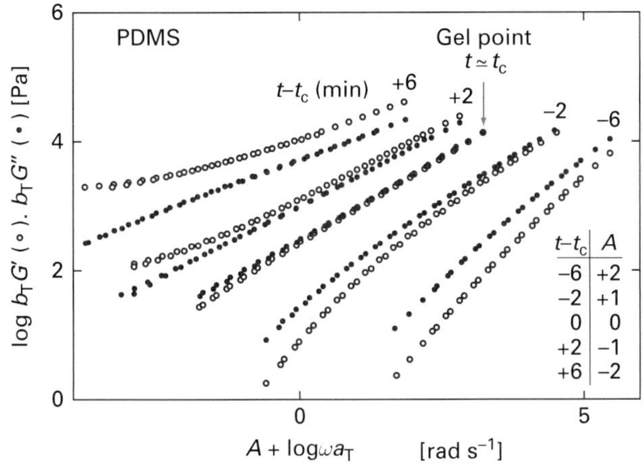

Figure 3.8 Reduced storage moduli for PDMS samples for which the reaction has been stopped at intermediate states of conversion. t_c is the instant of intersection of G' and G''. Curves are shifted horizontally to avoid overlapping. Reproduced from Winter and Chambon (1986) with permission of the American Institute of Physics for The Society of Rheology.

3.5.2 Experimental determination by the Winter–Chambon criteria

Nowadays it is accepted that rheological experiments are those most commonly employed in both chemical and physical gelation studies. It is frustrating, however, to appreciate that the most difficult task is understanding the evolution of viscosity or elasticity with respect to the relevant cross-linking parameter, even for the case of 'point-like', randomly distributed cross-links between flexible chains. Nor are dynamic measurements of the whole spectrum of the sample easy to perform. However, this approach, due to Winter and Chambon, has become much more familiar to scientists working on gelation, and their criteria are considered by many a useful method of determining when gelation occurs.

By stopping the chemical reaction of an end-linked PDMS reaction, Winter and Chambon (1986) were able to measure the frequency spectrum over a very large region (five orders of magnitude), as can be seen in Figure 3.8. Within a window of time around the point $t_{reac} = t_c$ there is a full superposition of G' and G'' spectra over the whole frequency domain, and very shortly after, at $t_c + 2$ min, G' exhibits a plateau at the lowest frequency domain, related to a finite relaxed modulus G_{eq}. Within the accuracy of the experiments, the definition of the gelation point, whether from the power-law dependence of the spectra of G' and G'' using the Winter–Chambon criteria or using percolation theory and exploring the low-frequency limit, seems to converge towards the same gelation time. Since this first publication, the frequency power-law criteria for the moduli have been used to define the connectivity transition in a number of cross-linking processes of polymer systems in solution, and the formation of an infinite network has been assumed to correspond to the Winter–Chambon definition. These criteria have been applied, with varying success, in processes of physical gelation, chemical gelation and inorganic gel formation.

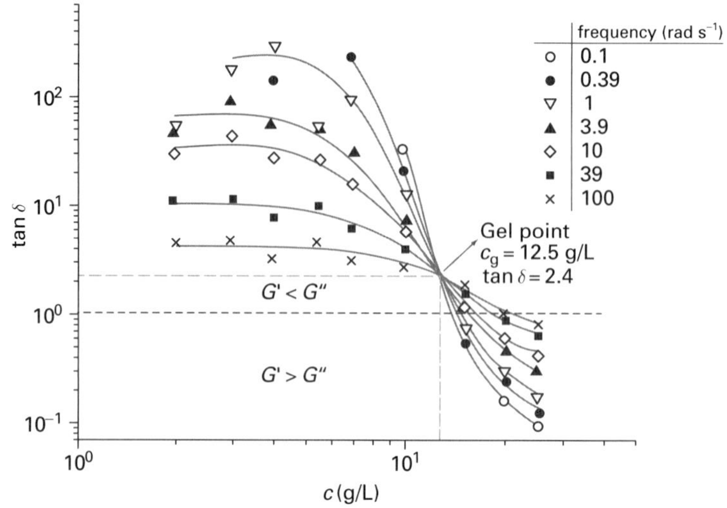

Figure 3.9 Winter–Chambon criteria applied to PVC gelation with concentration as gelation parameter (concentration at the gel point $c_g = 12.5$ g L^{-1}, angular frequencies 0.1, 0.39, 1, 3.9, 10, 39, 100 rad s^{-1}). Adapted with permission from Li *et al.* (1997) © 1997 American Chemical Society.

Working on PVC (poly(vinylchloride)) with various molecular masses in the presence of the plasticizer DOP (bis(2-ethylhexyl) phthalate), Li and Aoki (1997) studied the transition from sol to gel by applying these criteria. They took polymer concentration c as the parameter representing the change in connectivity of the solution. The exponent of the power law for G' and G'', $n = 0.75$, at the gel point indicates that, as shown in Figure 3.9, at a concentration of 12.5 g L^{-1} the medium is viscoelastic with $G'' > G'$ (as $\tan\delta = 2.4$, $G'' = 2.4G'$). The concentration at which G' overtakes G'' at the lowest available frequency (0.1 rad s^{-1}), $\tan\delta = 1$, corresponds to a higher concentration (close to 15 g L^{-1} from Figure 3.9).

It is difficult to define the state in which the system should be classified in this intermediate range of concentrations. Defining the medium as a gel while G'' is still higher than G' at any frequency investigated seems to contradict the usual definitions of a solid-like material. The mechanism of gelation for PVC is related to formation of crystallites in the syndiotactic polymer sequences, but the molecular structure of the crystallites is not determined. The authors assume that PVC clusters form before the gel point and that gelation corresponds to percolation of the clusters, but there is no direct evidence that, at their gel point, the clusters are indeed connected. The authors found gelation concentration to be inversely proportional to the molecular mass of the linear chains, and assumed that percolation between clusters is favoured by longer chains. Li *et al.* (1997), in looking for scaling exponents of percolation, interpreted the zero-frequency viscosity of the gelling solution and the shear modulus versus distance to the gel point as a function of $\varepsilon = (c - c_{gel})/c_{gel}$, although in general it is difficult to establish a simple relationship between c and p (Aoki and Kakiuchi, 1998). te Nijenhuis and Winter (1989) studied PVC gelation with various plasticizers and found the gel point

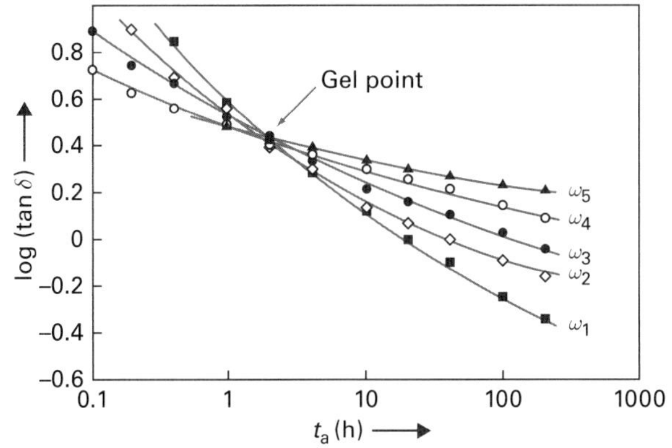

Figure 3.10 Winter–Chambon criteria for PVC–Reomol (Ciba Geigy) plasticizer, with ageing time t_a as gelation parameter at a fixed concentration of PVC: loss tangent against ageing time for angular frequencies $\omega_1 = 1.26$, $\omega_2 = 3.9$, $\omega_3 = 12.6$, $\omega_4 = 39$, $\omega_5 = 126$ rad s^{-1}, $T = 110$°C. Adapted with permission from te Nijenhuis and Winter (1989) © 1989 American Chemical Society.

with power-law criteria at $n = 0.8$, choosing as a parameter the time at a fixed concentration of PVC (9.9 wt%) at various temperatures; see Figure 3.10.

In the case of PVC gelation, therefore, authors working under various experimental conditions found a region where the Winter–Chambon criteria apply. This does not, of course, mean that these systems are gels, and it would be very interesting to validate these measurements by proving that the different systems exhibited the same state of connectivity and degree of crystallinity at the various locations of the measured gelation times.

Another example analogous to gel formation in synthetic polymer solutions is provided by an isotactic poly(propylene) (iPP) melt. There is no solvent in this case, and the melt starts to crystallize in supercooled conditions. The time evolution of tan δ versus the angular frequency ω is shown in Figure 3.11. The loss angle shows quite a different behaviour from the other figures: tan δ is only frequency-independent in a narrow frequency range ($10^{-2} < \omega < 2 \times 10^{-2}$ rad s^{-1}) marked by a horizontal line. The degree of crystallinity at the gel point for this sample is very low and the authors could not measure it. The question of the functionality of the cross-linking junctions is also raised in this investigation. The gel time for this sample decreases with the degree of supercooling associated with the higher degree of crystallinity. A low degree of supercooling allows the largest crystals to form with a low functionality of the crystals (number of tie chains per crystal) because of a larger extent of chain folding. The authors observed an apparent gel point with a low degree of crystallinity at low supercooling, meaning that the crystals would be more efficient in making the network than at high supercooling, but this contradicts current understanding of the crystal morphology. Here again the connectivity between the crystals was not established at the point associated with gelation.

A final example is found with a biopolymer gel, gelatin. Gelation was investigated by Peyrelasse *et al.* (1996) by taking temperature as the gelation parameter. The exponent was found to be $n = 0.62$ and the gelation temperature 36°C, independent of

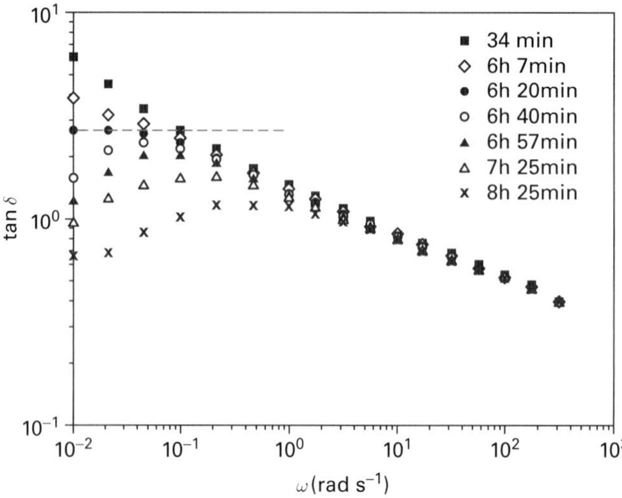

Figure 3.11 Winter–Chambon criteria for crystallization of iPPS: loss tangent against frequency at various times of crystallization ($T = 146$°C). Gelation time is marked by the horizontal dashed line. Reproduced from Schwittay *et al.* (1995) by permission of the Royal Society of Chemistry.

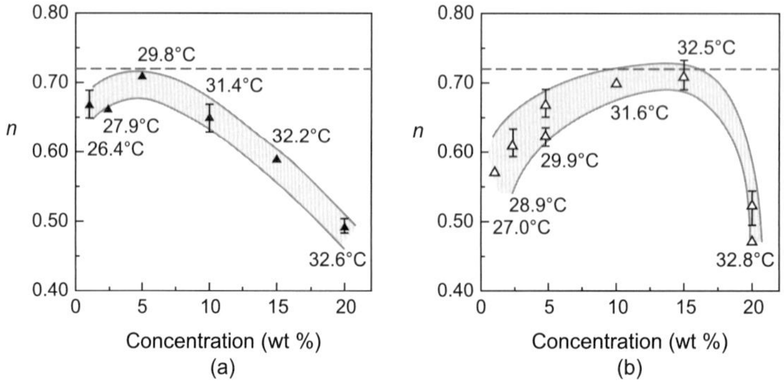

Figure 3.12 Winter–Chambon criteria: exponent n versus concentration for gelatin gels: (a) gelation; (b) melting. Corresponding temperatures for gelation or melting are also shown on the figures. Adapted from Michon *et al.* (1996) © Wiley-VCH Verlag GmbH & Co. KGaA. Reproduced with permission.

concentration, between 17% and 40% g cm^{-3}. Michon *et al.* (1996) reported different values for the exponent n during gelation and melting of gelatin gels, as shown in Figure 3.12 at various temperatures and concentrations, which contradicts the previous description of the gelation process.

Although most of the available data for gelation of physical or chemical gels gives exponents $n > 0.5$, one can see in Figure 3.12 that some values determined at high gelatin concentrations are below this value. The plots also indicate that the gelation temperature of gelatin solutions is a function of concentration.

It is known that the underlying mechanism of network formation in gelatin gels is the formation of triple helices of the collagen type. There is no equilibrium for the structure at a particular concentration and temperature once helices start to build. These experiments do not take into account the mechanism of gelation. Gelatin gelation is reported in detail in Chapter 7.

3.6 Eldridge–Ferry method

One of the longest-established ways of characterizing the behaviour of thermoreversible gels is due to Eldridge and Ferry (1954), who made a series of careful measurements of the melting temperature of gelatin gels as a function of polymer concentration and molecular mass. It should be made clear at the outset that the approach has been criticized by some theorists, but it is nevertheless still widely used, as are a number of variations and improvements. These are discussed in detail in the review by te Nijenhuis (1997).

Eldridge and Ferry (EF) decided to treat gel 'melting' as a first-order phase transition, in a manner similar to Flory's original treatment for crystalline melting of polymers. They had observed that a plot of ln c versus $1/T_{\mathrm{m}}$, where T_{m} is the gel melting temperature (K), was linear, as was a corresponding plot of ln (gelatin primary chain molecular mass) versus $1/T_{\mathrm{m}}$. This suggests that a van't Hoff law of the form

$$-\left[\frac{d\ln c}{dT_{\mathrm{m}}}\right]_{M_{\mathrm{w}}} = \frac{\Delta H_{\mathrm{m}}{}^{0}}{RT_{\mathrm{m}}{}^{2}}, \tag{3.29}$$

where ΔH_{m} is the melting enthalpy, could be applied. The method assumes an equilibrium between

2 cross-linking sites $\Leftrightarrow$ 1 cross-link,

although the nature of the equilibrium (in particular the relative rate of the back reaction) does not actually alter the tenets of the model. Here the sites are assumed to be tetrafunctional, but te Nijenhuis (1997) has generalized the model to the case of cross-link functionality f to give

$$-\left[\frac{d\ln c}{dT_{\mathrm{m}}}\right]_{M_{\mathrm{w}}} = \frac{\Delta H_{\mathrm{m}}{}^{0}}{(1-f/2)RT_{\mathrm{m}}{}^{2}}. \tag{3.30}$$

The limitations of this method are several. First, it assumes that the number of elastically active cross-links is a function of concentration but not of temperature, so it neglects wastage (which will increase at lower concentrations). Second, it requires that the number of unused junctions is also proportional to concentration, which only applies, for example for gelatin, if the ageing process has not proceeded very far. It also neglects the possibility of a critical gel concentration, which suggests that the plot cannot always be linear at lower concentrations. Despite all of these points, the method has great practical use, and is still widely applied in studies of thermoreversible gels, especially including work on PVA and gelatin. Much of this work has concentrated on evaluating 'best' values of the melting enthalpy, and these are discussed in the te Nijenhuis review (te Nijenhuis, 1997).

In 1972, Takahashi (1972) derived a different equation, based on the Flory copolymer melting treatment, to address the same problem. Rather than plotting ln c versus $1/T_m$, it uses Flory–Huggins coordinates, so that $1/T_m$ is now the abscissa and ln v_2N the ordinate, with v_2 the polymer volume fraction and N the degree of polymerization. While criticisms similar to those mentioned above for EF could be made about this treatment, it does appear to be quite successful in allowing the superposition of samples of differing molecular mass.

Yet another variation on the EF method, based on a more rigorous underlying theoretical model, is due to Tanaka and Nishinari (1996). They suggested that a plot of ln c versus $1000/T_m$ + ln M would also allow the apparent multiplicity of junction zones to be determined. Here the factor 1000 simply helps to represent the two plot components successfully, analogous to the scaling of the classical Zimm plot for light scattering.

The two slopes of this rectilinear Zimm type plot can be related respectively to the multiplicity and to the usual melting enthalpy. They describe the application of the method to a number of synthetic thermoreversible gel systems, including poly(styrene) in carbon disulphide (which gives a maximum in the EF plot) and PVA in water. In the latter case s was calculated to be in the range 2–3, with higher values, perhaps as expected, for the high-melting samples.

3.7 Critical gel concentration

One aspect of the model mentioned above is the critical gel concentration, here denoted c_0. That this exists follows from the existence of a percolation threshold, as described above and derived from both classical and non-classical formulations. Apart from this, it is all but self-evident that there has to be a certain ('critical') concentration of polymer present before a gel can be formed, and only a few approaches (Bremer *et al.*, 1993) have made the bold assumption that this concentration is effectively zero.

Using the equilibrium assumption of Eldridge and Ferry, Hermans (1965) assumed that there was a monomer ⇔ dimer (or link ⇔ two free sites) equilibrium with a site dissociation constant K^s. This could then be written in the form of an Ostwald dilution law,

$$K^s = \left[(1-p)^2/p\right] N_0 f,$$

where N_0 is the number of primary chains per unit volume, p is the percolation site occupancy and f is the functionality. In the limit of weak binding (where $K^s \gg 1$) and high functionality, f can be shown (Clark and Ross-Murphy, 1987) to give a straight proportionality, equivalent in our terms to

$$\frac{p}{p_c} = \frac{c}{c_0},$$

provided $p \gtrsim p_c$; and the critical concentration itself can be shown to be

$$c_0 = \frac{M_w}{K^s f^2}, \tag{3.31}$$

The assumption of high f is a very artificial limitation, even for the cross-linking of long primary chains, so Clark and Ross-Murphy (1985, 1987) derived a more general form valid for any functionality, which gives

$$c_0 = \frac{M_w(f-1)}{K^s f(f-2)^2}, \tag{3.32}$$

in which case

$$\frac{p}{p_c} \neq \frac{c}{c_0}.$$

On the basis of this approximation, and applying a polymer network branching model, the general relationship between gel modulus and concentration was also derived. In carrying out such a manoeuvre, there is no assumption that networks should be ideal rubbers, since the so-called rubber elasticity 'front factor' need not be close to unity, as it is for rubber networks (Chapter 4).

In this way the underlying network connectivity (percolation) can be factored out of the contribution to the modulus from each cross-link, which in turn will depend on the chain species, flexibility etc. It is via the system-independent connectivity part of the treatment that such essential features as critical concentration and the initial steepness of the concentration dependence are derived. The final – high – concentration modulus tends to reflect the modulus per cross-link terms. Clark (1993) showed subsequently that the form of the equations above still holds even if the system, rather than being in equilibrium, is under kinetic control.

The method has been quite successful, not only for cold-set gels such as gelatin (Chapter 7) but also for heat-set particulate and fibrillar gels such as those formed from β-lactoglobulin (Chapter 9). Not least, it has also served as a useful interpolation method in evaluating the modulus of mixed-gel systems (Chapter 10).

For thermoreversible gels, the critical concentration will obviously depend on temperature because, above the melting temperature, the critical concentration is nominally infinite. This means that there must be a maximum gelation temperature, and this relates to the induction time originally defined by te Nijenhuis, now more commonly referred to as the critical gelation time t_c, which also tends to infinity at this maximum gelation temperature. Various empirical models for the temperature dependence of c_0 and t_c are discussed by te Nijenhuis in his review (te Nijenhuis, 1997).

For some workers, c_0 is related to the overlap concentration c^* in solution, and some even treat the two as the same. That said, it seems that for physical gels there is little correlation between c^* and c_0. Indeed it has even been argued (Clark and Ross-

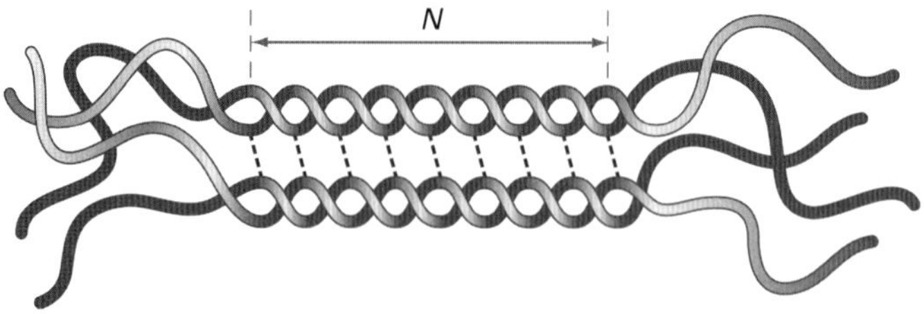

Figure 3.13 The zipper model: a single zipper with N links can be opened from either end. Helices may not necessarily be double helices. Here the double helices represent the association of some ordered structures.

Murphy, 1987) that if the cross-linking structure – however it is constituted – involves more than a few k_BT (the Boltzmann energy), then $c_0 < c^*$, and, insofar as it can be tested, this does seem to be the case. The experimental difficulty here is that establishing c^* for a system just as it undergoes a change in structure attendant with gelation is rarely straightforward.

3.8 Zipper model

Another very useful model which is applicable to the melting of all 'junction zone' type gels is the so-called 'zipper' model of Nishinari and co-workers (Nishinari *et al.*, 1990). Figure 3.13 presents the basic model.

They treated the dissolution of such gels as the opening of molecular zippers, consisting of N parallel links that can be opened from one end. First, we assume that the zipper can be opened from only the left end. Link i from the left can be opened only when the 0th, 1st, 2nd, . . ., $(i-1)$th links are all open. Each link has only two energy states: the energy of a closed link is assumed to be zero and that for the open link E. When one link is open, we assume that it has Ω orientations, corresponding to the rotational freedom of the links; that is, the energy level of this link is Ω-fold degenerate. Then, for a long enough chain, the partition function ζ for such a single zipper which can be opened from either end is

$$\zeta = \left[1 - (N+1)x^N + Nx^{N+1}\right]/(1-x)^2, \tag{3.33}$$

where the factor x is given by:

$$x = \Omega\exp(-E/k_BT) \tag{3.34}$$

and k_BT is the Boltzmann energy. The partition function for the whole system containing $\mathcal{N}$ zippers is then simply $\zeta^{\mathcal{P}}$, and from this the heat capacity of the whole system is directly found.

The advantage of this method is that the heat capacity for a gelling junction zone system can be calculated as a function of temperature and, even allowing for the number of parameters, a realistic fit can be obtained for a variety of systems, including PVA

(Chapter 8), κ-carrageenan (Chapter 5) and even mixed gels, such as konjac glucomannan and κ-carrageenan, where differential scanning calorimetry (DSC) melt traces are particularly broadened. The model predicts the gel–sol transition temperatures by means of the zipper mechanism and includes concentration effects. It also predicts a hysteresis between setting and melting temperatures.

3.9 Liquid crystal gels

One final aspect of the sol–gel transition which needs to be mentioned here involves those 'gels', which occur quite widely, formed from liquid crystalline materials. Liquid crystals show behaviour intermediate between that of a liquid and that of a crystalline solid: they will flow but, under flow and even statically, they tend to show birefringence, indicating an underlying crystalline type order.

Many of these materials are formed from low molecular mass surfactant systems, and are largely outside the scope of this monograph. However, lyotropic liquid crystal gels can also be formed from polymeric systems, particularly from high molecular mass 'rods' – by which we mean polymers of high persistence length. Such materials can form from many different molecular species, and some are discussed in other chapters, for example xanthan (Chapter 5) and protein fibrillar systems (Chapter 9). However, the best exemplar and among the most carefully studied, are the so-called Miller gels, formed from poly(γ-benzyl-α, L-glutamate) (PBLG) in various solvents including benzyl alcohol and dimethylformamide (DMF), studied by Miller and co-workers (Miller *et al.*, 1978; Tohyama and Miller, 1981). PBLG is helicogenic in these and some other solvents; that is to say it readily forms a single-chain α-helical conformer.

At this point, and particularly since this section is positioned in the sol–gel transition chapter, it is important to point out that these 'gels' are usually – rheologically at least – viscoelastic liquids, i.e. they have no equilibrium modulus, and can also sustain a steady shear flow, either at long times or at higher stresses and strains, without rupturing.

The formation of complex, optically anisotropic structures from rods was well known to Onsager in his 1949 paper (Onsager, 1949). He examined the behaviour of rod-like systems which mutually exclude one another's volume. The order–disorder transition is driven by a competition between orientational entropy (minimized by an isotropic distribution of rods) and translational entropy (minimized by a parallel array of aligned rods). He established that there can be a net reduction in overall entropy if a set of rods undergoes a transition to a state where they are aligned more or less parallel to one another. Then, above some critical volume fraction, an isotropic solution of rods will show a spontaneous phase transition to a nematic phase. This critical volume fraction depends on the ratio of length L to diameter D of the rod, and also on the precise form of the rod, but if $L >> D$ the approximate solutions give

$$\phi_{\mathrm{iso}} \approx \frac{3.3}{L/D}, \quad \phi_{\mathrm{nem}} \approx \frac{4.5}{L/D}. \tag{3.35}$$

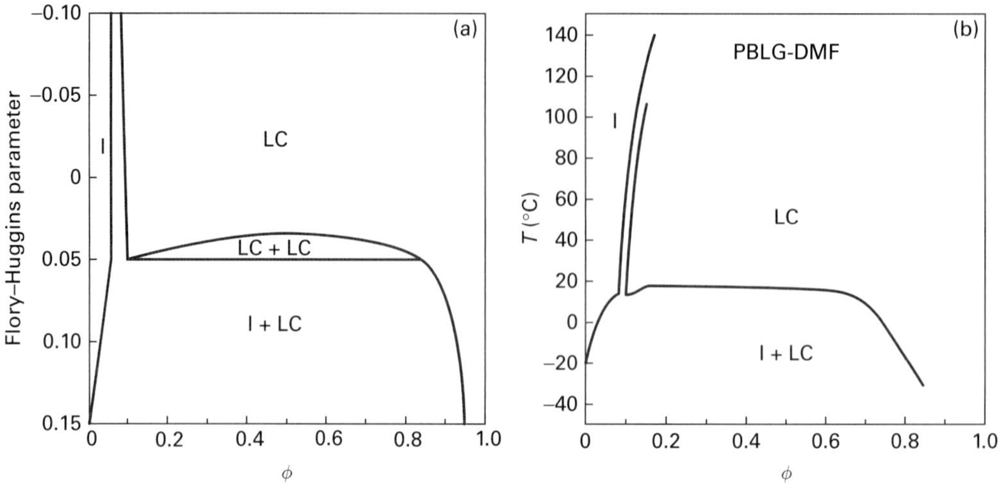

Figure 3.14 Flory's theoretical prediction for rods in solution (a), compared with experimental phase diagram for poly(benzyl L-glutamate) – dimethylformamide (PBLG-DMF) system (b). From Miller *et al.* (1978) with permission from John Wiley & Sons.

Here $\phi_{\rm iso}$ and $\phi_{\rm nem}$ are the respective critical volume fractions of the limit concentrations of isotropic and nematic phases. For volume fractions lying between $\phi_{\rm iso}$ and $\phi_{\rm nem}$ there is a transitional region which can display some nematic character.

Flory solved the same problem with a lattice model (Flory, 1956), and his solution predicts higher values of the two critical volume fractions:

$$\phi_{\rm iso} \approx \frac{8}{L/D}, \quad \phi_{\rm nem} \approx \frac{12.5}{L/D}. \tag{3.36}$$

Extensions such as those by Warner and Flory (1980) and Khokhlov and Semenov (1982) have included a degree of flexibility via, for example, the worm-like chain model. Such effects tend to increase the critical values of ϕ, by increasing the 'constant' in the numerators. Qualitatively all these predictions are useful, although they have proved to be quantitatively imprecise. This is partly because few macromolecules are even close to rod-like, and partly because the theories neglect the polydispersity of $M_{\rm w}$ (or length). Both factors act to increase the critical ϕs. However, the Flory theory makes the useful assertion that no isotropic–nematic transition can occur unless $L/D \gtrsim 6$. He was able to calculate a new phase diagram, distinguished by the so-called 'chimney region' shown in Figure 3.14.

Miller made a series of measurements at various concentrations and temperatures, and his results showed qualitative agreement with Flory's predictions, including an apparent 'chimney'. Sasaki extended Miller's work by investigating the rheology of a range of values for PBLG concentration and $M_{\rm w}$, particularly in benzyl alcohol (Sasaki *et al.*, 1983). These gels were extremely fragile and tended to flow at strains above a few percent. This and subsequent work strongly suggests that the gel structure is not linked directly to the tendency to form liquid crystals. Miller suggested that the gelation is a consequence of spinodal decomposition (Chapter 10). Moreover, the whole system is known to be under kinetic control. One aspect of interest here is that there is very little concentration dependence of the modulus, even close to the phase transition; some of the Miller data, for example, suggests that $G \sim c^{0.2}$. On this evidence alone it is reasonable to affirm that these 'gels' are not the result of a sol–gel percolation process.

Later work included that by Cohen and co-workers (Cohen and Talmon, 1987), who suggested that microfibrillar structures are formed by a non-specific cluster aggregation process, and by Donald and co-workers (Hill and Donald, 1989; Horton and Donald, 1991), who examined the systems by hot stage microscopy and concluded that the gel consisted of a weak network of solid PBLG crystals dispersed among the original chains. Shukla and co-workers (Shukla and Muthukumar, 1988) examined a range of concentrations and also established part of the phase diagram. Although the synthesis of PBLG-like derivatives and their exploitation in complex 'applications' such as the formation of nanotubes and channel cavities is still an active area of research, more fundamental work is somewhat scarce. Indeed, the most complete recent study appears to be that by Tadmor and co-workers (Tadmor *et al.*, 2002), who used SAXS and SANS. Their time-resolved measurements showed characteristics of a nucleation and growth process with an incubation period which was extremely dependent on the gelation temperature. They suggest that the structure is formed by homogeneous nucleation and one-dimensional growth of fibrils that then merge or branch to form the network structure.

To conclude, helical PBLG is an interesting physicochemical system, but of more general interest and applicability is the Flory type phase behaviour, and it is this that has been examined for other systems, and is discussed in later chapters.

3.10 Conclusions

Theoretical models for determining the gel point are based on the concept of connectivity between the polymer chains in the sol state or monomers undergoing polymerization. For a chemical gel it is possible to control the cross-linking reactions and unambiguously identify the moment when the network becomes insoluble. Provided the samples are handled with great care to avoid breakage, light scattering, size exclusion chromatography and intrinsic viscosity measurements can be performed on samples which have been diluted before the gel point is reached, or where the reaction has been stopped chemically. All types of rheological experiments can also be performed *in situ*, ensuring that, as far as can be determined, measurements remain in the linear regime. The only limitation comes from the instrumental precision and range of sensitivities.

When dealing with physical gelation, many more difficulties appear. The links are not point-like junctions, but contain local structures with many different characteristics, including local crystallization in synthetic polymers, multiple strand structures in biopolymers and micelle formation in copolymer solutions. The first question is whether these cross-links can be assumed equivalent to permanent links, i.e. whether the lifetime of any junction is long enough that the gel exhibits a permanent modulus, which does not relax during the time of observation; some physical gels can fulfil this requirement. The second difficulty is establishing the mechanism of macromolecular self-assembly, with respect to temperature, concentration variation or time, for instance. In this case there is no way of diluting the samples to characterize the clusters. The characterization has to be done *in situ* in the course of aggregation. For a kinetic process, the *time* is limited. When examining changes in rheological properties, it is not possible to explore any window of frequency; frequency has to be adapted to the time dependence of the gelation parameter, which is supposed to remain constant during any one experimental run.

Rheological methods have become among the most important and widely used for investigation of gelation processes. The use of the power-law criteria became very popular on this type of systems, albeit over a limited range of frequencies. In many but not all cases, these lead, at some 'reaction time', to parallelism of G' and G'' versus frequency in a double logarithmic scale. However, many cases reported correspond to a definition of the 'gel point' where $G'' > G'$ over the whole range of frequencies investigated. Clearly, this point should not be identified as the last step before formation of the infinite network; the next-closest step (one more link added to the largest cluster) should indeed correspond to a connected network, with an equilibrium modulus at low frequency. Sometimes this is the case, but in other cases it seems that the medium is not close to reaching a permanent modulus, and a larger concentration or a lower temperature must be used to step into the gel solid-like phase. We shall discuss these points further in the next chapters.

Despite this, percolation models and the Winter–Chambon criteria both seem to converge towards the same definition of the gel point, although how closely they agree is still a matter for debate. Certainly the Winter–Chambon definition of the gel point is open to criticism on a number of grounds: that it neglects the effect of topologically 'trapped' entanglements; that it assumes linear viscoelasticity even for extremely tenuous physical gels when the linear viscoelastic strain may tend to zero at the gel point; that systems are known that generate a Winter–Chambon type of mechanical spectrum even though they are patently not gels.

References

Aoki, Y., Kakiuchi, M., 1998. *Macromolecules* **31**, 8117–8123.

Bremer, L. G. B., Bijsterbosch, B. H., Walstra, P., van Vliet, T., 1993. *Adv. Colloid Interface Sci.* **46**, 117–128.

Carmesin, I., Kremer, K., 1988. *Macromolecules* **21**, 2819–2823.

Clark, A. H., 1993. *Polymer Gels and Networks* **1**, 139–158.

Clark, A. H., Ross-Murphy, S. B., 1985. *Br. Polym. J.* **17**, 164–168.

Clark, A. H., Ross-Murphy, S. B., 1987. *Adv. Polym. Sci.* **83**, 57–192.

Cohen, Y., Talmon, Y., 1987. *Ultramicroscopy* **23**, 251.

Colby, R. H., Rubinstein, M., Gillmor, J. R., Mourey, T. H., 1992. *Macromolecules* **25**, 7180–7187.

Daoud, M., 1979. *J. Phys., Lett.* **40**, L 201–205.

Daoud, M., Coniglio, A., 1981. *J. Phys. A: Math. Gen.* **14**, L301–L306.

de Arcangelis, L., 2003. *Comput. Sci. Eng.* **5**, 78–87.

de Gennes, P.-G., 1976. *J. Phys., Lett.* **37**, L1–L2.

de Gennes, P.-G., 1977. *J. Phys., Lett.* **38**, L355–L358.

de Gennes, P.-G., 1979. *J. Phys., Lett.* **40**, L197–L199.
Del Gado, E., de Arcangelis, L., Coniglio, A., 2002. *Phys. Rev. E* **65**, 041803.
Djabourov, M., Leblond, J., Papon, P., 1988. *J. Phys. France* **49**, 333–343.
Eldridge, J. E., Ferry, J. D., 1954. *J. Phys. Chem.* **58**, 992–995.
Farago, O., Kantor, Y., 2000. *Phys. Rev. E* **61**, 2478–2489.
Feng, S., Sen, P. N., 1984. *Phys. Rev. Lett.* **52**, 216–219.
Feng, S., Sen, P. N., Halperin, B. I., Lobb, C. J., 1984. *Phys. Rev. B* **30**, 5386–5389.
Flory, P. J., 1941. *J. Am. Chem. Soc.* **63**, 3083–3090.
Flory, P. J., 1942. *J. Phys. Chem.* **46**, 132–140.
Flory, P. J., 1953. *Principles of Polymer Chemistry.* Cornell University Press, New York.
Flory, P. J., 1956. *Proc. R. Soc. A* **234**, 73–89.
Flory, P. J., 1974. *Faraday Discuss.* **57**, 7–18.
Frisch, H. L., Hammersley, J. M., 1963. *J. Soc. Ind. Appl. Math.* **11**, 894–917.
Gauthier-Manuel, B., Guyon, E., Roux, S., Gits, S., Lefaucheux, F., 1987. *J. Phys. France* **48**, 869–875.
Ginzburg, V. L., 1960. *Phys. Solid State* **2**, 1824.
Gordon, M., Kajiwara, K., Peniche-Covas, C. A. L., Ross-Murphy, S. B., 1975. *Makromol. Chem.* **176**, 2413–2435.
Gordon, M., Ross-Murphy, S. B., 1975. *Pure Appl. Chem.* **43**, 1–26.
Gordon, M., Torkington, J., 1981. *Pure Appl. Chem.* **53**, 1461–1478.
Hammersley, J. M., 1957. *Proc. Cambridge Phil. Soc.* **53**, 642–645.
Hermans, J. R., 1965. *J. Polym. Sci., Part A: Gen. Pap.* **3**, 1859–1868.
Hill, A., Donald, A. M., 1989. *Liq. Cryst.* **6**, 93–110.
Horton, J. C., Donald, A. M., 1991. *Polymer* **32**, 2418–2427.
Jacobson, H., Beckmann, C. O., Stockmayer, W. H., 1950. *J. Chem. Phys.* **18**, 1607–1612.
Kantor, Y., Webman, I., 1984. *Phys. Rev. Lett.* **52**, 1891–1894.
Khokhlov, A. R., Semenov, A. N., 1982. *Phys. A (Amsterdam)* **112**, 605–614.
Li, L., Aoki, Y., 1997. *Macromolecules* **30**, 7835–7841.
Li, L., Uchida, H., Aoki, Y., Yao, M. L., 1997. *Macromolecules* **30**, 7842–7848.
Lusignan, C. P., Mourey, T. H., Wilson, J. C., Colby, R. H., 1995. *Phys. Rev. E* **52**, 6271–6280.
Lusignan, C. P., Mourey, T. H., Wilson, J. C., Colby, R. H., 1999. *Phys. Rev. E* **60**, 5657–5669.
Martin, J. E., Adolf, D., Wilcoxon, J. P., 1988. *Phys. Rev. Lett.* **61**, 2620–2623.
Michon, C., Cuvelier, G., Launey, B., Parker, A., 1996. *J. Chim. Phys. Phys-Chim. Biol.* **93**, 819.
Miller, W. G., Kou, L., Tohyama, K., Voltaggio, V., 1978. *J. Polym. Sci., Part C: Polym. Symp.* **65**, 91–106.
Nishinari, K., Koide, S., Williams, P. A., Phillips, G. O., 1990. *J. Phys. France* **51**, 1759–1768.
Onsager, L., 1949. *Ann. N. Y. Acad. Sci.* **51**, 627–659.
Patton, E. V., Wesson, J. A., Rubinstein, M., Wilson, J. C., Oppenheimer, L. E., 1989. *Macromolecules* **22**, 1946–1959.
Peyrelasse, J., Lamarque, M., Habas, J. P., Bounia, N. E., 1996. *Phys. Rev. E* **53**, 6126–6133.
Ross-Murphy, S. B., 1975. *J. Polym. Sci., Part C: Polym. Symp.* **53**, 11–22.
Ross-Murphy, S. B., 2005. *J. Macromol. Sci., Part B: Phys.* **44**, 1007–1019.
Sasaki, S., Tokuma, K., Uematsu, I., 1983. *Polym. Bull.* **10**, 539–546.
Schwittay, C., Mours, M., Winter, H. H., 1995. *Faraday Discuss.* **101**, 93–104.
Shukla, P., Muthukumar, M., 1988. *Polym. Eng. Sci.* **28**, 1304–1312.
Stanley, H. E., 1998. *Introduction to Phase Transitions and Critical Phenomena.* Oxford University Press, Oxford.

Stauffer, D., 1976. *J. Chem. Soc., Faraday Trans.* **72**, 1354–1364.
Stauffer, D., 1998. *Ber. Bun. Gesell. Phys. Chem.* **102**, 1672–1678.
Stauffer, D., Coniglio, A., Adam, M., 1982. *Adv. Polym. Sci.* **44**, 103–158.
Stepto, R. F. T., 1998. Non-linear polymerization, gelation and network formation, structure and properties. In Stepto, R. F. T. (ed.), *Polymer Networks: Principles of Their Formation, Structure and Properties*. Chapman and Hall, Glasgow, pp. 14–63.
Stockmayer, W. H., 1943. *J. Chem. Phys.* **11**, 45–55.
Tadmor, R., Khalfin, R. L., Cohen, Y., 2002. *Langmuir* **18**, 7146–7150.
Takahashi, A., 1972. *Polym. J.* **3**, 207–216.
Tanaka, F., Nishinari, K., 1996. *Macromolecules* **29**, 3625–3628.
te Nijenhuis, K., 1997. *Adv. Polym. Sci.* **130**, 1–252.
te Nijenhuis, K., Winter, H. H., 1989. *Macromolecules* **22**, 411–414.
Tohyama, K., Miller, W. G., 1981. *Nature* **289**, 813–814.
Vernon, D., Plischke, M., Joós, B., 2001. *Phys. Rev. E* **64**, 031505.
Warner, M., Flory, P. J., 1980. *J. Chem. Phys.* **73**, 6327–6332.
Winter, H. H., Chambon, F. J., 1986. *J. Rheol.* **30**, 367–382.

4 General properties of polymer networks

As noted in Chapter 1, there are essentially three types of polymer network structures: covalently cross-linked networks, in which the junctions are formed by chemical bonds; entangled networks (melts, concentrated solutions) in which the junctions are simply localized topological constraints; and physical networks where the junctions have complex structures, stabilized by secondary forces (such as hydrogen bonds, hydrophobic interactions or labile ionic complexes). The present volume is concerned predominantly with physicals gels and gelation. However, both from the historical viewpoint and from the overlaps in structure–property relationships which follow from the commonality of topology, it would be quite inappropriate to attempt to address the nature and science of physical gels without considering the many features of interest and theories that have been carried over, either modified or unmodified, from the older and scientifically more mature field of chemical gels and networks.

Interestingly, physical gels and networks have been known, at least through the application of their properties, for millennia. The early Egyptians were familiar with both starch and animal-based glues some 4000 years ago, while the technology of papermaking dates back to the Chinese Han dynasty, *c.*100 AD. By contrast, chemical networks have been known for only around two hundred years, starting with Mackintosh's early work (1823) on producing a waterproof fabric using natural rubber, and the subsequent progress by Goodyear in extending the applicability of natural rubber latex by the process of sulphur vulcanization, invented in 1839. The first synthetic resin was the phenol-formaldehyde system Bakelite, invented by Leo Baekeland in the first decade of the twentieth century. Subsequent work on urea-formaldehyde and polyester resins followed in the next 20 years.

Much of the above work is described in Stepto's 1998 monograph (Stepto, 1998). This volume describes how much of the simple chemistry was known well before its extension to useful cross-linked resins. What is of real significance is that much of this work predates Staudinger's macromolecular hypothesis of the early 1920s, and a real understanding of the nature of polymer networks did not become clear until Flory's work on non-linear polymerization, published in a series of classic papers in the late 1930s and early 1940s.

What this work also made clear for the first time was the similarities and differences between 'polycondensation' – now more correctly referred to as step-growth or end-linked polymerization, and involving low molecular mass species – and 'vulcanization', the cross-linking of pre-existing polymeric (i.e. high molecular mass) chains. Before we address these cases in detail, we must spell out that the major difference between polymer networks and polymer gels is simply the presence of a 'solvent' or swelling agent. As we will see below and elsewhere in this volume, the elastic constraints which follow from the three-dimensional network structure ensure that a polymer network can swell to a limited extent, but cannot dissolve. All viable theories of swelling are, in essence, concerned with the balance between dissolution (i.e. infinite swelling) and the abovementioned elastic constraints, which follow from macroscopic network connectivity.

4.1 Chemically cross-linked networks and gels

The earliest true science of polymer networks goes back to studies on the nature of rubber elasticity associated with such names as Gough, Joule and Kelvin in the nineteenth century, well before there was any appreciation of the underlying network structure (Treloar, 1975). Staudinger's demonstration that there was a class of materials which consisted of very high molecular mass species, both linear and non-linearly linked, as opposed to the original view that such species were simply colloidal aggregates, established the existence of polymer science. Subsequently it was appreciated that rubber elasticity is predominantly attributable to the reduction in entropy induced by the deformation of linked polymer chains, and many papers on the classical theory of rubber elasticity and swelling followed in the period 1936–1945, due to such seminal names as Kuhn, Flory, Treloar and their respective co-workers.

As mentioned in Chapter 3, the treatment of non-linear systems was a natural extension of work on linear polymerizations, which had been established by, among others, Carothers and co-workers in the development of the synthetic polyamides, now known as nylons, around 1934 at the DuPont Company.

4.1.1 Non-linear materials formed from the reaction of functional groups

The chemistry of these systems has already been mentioned, and studies have tended to follow the principles established in early work with reactant species – a multi-functional m-ol is reacted with, say, a multi-functional n-acid, e.g. pentaerythritol (functionality, $f_m = 4$) and the dibasic adipic acid ($f_n = 2$), as in a series of experiments by Stockmayer and Weil (Flory, 1953). Other systems include those from benzene triacetic acid ($f_n = 3$) and decamethylene glycol ($f_m = 2$) (BTA–DMG), described in Chapter 3.

Such materials, including such diverse systems as the phenol-formaldehyde, urea-formaldehyde, epoxy and alkyd resins, are of commercial interest. Epoxy resins are produced by reacting (typically) a diepoxide with a polyfunctional amine such as triethylenetetramine (TETA) ($f = 4$). The well-known Araldite® resin is an epoxy system which uses epichlorhydrin and bisphenol-A to produce bisphenol-A diglycidyl ether ($f = 2$), which is then 'cured' with TETA.

Such chemical networks are used as adhesives and/or rigid moulding resins by carrying out the reaction to high conversion, but such materials appear to be a long way away from what are normally considered to be polymer gels. Indeed many, such as moulding resins like Araldite, are deliberately prepared (or driven by appropriate processing) into the glassy state (Williams *et al.*, 1998). However, at degrees of conversion just greater than the corresponding gel point, i.e. when $p \gtrsim p_c$, they can be swollen, and then their properties are not dissimilar to those of any other swollen network, whether chemically or physically cross-linked, and so can be described, albeit with caution, as polymer gels.

Perhaps the most important materials in this category are the polyesters and polyurethanes (Stanford, 1998; Stepto, 1998). These form by the reaction of the constituents with (usually) the elimination of a simple product. For example, the production of polyesters involves the reaction of a polyacid with a polyol (often just a diol), with the elimination of water. Commercial polyurethanes involve the reaction of a diol such as ethylene glycol, a diisocyanate such as MDI (4,4′-diphenyl-methane diisocyanate) which gives a urethane (structurally –O(CO)NH–) linkage, and a polyfunctional additive such as a hydroxyl-functionalized polyether (Stanford, 1998). In all cases a catalyst and high-temperature/pressure processing is required to produce an acceptable commercial product. For example, polyurethanes are commonly produced by the reactive injection moulding (RIM) process, while polyurethane foams are producing by exposing the reactive urethane groups to water or steam.

4.1.2 Non-linear materials from preformed polymer chains

These systems include the traditional elastomers, both natural and synthetic rubbers (Erman and Mark, 1997). The vulcanization of natural rubber, a process some two centuries old, involves heating the natural rubber latex extracted from the tree – essentially *cis*-poly(isoprene) – with elemental sulphur. The sulphur forms cross-links, normally assumed to be of the form –S–S–S–, between adjacent polymer chains, a process commonly referred to as curing; see Figure 4.1. Provided it has more than the necessary number of intermolecular cross-links, the newly cross-linked rubber becomes a solid (albeit a viscoelastic solid), whereas prior to this it was a viscoelastic fluid, so curing a rubber ('gelling' it) converts the material from liquid into solid.

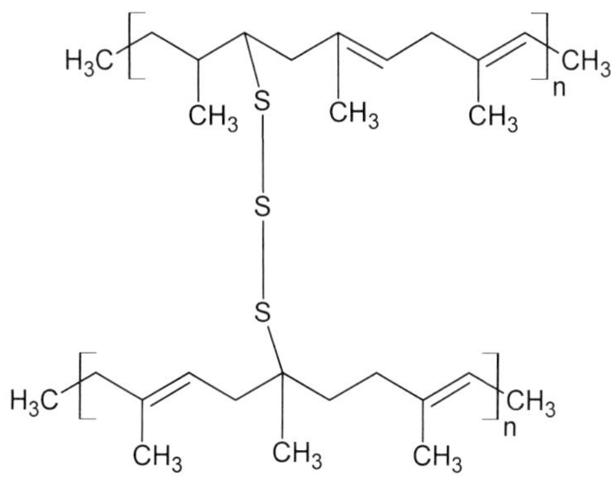

Figure 4.1 Sulphur vulcanization of natural rubber, *cis*-poly(isoprene).

Of course natural rubber is now used to a much more limited extent than previously; instead, synthetic rubbers based on poly(butadiene), styrene–butadiene or silicone chemistries are commonly employed. For these the cross-linking chemistry is different, but is still commonly known as vulcanization. Such non-traditional vulcanizing agents include peroxides and urethane cross-linkers. Vulcanization, whether by sulphur or other agents, is only one possible means of cross-linking preformed chains. Conventional free-radical-initiated polymerization of vinyl polymers is very commonly used for linear chain systems, but incorporation of small amounts of difunctional monomers can also produce a cross-linked system. For example, the addition of divinylbenzenes to styrene can soon produce a rigid resin, but the process can be controlled by suspension polymerization, by adding a non-solvent such as water. Physical methods for cross-linking, such as radiation, can also be employed; the work by Charlesby and co-workers (Charlesby and Swallow, 1959) pioneered this approach in the 1950s. Finally, many commercial ion-exchange resins are formed by reacting difunctional styrene with divinylbenzenes and then modifying the polymer by adding the appropriate ionic groups.

For control and flexibility of material properties, silicones have a number of advantages over vinyl polymers, and polydimethylsiloxane (PDMS) elastomers have proven invaluable in academic studies because they can be persuaded to form elastomers by end-linking, a process which allows far greater control and quantitative conversion degrees (Erman and Mark, 1997).

The critical degree of cross-linking is approximately equal to $1/n_0$, where n_0 is the number of potential cross-linking sites, which is in turn related to chain length. The exact proportionality actually depends upon the primary chain molecular mass distribution; for a monodisperse system it will be as stated, but for a most probable or Flory distribution it tends towards $2/n_0$. However, in any practical process, the majority of cross-links formed are not intermolecular, but are wasted in forming various defects such as loops or cycles. At the same time, other network defects such as entanglements can also contribute, so predicting the exact properties of an elastomeric network is not nearly as simple as, for example, the Flory gel point estimate (3.3) might suggest. This has a number of implications when we come to consider swelling theories.

4.1.3 Poly(acrylamide) and poly(NIPAm) gels

Despite the history of polymer networks discussed above for bulk systems, one of the most important classes of chemical gels are those derived from poly(acrylamide). These are widely used in a number of applications, most particularly in the technique of poly(acrylamide) gel electrophoresis (PAGE). Early-generation DNA sequencing techniques used poly(acrylamide) gels to separate DNA fragments differing in length by a single base-pair, so the sequence could be read. Although now largely replaced in this application by agarose gels (see Chapter 7), PAGE is still used in some applications (Stryer, 1981).

A poly(acrylamide) gel is prepared by mixing acrylamide and a small amount of an analogous branching component, typically the bifunctional form bisacrylamide, with a persulphate, usually ammonium persulphate, and tetramethylethylenediamine (TEMED)

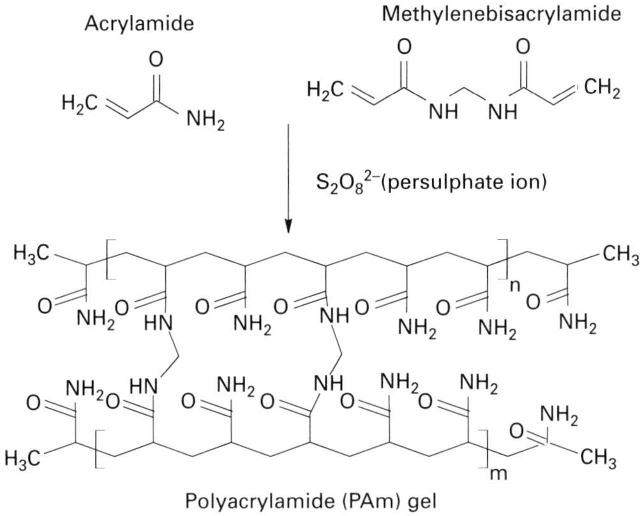

Figure 4.2 PAm gel synthesis.

as the free radical polymerization initiator components; see Figure 4.2. Despite the somewhat cluttered chemistry, a fine transparent gel can be formed. By applying an electric field across the gel, typically swollen with buffer, proteins and nucleic acids can be separated and visualized by appropriate staining.

Poly(acrylamide) – polyAm or PAm – polymers are hydrophilic, but by making the monomer core more hydrophobic, by N-isopropyl substituting the acrylamide monomer to give N-isopropylacrylamide (NIPAm), new materials can be formed with intriguing properties. Pure poly(NIPAm) gel materials are sufficiently hydrophobic that they tend to phase-separate on heating, and tuned temperature-sensitive swelling/de-swelling can be achieved by copolymerizing acrylamide, N-isopropyl acrylamide and bisacrylamide in various proportions. One obvious aim of such work is to control the swelling or de-swelling so it occurs around either room or body temperature, in so-called thermo-responsive gels.

4.1.4 Copolymer networks

Some copolymer networks have already been mentioned, but great control and flexibility of final properties can be gained by copolymerizing, say, urethane and urea with a poly(ethylene oxide) and poly(propylene oxide) prepolymer referred to as PEO–PPO to produce segmented copolymers (Stanford, 1998). The urethane–urea segments tend to be 'hard' and the POE–PPO segments 'soft', so material properties can be fine-tuned. Added to this, as is now well known, such segmented copolymers can undergo phase separation; these systems often have complex morphologies, as discussed in Chapter 6.

4.2 Theories of rubber elasticity

As mentioned earlier, the decrease in conformational entropy of a stretched rubber gives rise to an elastic restoring force; this force is, of course, the signature property of a rubber. The classical theory of rubber elasticity (Flory, 1953; Treloar, 1975) is now some 70 years old (three of the most important papers were published in the same year, 1943), so it is not surprising that it has been subsequently challenged, extended and improved (Erman and Mark, 1997; Mark and Erman, 1998). That said, for elastomeric systems, and where exact quantitative agreement between theory and experiment is not required, its beauty and simplicity still makes it a useful starting point for any more detailed discussion.

In the original theories (mostly pre-1975) the exact numerical factors depend upon whether the derivation assumes 'phantom' networks, in which chains have no excluded volume and so do not 'see' one another, or 'affine' networks, so that network junction points move affinely following macroscopic deformation.

Pre-1975 theories give the elastic free energy as

$$\Delta F_{\mathrm{el}} = ak_{\mathrm{B}}T(\Lambda_x{}^2 + \Lambda_y{}^2 + \Lambda_z{}^2), \tag{4.1}$$

where the Λs refer to macroscopic deformation ratios in the x, y and z directions, and a is a 'front factor' which depends on the particular assumption used. In this way the stress–strain behaviour of an elastomeric network can be calculated, and the stress is found to be proportional to $(\Lambda-\Lambda^{-2})$, where Λ^3 is simply the product $(\Lambda_x \Lambda_y \Lambda_z)$ (Treloar, 1975).

What the theory also produces is a relationship between the equilibrium elastic modulus – the ratio of the retractive force per area, i.e. stress, per unit strain – in terms of the number of cross-links (more exactly the number of junction points and chains in the network). In one version of the theory, there is a quantitative relationship between the modulus of the network (in this case the equilibrium elastic shear modulus G) and the number of elastically active network chains (EANCs) (Gordon and Ross-Murphy, 1975).

This version of the theory gives

$$G = a\nu_{\mathrm{e}}k_{\mathrm{B}}T. \tag{4.2}$$

Here ν_{e} is the number density of EANCs, and is dependent upon the degree of conversion p and the functionality f. The parameter a is again the rubber front factor, and $k_{\mathrm{B}}T$ is the usual Boltzmann energy term. More particularly in this formulation, ν_{e} will be zero before the gel point, defined as in (3.3), and so, up to this point, $G = 0$. An EANC is defined as any sequence of cross-linked units joining two units known as 'ties', and a tie is any unit which has at least three non-extinct cross-links, i.e. is 'tied' into the infinite gel molecule by at least three separate paths to infinity (see Figure 3.2). The value of ν_{e} is then equal to half the average number of non-extinct cross-links per unit present, because each EANC has two ends. The average is calculated excluding contributions from units with only one or two non-extinct cross-links, using methods described elsewhere. (Note that in an alternative form of this equation, due originally to Flory (Erman and Mark, 1997), the relevant parameter is M_{c}, the molecular mass between cross-links, rather than ν_{e}.)

Simple and beautiful though this approach is, it became clear quite soon after its development in the 1940s that a number of corrections would need to be applied. This is particularly an issue when applying the theory of ideal ('virtual') networks to real systems. Corrections include those mentioned above, which compensate for various more specific forms of 'wastage', and also those arising from topological entanglements, including ones which have become permanently 'trapped' during the cross-linking process. Progress in improving the theory was relatively slow until the 1970s, because the chemistry of network formation was not easily controlled, and so model samples were not available. However, when samples of networks formed from end-linking PDMS were prepared and tested, newer theories began to appear, including those due to Flory and Erman, which included contributions from network junction constraints. Detailed discussions of these approaches lie outside the scope of this volume, but are well argued in the monograph by Erman and Mark (1997).

Almost at the same time, Edwards and his co-workers (Doi and Edwards, 1986; Edwards and Vilgis, 1988) published a number of papers that included the effects of entanglements, in the so-called tube or slip-link model. Detailed observation of networks during deformation also became possible using the newly developing technique of neutron scattering. Such modifications to the theory become important when considering large degrees of swelling, discussed in the next section, since the assumptions behind the classic theory include that of Gaussian chain behaviour, which becomes increasingly unlikely at high degrees of swelling. Some workers have assumed limited extensibility Langevin chain behaviour. These Langevin chains (strictly chains whose end-to-end distribution follows the inverse Langevin function approximation) have the property that they behave like freely jointed chains at low deformation but, unlike the Gaussian function, have a realistic and finite large-deformation limit, so a chain cannot be longer than its maximum contour length. However, attributing deviations solely to finite-chain effects has proved difficult because, for real networks, deformation-induced order (crystallization, even liquid crystal formation) can also contribute to deviations from ideal behaviour.

Finally we need to reassert one of the conclusions of Chapter 3, that the equilibrium modulus of any gel depends essentially on the product of two factors – the number of 'cross-links' or 'junction zones' per unit volume, which in turn is related to the percolation degree of conversion parameter p, and the contribution of these to the overall modulus. For idealized 'entropic' or ideal rubber-like networks the latter term is small, of order unity. For more rigid network structures such as agarose, elasticity is very unlikely to be entropic. In other words, the contribution per cross-link comes from bending and twisting contributions, which raise the enthalpy of the system. Consequently we adopt the term 'enthalpic' and assign its contribution to a generalized (i.e. non-ideal rubber-like or entropic) front factor, which for agarose, for example, is estimated to be $\gtrsim 10$ (Clark and Ross-Murphy, 1987). That said, separating these two contributions is not trivial, and for these physical gels requires a number of additional assumptions, which are discussed throughout this book.

4.2.1 Reel chain models

For thermoreversible gels, which are a major topic of this volume, Nishinari *et al.* (1985) formulated an approach to the modulus–temperature relationship (a direct proportionality for the classical networks above) by assuming that the gel consists of a network of Langevin chains. In this so-called 'reel-chain' model, both ends of a Langevin chain are bound into a cross-link (or junction zone) region, but can be progressively released on increasing the temperature by 'sublimation', as if chains were being pulled off a reel. Using this model, Nishinari and co-workers were able to calculate the temperature dependence of the modulus (in particular the static Young's modulus E) and compare results with their own data for PVA gel networks. They found that the temperature dependence of E was a sensitive function of a constraint release parameter, defined as the ratio of the number of chains released to the (RMS) end-to-end distance of the chain. They also found that E had a maximum value as a function of temperature, first increasing and then decreasing.

Such reel-chain models have been published by others, including Higgs and Ball (1989), who drew on the NKO model (Nishinari *et al.*, 1985) and also investigated so-called rod-chain models, which apply before the unwinding or 'sublimation' occurs. Using their model, they concluded that, rather than approaching its extension limit and paying a high free-energy cost due to reduction in entropy, a Langevin chain will tend to pull out another chain segment from the junction zone. Consequently, as long as a Langevin chain does not approach its extension limit, it behaves in the same way as a Gaussian chain, so the effects of limited extensibility are somewhat outweighed by the unwinding of the junction zone.

4.3 Swelling of gels

While the various theories of rubber elasticity allow us to calculate the retractive force on a polymer network, the swelling of a polymer network or gel is assumed to be governed by a solvent-induced expansion force (or, more formally, free energy) involving osmotic and, for polyelectrolyte systems, ionic contributions. The net result is that a gel will tend to swell to an equilibrium degree which depends upon its nature, the amount of cross-linking, the nature of the solvent, the temperature and other factors. For a given network, these factors allow the equilibrium degree of swelling to be altered, but the essential point is clear: to achieve a high degree of swelling, the retractive force must be small, so the degree of cross-linking must itself be not much more than the critical amount.

Since absolute degrees of swelling are, in the context of the present volume, of less significance than the changes that can be wrought by changes of temperature, solvent etc., we will discuss swelling below in terms of the simplest such model, that of Flory and Rehner (Flory, 1953). This is now very ancient but, like the Flory–Stockmayer theory, can be regarded as a useful and easily comprehensible zeroth-order approximation, upon which we can hang future developments and improvements.

As stated above, when a chemically cross-linked gel is immersed in an excess of liquid and allowed to equilibrate, the size of the sample may increase, decrease or remain constant; see Figure 4.3. This effect depends strongly upon the 'thermodynamic quality' of the liquid for the gel and the degree of cross-linking. If a hypothetical cube of gel (with sides of length l) is immersed in a good solvent the volume will increase from an initial value of V_0 to V. The swelling ratio q is defined as V/V_0. For a cube, V_0 is equal to l^3, and

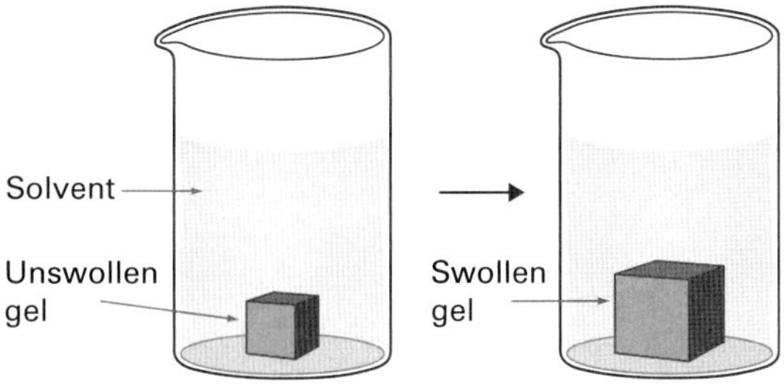

Figure 4.3 Swelling and de-swelling of a gel cube.

since the gel is assumed to swell affinely (that is, each side deforms proportional to the macroscopic strain ε), $V = (\varepsilon l)^3$, so that the degree of swelling q is also equal to ε^3. (For very high degrees of swelling, this ratio could also be defined in terms of the mass of gel.)

The factors which affect swelling also need to be considered. The excluded-volume effect determines that the average coil dimensions tend to increase, in order to maximize the number of polymer segment–solvent interactions. For the bulk (unswollen) gel, the chains will have approximately unperturbed dimensions, so when immersed in solvent the polymer will swell by an amount depending on the polymer molecular mass M_w:

$$\langle R_g^2 \rangle / \langle R_g^2 \rangle_0 \approx M_w^{1/5}. \tag{4.3}$$

In a poor solvent the polymer will appear to shrink, and eventually

$$\langle R_g^2 \rangle \approx M_w^{2/3}. \tag{4.4}$$

The exponents 0.2 and 2/3 follow from excluded-volume theory (Yamakawa, 1971). Obviously the overall increase (or decrease) in dimensions will depend upon the original dimensions; and the higher the molecular mass, the greater the effect. Indeed, for the gel, the nominal overall weight average molecular mass M_w is infinite, so an infinite degree of swelling would be predicted. This does not happen because the tendency to swell is counteracted by the elastic restoring force from rubber elasticity theory.

The total swelling pressure Π can be written more formally as

$$\Pi = \sum_{i=1,4} \pi_i, \tag{4.5}$$

where the individual pressure terms are

$$\pi_i = \frac{\Delta F_i}{\phi_1}$$

and ϕ_1 is the volume fraction of solvent. Here, for example, ΔF_1 is the free energy for swelling and π_1 is the excluded-volume (or mixing) term from Flory–Huggins theory, with an associated polymer–solvent 'goodness' parameter χ . π_2 of opposite sign reflects the change in configurational free energy with swelling (the rubber elasticity term).

In Flory's original approach (Erman and Mark, 1997), the Flory–Huggins mixing terms give

$$\Delta F_1 = RT(n_1 \ln\phi_1 + n_2 \ln\phi_2 + \chi_1\phi_1 v_2). \tag{4.6}$$

Here n_1 is number fraction of solvent molecules and ϕ_2 is the volume fraction of the network, with $\phi_1 = (1 - \phi_2)$. The elastic term can be written in a number of ways, depending upon the network model, but is related to the number of elastically effective chains and the volume deformation (the swelling ratio) (Dusek and Prins, 1969; Schroder and Oppermann, 1996).

Further terms which may be added include π_3, which is a measure of the difference in osmotic pressure between the gel and the solution. This includes the Donnan contribution from the mixing of ions with the solvent, where it is assumed that a polyelectrolyte gel acts like an osmotic semi-permeable membrane. If the polymer chains have, say, a net negative charge, there will be a net accumulation of mobile cations in the 'compartment' containing the gel, which gives rise to this free energy contribution. Finally, a term π_4 comes from the free energy of electrostatic interactions. For neutral gels in non-polar solvents, the original Flory–Rehner case, only the first two of these terms are required. The osmotic swelling force depends on the factor $\chi - 0.5$, reflecting solvent quality. As already mentioned, the retractive force is in turn proportional to the number density of cross-links (more formally, to the number density of elastically active network chains). Thus for higher degrees of cross-linking, the net elastic force is higher (the shear modulus is higher) so the equilibrium degree of swelling is lower; conversely, the lower the degree of cross-linking, the greater the degree of swelling.

The previous comments assume that the polymer and solvent are not carrying any charges. However, because of the polyelectrolyte effect, the need to reduce charge–charge interactions along the chain contour, a polyelectrolyte gel will swell to a large extent in pure water, or in low ionic strength electrolyte solutions (Chapter 5). At high electrolyte concentrations the gel will tend to de-swell, because of the dependence of chain dimensions on $I^{-1/2}$, where I is the ionic strength. This effect can be amplified by adjusting the mix of ionic species.

Overall, this means that the change in chemical potential due to swelling can be obtained by evaluating the derivatives of the free energies with respect to n_1. Swelling equilibrium is then established for polyelectrolyte gels when the chemical potential of the solvent in the gel and of the surrounding solvent are the same; the subscripts 1 to 4 refer respectively to the mixing, elastic, osmotic (including Donnan) and electrostatic contributions:

$$\Delta\mu_{1,\mathrm{eq}} = \Delta\mu_{1,1} + \Delta\mu_{1,2} + \Delta\mu_{1,3} + \Delta\mu_{1,4}, \tag{4.7}$$

where the mixing term is simply the classic Flory–Huggins contribution,

$$\Delta\mu_{1,1} = RT\left(\ln(1 - \phi_2) + \phi_2 + \chi\phi_2^2\right). \tag{4.8}$$

The other terms are more complicated to evaluate, and there are several alternate forms in the literature (Erman and Mark, 1997). In terms of overall and readily measurable swelling parameters, it can be more useful to consider swelling in terms of its overall degree. For example, in many experiments what is actually determined is the mass m of swollen gel relative to the mass m_0 of the initial un-swollen system, and the swelling ratio q is simply m/m_0. Depending on the specific assumptions, a double logarithmic plot of the modulus of the gel against the degree of swelling should have a slope of $-1/3$ for an ideal Gaussian network. For chemical gels, however, any upward deviation from this $-1/3$ law at low swelling degrees can be interpreted in terms of constraints on the chain motion becoming more significant when the network volume fraction increases. For physical gels, this becomes more complicated because the modulus has its own concentration dependence, and the degree of swelling is nothing more than the reciprocal of the volume fraction of gel in the network, i.e. effectively this concentration dependence.

Finally, we comment that the simple separability of the retractive and swelling force terms is still a controversial topic. We will not discuss this further, but merely reiterate that we have used this simple form of the swelling equations for the purpose of clarity.

4.3.1 Discontinuous swelling

The explanation of the swelling and de-swelling of polymer gels given above was accepted until about 20 years ago (and remains true in the majority of cases). However, Tanaka and co-workers then reported the behaviour of partly hydrolysed PAm gels (Tanaka, 1981; Osada and Ross-Murphy, 1993). In a mixed solvent of good (water) and poor (acetone) thermodynamic quality these are weakly polyelectrolytic in nature. As small amounts of acetone are added to a PAm gel, the gel de-swells. Moreover, for certain samples corresponding to certain amounts of alkaline hydrolysis, the swelling become discontinuous: there is a sudden de-swelling over a very narrow range of acetone–water compositions. This occurs because the net elastic, polymer–solvent affinity and electrolyte contributions (introduced above) produce, instead of a continuous temperature (or acetone–water composition) swelling 'phase diagram', one which is discontinuous.

The explanation based on polymer–solvent affinity and electrolyte terms neglects specific Donnan contributions. However, the higher the ionic concentration outside the membrane (gel), the lower the concentration difference of mobile ions between the two sides of the membrane. The concentration of mobile ions depends upon the ionic strength, the polymer concentration and the effect of (Manning) counterion condensation (Chapter 5). For example, Moe and co-workers (Moe *et al.*, 1993) have shown that the contribution of the Donnan term to the swelling of polyelectrolyte gels is actually the most significant one, and can lead to very high degrees of swelling (and de-swelling).

Whichever factor causes the gel collapse in the acetone–water system, the consequences are profound, since changes in swelling can be induced by appropriate changes in temperature, pH, applied voltage etc., depending upon the precise chemical composition of the gel–solvent system. Figure 4.4 illustrates the simple case of continuous swelling, and contrasts this with the case of discontinuous swelling.

As mentioned above, so-called thermoresponsive gels can be formed by copolymerizing acrylamide and NIPAm. That this can be done is not in dispute, but whether this

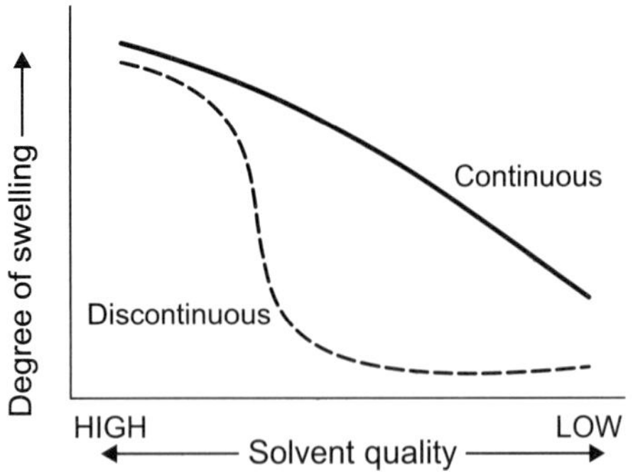

Figure 4.4 Continuous and discontinuous swelling of a gel, induced by change of solvent quality.

alone justifies the many synthetic papers which have appeared is less clear. The chemistry of formation of such PAm gels is straightforward, at least in the laboratory, but it is far from simple in terms of theories of polymer network formation. In almost all cases, the proportion of intermolecular cross-links is completely unknown, but is certainly much lower than the total number of cross-links formed, so the majority are actually 'wasted' in various chemical and physical side-reactions. The approach that many authors have adopted is simply to show that, if they add more cross-linker, the Flory 'molecular mass between cross-links' parameter M_c, mentioned above, tends to decrease. However, the latter is itself a less than useful factor since it has little physical meaning, except that it naturally decreases when the swelling ratio falls, so this becomes a circular argument. Added to this is the fact that no correction tends to be made for the wastage, so quoted M_c values are usually far too low to have any significance.

4.3.2 Kinetics of swelling

The diagrams above show the equilibrium behaviour on swelling or de-swelling, but do not, of course, say anything about the (de)swelling rate. In practice this is mainly under osmotic control so, for a typical gel, it may take hours to occur, even for small – say 1 mm – samples. Since for a spherical gel sample the rate of swelling is approximately inversely proportional to the square of the radius (Tanaka and Fillmore, 1979), for working devices micro- or nano-scale samples are required, so reducing the size of the gel particles from 1 mm to 10 μm will reduce the time from, say, 3 h to around 1 s. In other words, the rate of swelling can be accelerated simply by chopping the macroscopic gel into micro-particles.

4.3.3 Smart gels

The exploitation of swelling phenomena owes much to the work of the late Toyoichi Tanaka (Tanaka, 1981; Shibayama and Tanaka, 1993). Drawing on, but extending, work in the review by Dusek and Prins (1969), he was able to establish the conditions leading to discontinuous swelling. This immediately suggests a way to construct practical devices in which the so-called volume phase transition can be induced, say by changes in temperature (or voltage, pressure etc.). The kinetic aspects can be overcome by combining micro-gel particles, and a number of potential applications of so-called 'smart' or 'intelligent' gels can be formulated.

A thermodynamic system capable of transforming chemical energy directly into mechanical work is referred to as a chemomechanical system. The history of such materials goes back to such distinguished names in polymer science as Kuhn and Katchalsky in the early 1950s. All living organisms move by the isothermal conversion of chemical energy into mechanical work, as exemplified by muscle, flagellar and ciliary movements. Such highly efficient energy conversion systems can be realized, at least potentially, in synthetic polymer gels, which expand and contract upon changing their solubility and/or degree of ionization in response to an external stimulus in the form of thermal, chemical or electrical energy (Kaneko *et al.*, 2002).

As mentioned, hydrogels prepared from poly(NIPAm)–PAm which show so-called lower critical solution temperature (LCST) behaviour (see Chapter 10) in aqueous solution are thermoresponsive, and expand/contract below/above the phase transition temperature. Another type of gel containing ionic groups can be actuated isothermally by an electric field. When the gel is negatively charged, it swells near the anode and contracts near the cathode. The contraction rate is proportional to the external electric current, and changes in volume of, say, ±50% are quite usual. Such a polyelectrolyte gel can be made to bend backwards and forwards by the application of an alternating electric field. Here water and ions migrate towards the electrode, bearing a charge opposite in sign to the net charge in the gel, and this coupling of electro-osmosis and electrophoresis is thought to be responsible for the observed chemomechanical behaviour (Osada and Ross-Murphy, 1993). For a long time development of such devices seemed not to progress beyond lab 'toys', but work by, for example, Siegel and co-workers (Siegel *et al.*, 2004) has led to novel swelling/de-swelling of glucose-binding hydrogels (Kataoka *et al.*, 1998), for use in oscillatory feedback insulin delivery systems.

4.4 Transient networks

The dynamical mechanical moduli of covalent networks have been extensively studied in relation to the mechanical properties of rubbers and elastomers; the chemistry has been outlined above. A second type of network, entangled networks, has also been extensively studied in relation to the flow characteristics of polymer melts (Graessley, 1974; Doi and Edwards, 1986). The various mechanisms for stress relaxation and flow of melts and semi-dilute polymer solutions are now well understood. In the widely accepted 'reptation' model, each chain acts as if it is surrounded by a restrictive virtual tube formed by all the other chains in the matrix. This modifies – restricts – a number of transport properties. The effect of entanglements on the relaxation of polymer chains is seen in the spectrum of shear modulus versus frequency, which shows a plateau which grows with the molecular mass (more commonly referred to in this area as molecular weight). The plateau value is called the entanglement plateau, and appears to correspond to a molecular mass at which the zero-shear viscosity begins to rise as $M^{3.4}$. The plateau value is related to the density of entanglements and it defines the molecular mass M_e between entanglements. M_e tends to be specific to each polymer type but is typically in the range $5–20 \times 10^3$ g mol^{-1}. Entanglements are present in melts and solutions when the product of normalized concentration (assumed to be 1 in the melt) and molecular mass is $>M_e$.

When dealing with physical gels, the local structures created by cross-linking mechanisms are complex and depend on the type of local association of the chains (crystallization, secondary structure formation, ionic interactions). The junctions are multi-functional, although the functionality is not a fixed number and the junctions either have finite lifetimes or can be disrupted by changes in temperature, solvent quality etc. As well as those association mechanisms which occur in nature, some polymers have been specially designed to create labile network structures under controlled conditions. These are the so-called associative polymers, of which the most common are the telechelics, polymers that have been end-functionalized. Telechelic polymers have generated much scientific activity and many practical applications. Yet another type of system forming transient networks is made up of flexible, entangled chains bearing a fixed number of 'stickers' at fixed positions along the chains, which can associate reversibly.

In the next sections we present three theoretical models which are generally used to describe the rheological properties of the above transient networks.

4.4.1 Model polymers with sticky end groups

One class of system that has attracted widespread interest is 'associative thickeners', and they have found applications in many surface coating formulations. These are solutions of flexible, water-soluble chains which bear associating groups at both ends. A well-known example is poly(ethylene glycol) (PEG) chains extended by diisocyanates and end-capped with long-chain alkanols, also called hydrophobic ethoxylated urethane (HEUR) thickeners (Lundberg *et al.*, 1991). The composition of a HEUR thickener is shown in Figure 4.5.

With increasing chain length (molecular mass), two distinct regimes have been considered. In the un-entangled network, the molecular mass M between the temporal junctions is smaller than the entanglement molecular mass M_e of the polymer. Here the major part of the stress is sustained by the elastically active chain strands connecting the junctions. In the second case, where when $M > M_e$, localized entanglements play a role similar to cross-linked junctions.

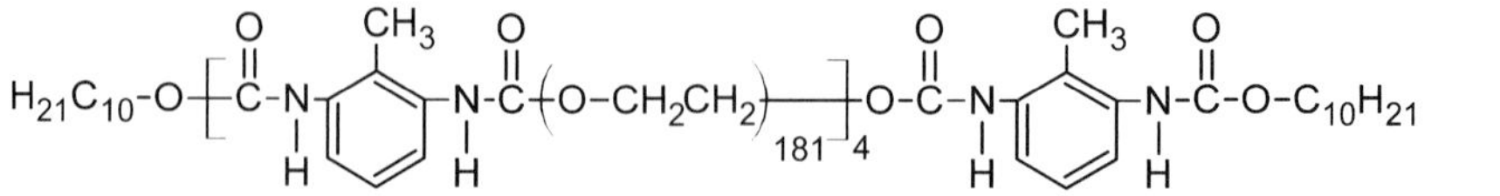

Figure 4.5 Structure of a HEUR thickener (Lundberg, 1991). Reproduced with permission of the American Institute of Physics for The Society of Rheology.

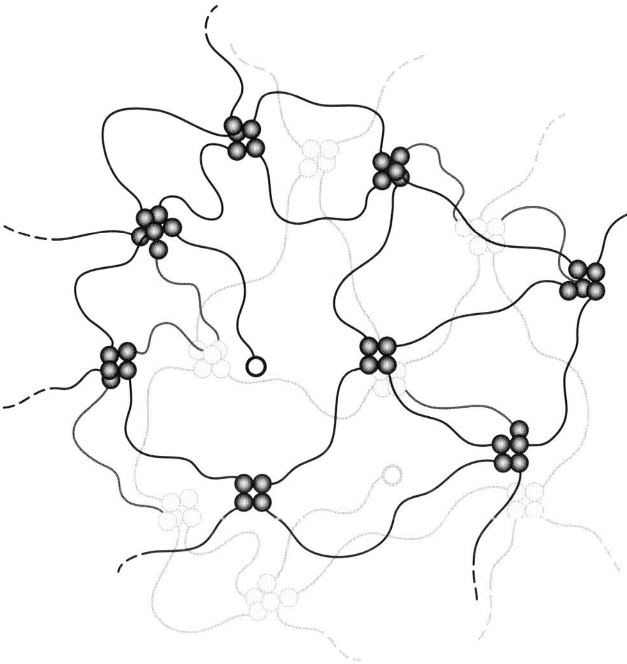

Figure 4.6 Network made of polymer chains with sticky end groups.

A schematic view of the network structure is shown in Figure 4.6. A small number of end-chains associate in micelles with an aggregation number of the order of 10 or less. This association mechanism is reversible, but the chains form viscoelastic solutions because the binding energy of a hydrophobe in a micelle is small enough that it can disentangle itself randomly, and at a finite rate. This is a reversible, non-permanent or transient network.

The first systematic study on reversible networks was carried out in 1946 by Green and Tobolsky (1946), in which the stress relaxation in rubber-like polymer networks was treated by the kinetic theory of rubber elasticity, extended to allow for the creation and annihilation of junctions during network deformation. The theory predicts a high-frequency storage modulus given by

$$G_\infty = v_e k_B T, \tag{4.9}$$

where v_e is again a number density of elastically active chains. The resultant Newtonian (shear-rate–independent) shear viscosity is

$$\eta(\dot{\gamma}) = \eta(0) = \tau_0 \, G_\infty, \tag{4.10}$$

where, in the present context, the relaxation time τ_0 is the reciprocal of the chain-end disentanglement rate. This model does not predict the shear thinning or shear thickening effects observed with HEUR solutions; to include such effects, an additional assumption is introduced to relate the disentanglement rate to the shear rate.

Although other theories were developed in the intervening years, that presented here (and referred to as the TE model) was proposed in a series of papers in 1992 by Tanaka and Edwards (1992a, 1992b, 1992c). This model deals with flexible polymers with a uniform molecular mass M, each carrying associative groups at both ends, as represented schematically in Figure 4.6, and deals with the un-entangled regime $M << M_e$.

The junctions of the network are created by the 'sticky' functional groups attached at the ends of the chains. There are two kinds of chains in the network: elastically active chains and dangling chains, the latter being chains with only one end attached to a junction, shown by open circles in Figure 4.6. The TE model introduces a potential barrier W for the dissociation of an end reactive group from a junction. One end can dissociate from a junction either by its own thermal motion or by being pulled by the chain attached to it. The probability for an isolated end group (with no chain connected to it) to dissociate is $\exp[-(W/k_B T)]$ and hence the dissociation rate of the reactive group is given by

$$\beta_0 = \omega_0 \exp(-W/k_B T), \tag{4.11}$$

where ω_0 is a characteristic frequency of thermal vibration of the reactive group, estimated to be of the order of 10^8–10^9 s^{-1}.

Because the reactive group is attached to a chain, an additional force works in the direction of the other end of the chain, and the potential barrier for the attached end to dissociate is reduced, owing to the elasticity of the chain. The TE model introduces a disentanglement rate $\beta(r)$ which depends on the end-to-end distance of the attached chain:

$$\beta(r) = \beta_0 \exp \frac{3r}{N a_s}, \tag{4.12}$$

where N is the number of monomers (residues) in the chain and a_s is the monomer size. This correction accounts for the non-Newtonian behaviour in steady-state shear, because the chains can stretch under shear.

As well as this, a dangling chain can also capture one of the junctions in its neighbourhood by its reactive end, with a probability per unit time which is the recombination rate p_r. In the TE model, p_r is given by

$$p_r \approx \beta_0 \exp(B/k_B T), \tag{4.13}$$

where B is the binding energy.

When n is the total number of chains per unit volume, the equilibrium number of active chains ν_0 under no external force is given by

$$\frac{\nu_0}{n} \approx \frac{\exp(B/k_B T)}{1 + \exp(B/k_B T)}. \tag{4.14}$$

The TE model predicts the linear viscoelastic response of the transient networks as a function of the angular frequency at various temperatures. It shows that the modulus–frequency curves at any temperature T can be superposed on the curve at the reference temperature T_0 if it is vertically and horizontally shifted according to the following expressions for the horizontal shift factor a_T and vertical shift factor b_T:

$$a_T \equiv \frac{\beta_0(T_0)}{\beta(T)} = \exp\left[-\frac{W}{k_B}\left(\frac{1}{T_0} - \frac{1}{T}\right)\right], \tag{4.15}$$

$$b_T \equiv \frac{\nu_0(T_0)T_0}{\nu_0(T)T}. \tag{4.16}$$

The fact that the frequency shift factor depends exponentially on the reciprocal of the temperature indicates that the linear viscoelasticity is dominated by the activation process of the junction dissociation.

The model predicts the viscosity at a stationary shear rate $\eta_{sta}(\dot{\gamma})$ and shows that the characteristic shear rate for observing the shear thinning of the viscosity – determined, for instance, by $\eta_{sta}(\dot{\gamma}) = \eta(0)/2$ – is very sensitive to the choice of the disentanglement function $\beta(r)$. It also shows that the Cox–Merz rule (Chapter 2) is not obeyed in such systems. Figure 4.7 shows this deviation, with dynamic viscosity falling below the steady shear value.

This model successfully describes (i) the temperature scaling of the viscoelastic moduli, (ii) the single relaxation time which controls the viscoelastic response, through the chain breakage rate, (iii) the observation of a non-Newtonian behaviour (shear thinning and shear thickening) and (iv) the breakdown of the Cox–Merz rule.

Rheological measurements of HEUR type systems show good agreement with these predictions. However, the concentration dependence of the high-frequency storage modulus G_∞ in these systems does not follow the prediction of the TE model, which

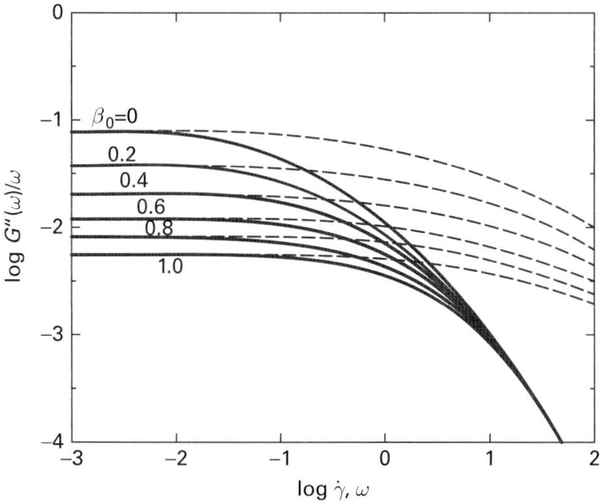

Figure 4.7 Deviation from the Cox–Merz rule in associating polymers derived from the TE model. Comparison between $G''(\omega)/\omega$ versus ω (solid line) and $\eta_{sta}(\dot{\gamma})$ versus $(\dot{\gamma})$ (broken line). The parameter β_0 is varied from curve to curve. From Tanaka and Edwards (1992a) © 1992 Elsevier.

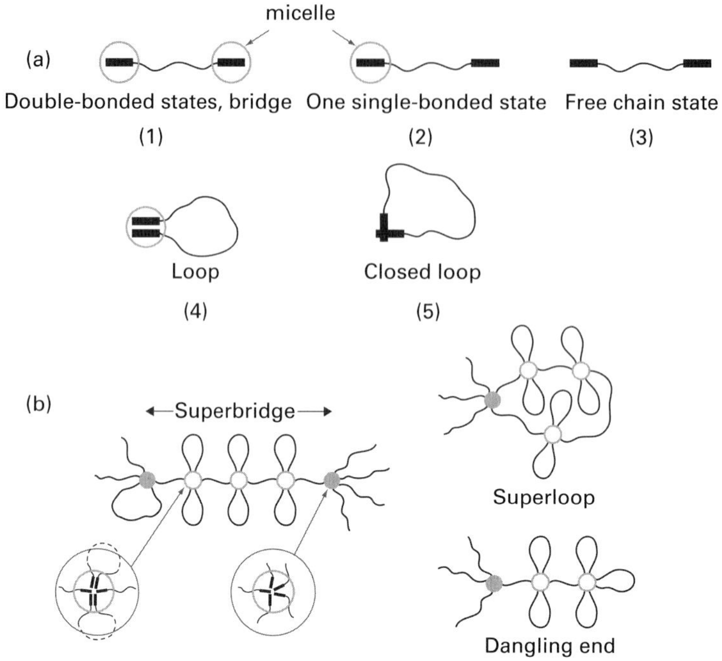

Figure 4.8 Various possible states of association of telechelic polymers: (a)(1) double-bonded state, bridge; (a)(2) one single-bonded state; (a)(3) free chain state; (a)(4) loop; (a)(5) closed loop; (b) more complex topologies: micelles with a cross-link functionality greater than 2 are shown as filled discs. Adapted from Annable *et al.* (1993) with permission of the American Institute of Physics for The Society of Rheology.

suggests a linear increase with concentration, or $G_\infty/(nk_BT)$ to be 1 at all concentrations (in the limit $W >> k_BT$ there are few free stickers). It was observed experimentally that $G_\infty/(nk_BT)$ varied strongly with polymer concentration and was always smaller than 1. Therefore, it has to be assumed that the topology of the network changes with concentration. This important feature was pointed out by Annable *et al.* (1993), who suggested that the fundamental process underlying the concentration dependence experimentally observed for G_∞ is the entropically driven transformation from a network composed mainly of loops, at low concentration, to one containing mainly links, at high concentration ('ring–chain competition').

In dilute solutions, various types of associations can be generated, as shown in Figure 4.8. Most of the structures represented are not effective cross-links, but with an increase in the polymer concentration the loops transform into bridges, giving rise to a strong initial concentration dependence of the modulus, viscosity and relaxation time. Thus the TE transient network theory represents a limiting case for HEUR solutions at high concentrations. Annable *et al.* (1993) proposed a statistical mechanical model, supported by Monte Carlo simulations which show the gradual transition from micelles built mainly from loops to micelles linked by bridging chains.

Figure 4.9 shows semi-quantitative agreement between the experimental results and the simulation for the high-frequency modulus derived by Annable *et al.* from their

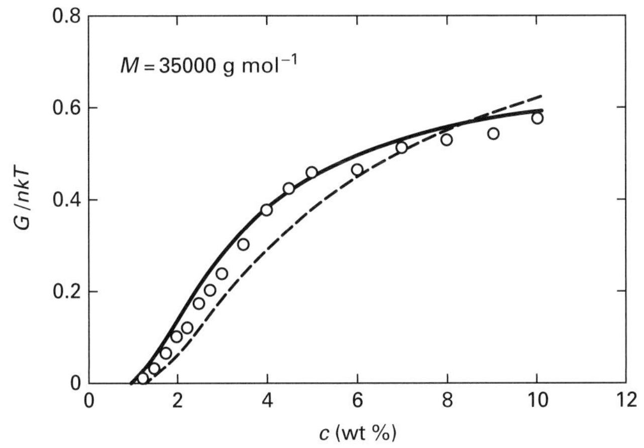

Figure 4.9 Comparison between theory and experiment for the fraction of elastically active chains, assuming an end-capping efficiency of 70% (dashed line) or 100% (continuous line) for C_{16} end-cap chains and a molecular mass of 35 000 g mol^{-1}. Adapted from Annable *et al.* (1993) with permission of the American Institute of Physics for The Society of Rheology.

statistical mechanical model. It is seen that $G_\infty/(nk_BT)$ does not reach the limit 1 even at high concentration, but is close to that found in experiments.

4.4.2 Flowers connected by bridges (SJK model)

The topology of associating polymer solutions with end capped associative groups was further analysed theoretically by Semenov *et al.* (1995; SJK model), who explored various regimes. They pointed out other important features related to micelle formation in the associative polymers. Telechelic polymers form micelles which consist of a compact hydrophobic core, surrounded by a corona of the long, soluble parts of the chains that form loops. This structure is similar to block copolymer micelles. The aggregation number (or number of chains per micelle) N_{agg}, determined experimentally, is in general between 5 and 50. This number is rather insensitive to concentration, the molecular mass of the soluble parts and other parameters. The inner structure of a micelle is similar to that of a star polymer, and the authors consider that the properties of star polymers in solution in good solvents also apply to the micelles. While star polymers in good solvents always repel each other, the important difference with telechelic polymers is the possibility of bridging that gives rise to attraction between micelles. The attraction between micelles can be large enough that the micelles phase separate and form a macro-phase of densely packed 'flowers', where neighbouring micelles are connected by bridges. A temporary, reversible network is formed by the connected flowers, as shown in Figure 4.10.

4.4.2.1 Elastic properties of the network of flowers (SJK model)

The elastic properties of the network of flowers are interpreted in the SJK model by analogy with grafted polymer layers on flat surfaces, in good solvents – the so-called 'brushes'. When two brushes come into contact the chains are less and less stretched

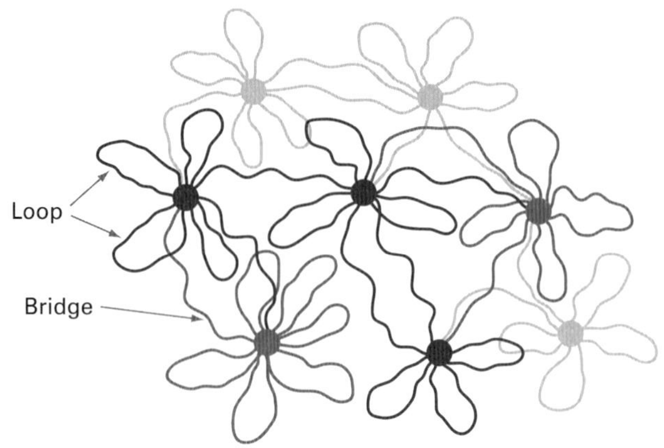

Figure 4.10 Transient network of flowers connected by bridges. Adapted from Semenov *et al.* (1995) © 1995 American Chemical Society.

close to the brush edge and, by interpenetration of the chains, become similar to the conformation of a semi-dilute solution. The distance of interpenetration is of the order of the correlation length of the semi-dilute solution. When brushes are made of telechelic polymers, each telechelic chain is a dense, highly stretched brush, and can be considered as two half-chains. However, the possibility of bridge formation in telechelic brushes gives rise to an additional attraction.

In the micellar structure, the same mechanism of interaction takes place: the coronas of the two micelles are deformed in the overlapping region, to an extent that depends on the distance between their centres, and if the aggregation number is assumed to be constant, a large energy of attraction is derived by analogy with the planar surfaces.

The attractive free energy $\Delta F_{\text{attract}}$ between two flowers depends on the number of bridges in the 'cone of contact'. This is much larger than $k_B T$, and is estimated in the model to vary as

$$\Delta F_{\text{attract}} \sim N_0 \sim N_{\text{agg}}{}^{0.3} k_B T, \tag{4.17}$$

where N_0 is the number of bridges between micelles. This large attractive energy leads to a second virial coefficient which is negative, meaning that the system phase separates into a 'liquid of micelles' while the second phase is a correspondingly very dilute solution of micelles. The flow properties of the 'liquid of micelles' phase are those of a transient gel network or a physical gel, since micelles are connected by multiple bridges, but the chains can also disentangle from one micelle and stick to another one.

The important factor which controls the rheology of the solutions is the monomer concentration c_m. At a concentration c_{min} the distance between neighbouring micelles corresponds to the minimum of the attraction energy. When the monomer concentration c_m is large enough ($c_m > c_{min}$), a micellar gel can form. Under compression or under shear, the free energy is increased because the distance between micelles is changed.

Three regimes can be analysed: non-compressed micelles for $c_m < c_{min}$, weakly compressed for $c_m \sim c_{min}$ and strongly compressed micelles for $c_m >> c_{min}$. For each regime

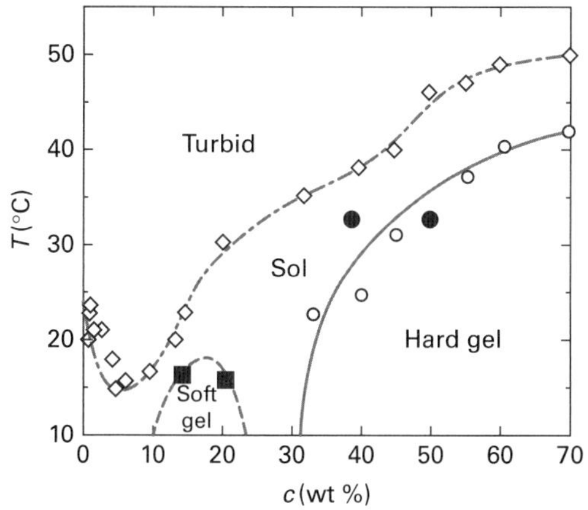

Figure 4.11 Phase diagram of aqueous solutions of $S_5E_{45}S_5$. Open symbols: tube inversion; solid symbols: rheology. Adapted from Ricardo *et al.* (2004) © 2012 American Chemical Society.

the authors predict the compressional and shear moduli. The moduli vary with different power laws for (a) the aggregation number in the micelles, (b) the polymer chain length and (c) the degree of compression $X = (c_m/c_{min} - 1)$.

Experimental work on telechelic polymers and other associative type polymers is presented in detail in Chapter 6. Light and neutron scattering, rheological experiments and optical observations clearly show the tendency for phase separation of the telechelic polymers and the formation of networks of micelles. As an example, Figure 4.11 shows results for solutions of a triblock copolymer of ethylene oxide and styrene oxide, $S_5E_{45}S_5$, where E denotes (OCH_2CH_2) and S denotes ($OCH_2CH(C_6H_5)$) (Ricardo *et al.*, 2004). The phase separation corresponds to the cloudy region. It starts around room temperature and appears when the sample is heated. In the homogeneous phase, at increasing polymer concentrations one observes various states of aggregation from soft gels (micellar liquids) to hard gels (closely packed micelles).

4.4.2.2 Zero-shear viscosity (SJK model)

Zero-shear viscosity was calculated for various concentration regimes using (4.10). Since this is the product of the shear modulus and the characteristic time of dissociation of the micelles, it is important to determine the characteristic relaxation times in the various states, according to c_m.

If the fraction of bridges is small ($c_m << c_{min}$), the authors show that the characteristic time of relaxation is the time of transformation of a loop into a bridge τ_1 or back, and is controlled by the expulsion of one adsorbing end group from the core of the micelle, as in (4.11) of the TE model:

$$\tau_1 \sim \tau_0 \exp(B/k_B T), \tag{4.18}$$

where τ_0 is the microscopic exchange time for an end group, determined by its size and the solvent viscosity.

The zero-shear viscosity is expected to increase strongly at $c_m \sim c_{min}$. In weakly compressed gels ($c_m > c_{min}$), the authors assume that the deformation of micelles is achieved through a hopping mechanism similar to the one adopted in microscopic 'free volume' theories of simple liquids. The creation of a void is necessary for a micelle to jump into a new position. The process of hole formation implies an effective energy of deformation of the micelles, where some of the bridges between micelles are stretched while the micelle near the hole is compressed. The deformation barrier U associated with a hole, for a weakly compressed gel, can be calculated ignoring the bridges. It is a function of the degree of compaction X and of N_{agg}. The activation energy of this process is governed by the stress relaxation time τ^*.

For a strongly compressed gel ($c_m >> c_{min}$), the mechanism of the elementary jump of a micelle is different. The energy of vacancy is very high, but there is no need to create a hole because the micelle can creep to another position without changing its volume, by a motion which requires an (albeit considerable) deformation in shape of the given micelle and of its neighbours. The corresponding free energy is (for shear amplitudes of the order of 1)

$$U \sim G R_m^{\ 3}. \tag{4.19}$$

Here G is again the elastic shear modulus and R_m is the micellar size.

The stress relaxation time depends crucially on whether the deformation barrier U is larger than the bridge exchange barrier B. If $U > B$, the activation state is controlled by U (micelles without bridges, with micelles and neighbours changing shape):

$$U^* \sim \Delta F_b + U, \tag{4.20}$$

where ΔF_b, the free energy of de-bridging of a given micelle from its neighbours, is proportional to N_0, the number of bridges per micelle.

On the other hand, when $U < B$,

$$U^* \sim \Delta F_b + B. \tag{4.21}$$

Finally, the stress relaxation time τ^* is estimated as

$$\tau^* \sim \tau_0 \exp(U^*/k_B T). \tag{4.22}$$

The zero-shear viscosity (4.10) of the solution increases, and the most important contribution to the viscosity comes from the energy barrier U^* and therefore the change in the characteristic relaxation time τ^*.

Overall, the SJK model provides a more detailed analysis of the mechanism responsible for the elasticity and the large viscosity increase in micellar gels of telechelic polymers, via attractive interactions from bridging chains. Micelles are compared to brushes that can slightly interpenetrate in the gel state. However, the predictions of the model, such as the power-law dependence of the moduli with the aggregation number or the polymer length, are difficult to validate experimentally.

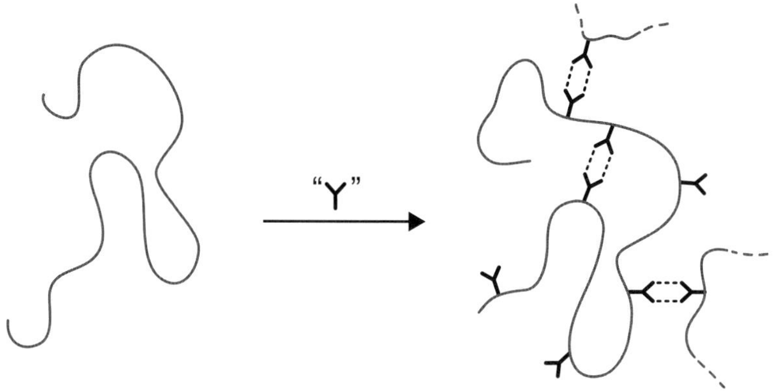

Figure 4.12 Schematic representation of the introduction of specifically interacting groups to a linear chain. Adapted from de Lucca Freitas and Stadler (1987) © 1987 American Chemical Society.

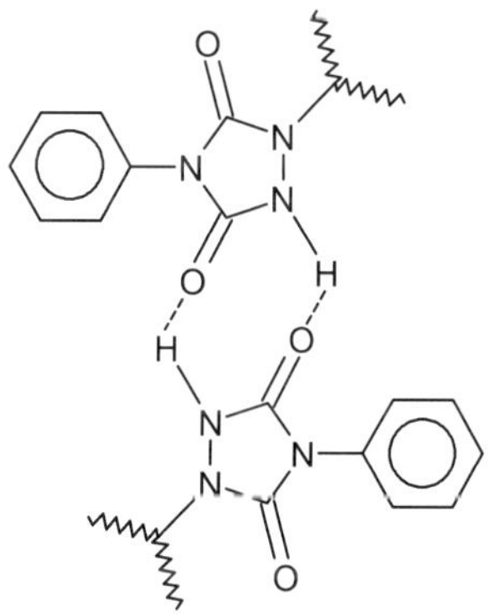

Figure 4.13 Hydrogen bond complexes between two urazole groups. Adapted from de Lucca Freitas and Stadler (1987) © 1987 American Chemical Society.

4.4.2.3 Entangled sticky chains (LRC model)

There is a third model for entangled melts or concentrated solutions of polymers, which also aims to relate the microscopic lifetime of the junction to the macroscopic stress. An early example of this class of materials is the poly(butadiene) chains with attached urazole groups studied by de Lucca Freitas and Stadler (1987), designed to form binary hydrogen bonded cross-links of this type, as shown in Figures 4.12 and 4.13.

Data for storage modulus versus reduced frequency for melts of sticky poly(butadiene) chains with various numbers of urazole groups are shown in Figure 4.14. The added urazole groups dramatically change the shape of the curve and the transition to the terminal behaviour ($G' \sim \omega^2$) becomes more gradual.

The model by Leibler, Rubinstein and Colby (LRC) (Leibler *et al.*, 1991) deals with the situation where the chains have many temporary cross-links, with S units attached to each chain. These units are called 'stickers' and can associate to form reversible cross-links. The stickers are at fixed positions.

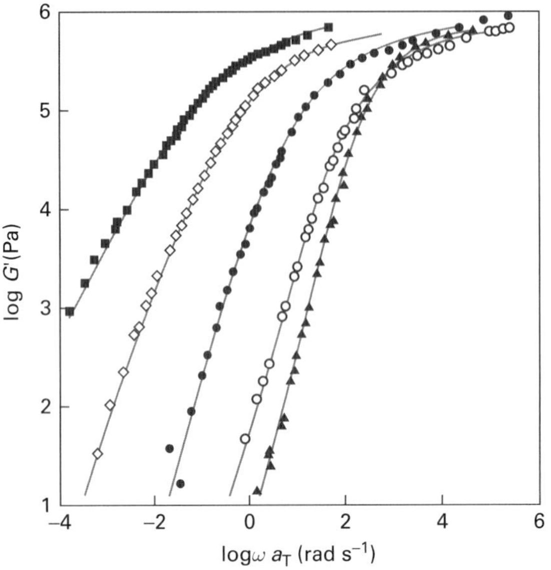

Figure 4.14 Master curves for storage modulus for a series of poly(butadiene)s with various degrees of modification x (mol%): $x = 0$ (▲), 0.5 (○), 2(●), 5 (◊), 7.5 (■). Reduced temperature is 0°C; molecular mass $M_n = 26\,000$ g mol^{-1}. Adapted from de Lucca Freitas and Stadler (1987) © 1987 American Chemical Society.

Let p' be the average number of stickers which are closed and τ the average lifetime of a sticker in the associated state. For lifetimes longer than τ (frequencies $< 1/\tau$, where τ is the average time of a sticker in the associated state), the dynamics of the network changes as stickers detach from one tie point and reassociate at another tie point.

The chains can diffuse in the reversible network and the stress can relax. The model is based on the topology of the chain in a confining tube. The chains diffuse by a reptation mechanism (Doi and Edwards, 1986). Three important relaxation times must be considered: τ_e, the Rouse time of an entanglement strand; τ, the average time of a sticker in the associated state; and T_d, the time required for a chain to escape from its tube by curvilinear diffusion.

There are four time regimes that are important for the stress relaxation modulus of reversible gels. At times shorter than τ_e, relaxation is indistinguishable from that in the polymer without stickers.

On time scales $\tau_e < t < \tau$ there is a rubbery plateau analogous to the one observed in permanently cross-linked networks or entangled melts, with a modulus which has contributions from both cross-links and entanglements. The plateau modulus G_1 is then the product of the number density of elastically active network strands and the stored energy per strand, $k_B T$. If one makes the assumption that the number densities of entanglements and cross-links are additive, then G_1 can be written as

$$G_1 \cong cRT\left[\frac{p'}{N_S} + \frac{1}{N_{\text{ent}}}\right], \tag{4.23}$$

where c is the number concentration of monomers, N is the number of monomers of the chain, N_{ent} is the number of monomers in an entanglement strand and N_S is the average number of monomers along the chain between the S stickers:

$$N_S = \frac{N}{S+1}. \tag{4.24}$$

On time scales $t > \tau$, the stickers open, the stress due to the stickers relaxes and the modulus drops to the level G_2 of the identical linear entangled chains without stickers:

$$G_2 \cong cRT\left[\frac{1}{N_{\mathrm{ent}}}\right]. \tag{4.25}$$

The second plateau G_2 persists until it reaches the terminal relaxation time T_{d} of the reversible network, given by

$$T_{\mathrm{d}} \cong \left(\frac{N}{N_{\mathrm{ent}}}\right)^{1.5} \frac{2S^2\tau}{1 - 9/p' + 12/p'^2}. \tag{4.26}$$

The predictions of the LRC model are summarized in Figure 4.15. T_{d} is enhanced significantly relative to that of unmodified polymers T_{d}^0 because, for $\tau > \tau_{\mathrm{e}}$, the stickers retard the reptation process. At the same time, the $G''(\omega)$ spectrum shows two maxima, one at high frequency, corresponding to $1/\tau$ and independent of the degree of substitution, and the other at $1/T_{\mathrm{d}}$ for the longest relaxation time, and which depends on p'.

The prediction for the zero-shear viscosity from the sticky reptation model is given by

$$\eta = G_2 T_{\mathrm{d}} = \frac{cRT}{N_{\mathrm{ent}}}\left(\frac{N}{N_{\mathrm{ent}}}\right)^{1.5} \frac{2S^2\tau}{1 - 9/p' + 12/p'^2}. \tag{4.27}$$

Experimental evidence for this model is discussed by Candau *et al.* (1998), Regalado *et al.* (1999) and Feldman *et al.* (2009), according to various chain architectures. The

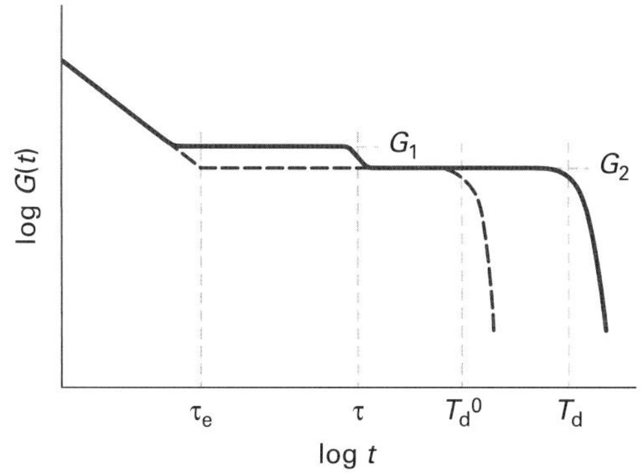

Figure 4.15 Time-dependent relaxation moduli of reversible networks of linear chains with stickers (solid line) and without stickers (dashed line). Adapted from Leibler *et al.* (1991) © 1991 American Chemical Society.

$G''(\omega)$ spectra of telechelic ionomers also show the presence of double peaks (Agarwal and Lundberg, 1984). For poly(isoprene) ionomers with carboxylate, amine and zwitterion end groups, the shift factors for the frequency dependence of the $G'(\omega)$ spectrum could reach very high values (10^5–10^7) at room temperature, before the network eventually relaxed (Fetters *et al.*, 1988).

4.5 Conclusions

For permanent rubber-like polymer networks, the science and the development of theory can be said to be quite mature. The essential assumption of classical rubber elasticity, that the retractive forces are entropic, is well accepted, and development of the theory is essentially now concerned with second-order effects. There still remain some controversies, but they are well discussed in texts such as the monograph by Erman and Mark (1997), and we do not consider them further in this volume.

By contrast, for temporary networks – the topic of major interest to us – it appears from both the models and the experiments that the rheological response of linear polymers melts and concentrated solutions can be considerably modified by the introduction of associating groups which can be either end-capped or randomly distributed along the chains. The viscoelasticity of such systems depends on the structure of the network (for example, low functionality junctions, micelles and entanglements) and on the lifetime of the clusters formed by the associating groups. Most studies have been devoted to telechelic materials, some on polymers with only one functionalized chain end. More detailed results and comparisons with theory will be discussed in Chapter 6 for triblock copolymers.

References

Agarwal, P. K., Lundberg, R. D., 1984. *Macromolecules* **17**, 1918–1928.
Annable, T., Buscall, R., Ettelaie, R., Wittlestone, D., 1993. *J. Rheol.* **37**, 695–726.
Candau, F., Regalado, E. J., Selb, J., 1998. *Macromolecules* **31**, 5550–5552.
Charlesby, A., Swallow, A. J., 1959. *Ann. Rev. Phys. Chem.* **10**, 289–330.
Clark, A. H., Ross-Murphy, S. B., 1987. *Adv. Polym. Sci.* **83**, 57–192.
de Lucca Freitas, L. L., Stadler, R., 1987. *Macromolecules* **20**, 2478–2485.
Doi, M., Edwards, S. F., 1986. *The Theory of Polymer Dynamics*. Clarendon Press, Oxford.
Dusek, K., Prins, W., 1969. *Adv. Polym. Sci.* **6**, 1–102.
Edwards, S. F., Vilgis, T. A., 1988. *Rep. Prog. Phys.* **51**, 243–297.
Erman, B., Mark, J. E., 1997. *Structure and Properties of Rubberlike Networks*. Oxford University Press, New York.
Feldman, K. E., Kade, M. J., Meijer, E. W., Hawker, C. J., Kramer, E. J., 2009. *Macromolecules* **42**, 9072–9081.
Fetters, L. J., Graessley, W. W., Hadjichristidis, N. *et al.*, 1988. *Macromolecules* **21**, 1644–1653.
Flory, P. J., 1953. *Principles of Polymer Chemistry*. Cornell University Press, New York.
Gordon, M., Ross-Murphy, S. B., 1975. *Pure Appl. Chem.* **43**, 1–26.
Graessley, W. W., 1974. *Adv. Polym. Sci.* **16**, 1–179.

Green, M. S., Tobolsky, A. V., 1946. *J. Chem. Phys.* **14**, 80–92.

Higgs, P. G., Ball, R. C., 1989. *Macromolecules* **22**, 2432–2437.

Kaneko, D., Gong, J. P., Osada, Y., 2002. *J. Mater. Chem.* **12**, 2169–2177.

Kataoka, K., Miyazaki, H., Bunya, M., Okano, T., Sakurai, Y., 1998. *J. Am. Chem. Soc.* **120**, 12694–12695.

Leibler, L., Rubinstein, M., Colby, R. H., 1991. *Macromolecules* **24**, 4701–4707.

Lundberg, D. J., Glass, J. E., Eley, R. R., 1991. *J. Rheol.* **35**, 1255–1274.

Mark, J. E., B. Erman, 1998. Molecular aspects of rubberlike elasticity. In Stepto, R. F. T. (ed.), *Polymer Networks: Principles of Their Formation, Structure and Properties*. Chapman and Hall, Glasgow, p. 215.

Moe, S. T., Skjåk-Bræk, G., Elgsaeter, A., Smidsrød, O., 1993. *Macromolecules* **26**, 3589–3597.

Nishinari, K., Koide, S., Ogino, K., 1985. *J. Phys. France* **46**, 793–797.

Osada, Y., Ross-Murphy, S. B., 1993. *Sci. Am.* **268**, 82–87.

Regalado, E. J., Selb, J., Candau, F., 1999. *Macromolecules* **32**, 8580.

Ricardo, N. M. P. S., Honorato, S. B., Yang, Z. *et al.*, 2004. *Langmuir* **20**, 4272–4278.

Schroder, U. P., Oppermann, W., 1996. Properties of polyelectrolyte gels. In Cohen-Addad, J. P. (ed.), *Physical Properties of Polymeric Gels*. John Wiley & Sons, Chichester, pp. 19–38.

Semenov, A. N., Joanny, J.-F., Khokhlov, A. R., 1995. *Macromolecules* **28**, 1066–1075.

Shibayama, M., Tanaka, T., 1993. *Adv. Polym. Sci.* **109**, 1–62.

Siegel, R. A., Gu, Y. D., Baldi, A., Ziaie, B., 2004. *Macromol. Symp.* **207**, 249–256.

Stanford, J. L., 1998. Reactive processing of polymer networks. In Stepto, R. F. T. (ed.), *Polymer Networks: Principles of Their Formation, Structure and Properties*. Chapman and Hall, Glasgow, pp. 125–186.

Stepto, R. F. T. (ed.), 1998. *Polymer Networks: Principles of Their Formation, Structure and Properties*. Chapman and Hall, Glasgow.

Stryer, L., 1981. *Biochemistry*. W. H. Freeman, San Francisco.

Tanaka, F., Edwards, S. F., 1992a. *J. Non-Newtonian Fluid Mech.* **43**, 247–271.

Tanaka, F., Edwards, S. F., 1992b. *J. Non-Newtonian Fluid Mech.* **43**, 273–288.

Tanaka, F., Edwards, S. F., 1992c. *J. Non-Newtonian Fluid Mech.* **43**, 289–309.

Tanaka, T., 1981. *Sci. Am.* **244**, 124–138.

Tanaka, T., Fillmore, D. J., 1979. *J. Chem. Phys.* **70**, 1214–1218.

Treloar, L. R. G., 1975. *The Physics of Rubber Elasticity*. Clarendon Press, Oxford.

Williams, M. A. K., Keenan, R. D., Halstead, T. K., 1998. *Magn. Reson. Chem.* **36**, 163–173.

Yamakawa, H., 1971.*Modern theory of Polymer Solutions*. Harper and Row, New York.

5 Ionic gels

5.1 Introduction

In this chapter, we consider ionic polysaccharides and the gels formed from them, since these form by far the majority of such physical gels of commercial and practical interest. A polyelectrolyte consists of a macroion (a macromolecule carrying covalently bound anionic or cationic groups) and sufficient low molecular mass counterions to assure electroneutrality.

The term 'ionic gels' is, of course, well known from those chemically cross-linked polyelectrolytes which form superabsorbent gels (particularly of acrylamides, acrylic acid etc.), and the swelling of these was examined in detail by the late Toyoichi Tanaka and co-workers, who quantified the very spectacular degree of swelling of the networks in pure water, their lesser swelling in electrolyte solutions and their de-swelling in poor solvents (see Chapter 4). More detailed discussion of these systems lies outside the scope of this volume.

However, of direct relevance here are the major anionic polysaccharide gels formed from the carrageenans, gellans, alginates and pectins. These are well known in a number of applications (including food, cosmetics, personal products and biomedical applications). They exhibit considerable variability in their properties, particularly in relation to the ionic content of the aqueous solution, and have been examined in the numerous investigations discussed here. The choice of ions allows changes in the mechanism and the structure of the network, thus allowing a tuning of the gel 'stiffness' and/or temperature of gelation. In these gels, in addition to hydrogen bonds and hydrophobic interactions, electrostatic forces and ordered secondary structures play an important role in gelation. These are the main features distinguishing these ionic gels from other physical gels. In this chapter we also introduce the cation polysaccharide chitosan. On its own, this system does not gel, but it can do so by interacting with other, particularly anionic, polymers; such systems are discussed in Chapter 11.

For all such naturally occurring polymers, changes in the molecular structure and composition according to the origin of the raw material – and, just as crucially, in the extraction regime – not only are sources of difference in the properties observed but also introduce extra complexity into their study. This chapter first presents a brief survey of overall polyelectrolyte behaviour, before going on to discuss the properties and mechanisms of gelation of these polymers.

5.2 Molecular characteristics of polyelectrolytes

In principle, any macromolecule can be transformed into a polyelectrolyte by covalent attachment of ionic groups to the polymer backbone. For example, certain essentially non-ionic polysaccharides such as cellulose and starch can be transformed into polyelectrolytes by chemical modification (functionalization).

Using the intrinsic equilibrium constant pK, ionic groups are classified as 'weak' or 'strong', depending respectively on whether they do or do not have a pK in the range 0–14, so that they can undergo proton exchange reactions at experimentally relevant pH values. Polymers with weak ionizable groups ($0 < pK < 14$) can adjust their average degree of dissociation to the physicochemical conditions, and the charge on individual groups can fluctuate. Polymers with strong groups ($pK < 0$; $pK > 14$) have fixed charges and charge distributions along the chain, whatever the pH (Stuart *et al.*, 2005).

A *polyampholyte* is a polyelectrolyte which has both anionic and cationic groups covalently bound to the macromolecule, and proteins having both $-NH_2$ and –COOH groups belong to this category. The $-NH_2$ groups are converted to $-NH_3^+$ at low pH (high hydrogen ion concentrations), while –COOH groups are coverted to $-COO^-$ at high pH (low hydrogen ion concentrations). At an intermediate condition, these two ions are present in equal amounts. This is the so-called *isoelectric point* (pI), where the viscosity of the solution shows a minimum.

The conformation of such charged polymers in solution depends strongly on the distribution of charged monomers along the polymer backbone and their environment. If these groups are weak acids or bases, the net charge of a polyampholyte in aqueous solution can be changed by varying the pH. In the vicinity of the isoelectric pH, the polymers are nearly charge-balanced and exhibit the unusual properties of polyampholytes. At high charge asymmetry (far above or far below the isoelectric pH), these polymers demonstrate polyelectrolyte-like behaviour. Dobrynin *et al.* (2004) have summarized solution properties of such amphoteric polymers and their interactions with surfaces and polyelectrolytes.

In addition to the acid and base strength of the ionic site, the average distance between adjacent anionic or cationic charges along the chain contour is an influential parameter governing overall polyelectrolyte behaviour. The *charge density* represents the spatial distribution of electric charge, and is usually defined as the average number of ionic sites per monomer unit. The regularity of the distribution of ionic sites along the chain is also important in determining solubility. Besides the acid and base strength and the charge density, the location of charged sites within the macroion is also an important factor in determining its behaviour.

Specific low molecular mass counterions have an important influence, especially on solubility and structure formation. The counterion is not just an anonymous particle maintaining the electroneutrality, but has specificity (Dautzenberg *et al.*, 1994). Consequently the final properties of gels discussed in this chapter often depend on the cation(s) employed.

5.3 Polyelectrolyte theories

Some concepts necessary for describing characteristic behaviour of polyelectrolyte solutions are given here. The Debye length λ_D, which represents a screening length, giving a measure of the range of the potential around a point charge surrounded by other point charges in such a way as to 'screen' interactions, is written as

$$\lambda_D = \left(\frac{\varepsilon_0 \varepsilon' k_B T}{2e^2 I}\right)^{1/2}, \tag{5.1}$$

where $I = \frac{1}{2}\sum z_i^2 n_i^0$ is the ionic strength (note that $\sum z_i n_i^0 = 0$ because of electroneutrality), e is the elementary charge, ε' is the dielectric coefficient, ε_0 is the dielectric constant *in vacuo*, $n_i(r)$ is the local ion concentration, z_i is the valence of the ith species and n_i^0 is the corresponding bulk concentration.

However, in many polyelectrolyte solutions the effect of counterions, always present because of the requirement for electroneutrality, cannot be reduced to straightforward Debye–Hückel screening. In practice, it can be energetically favourable to have a proportion of the counterions located in the vicinity of or 'at' the surface of the macroion. Then, the apparent charge of the polyelectrolyte will be reduced. The term *counterion condensation* was used by Oosawa (1971) and Manning (1988) and reflects the binding of counterions to the charged chain. However, it is not appropriate to speak simply of 'bound' and 'free' ions, because even diffuse ions are bound, at least in a thermodynamic sense.

What is the implication of chain stiffness? Even a very flexible polyelectrolyte has a tendency to 'stiffen' at low ionic strengths in order to maximize the separation of ionic groups attached along the chain contour. The elastic and electrostatic contributions to the free energy of chain curvature have been calculated for a polyelectrolyte chain of contour length $L = Na$ (N is the degree of polymerization and a is the length of a monomer unit) consisting of Z units of elementary charge e separated by a contour distance $b = L/Z$. The repulsion between charges was described by a screened Debye–Hückel potential with screening length λ_D. From the free energy, the following expression for the total persistence length l_p of long chains was derived (Odijk, 1977; Skolnick and Fixman, 1977):

$$l_p = l_{pi} + l_{pe}, \tag{5.2}$$

with

$$l_{pe} = \lambda_B \lambda_D{}^2 / \left(4b^2 \xi_M{}^2\right), \tag{5.3}$$

where l_{pi} is the intrinsic persistence length in the absence of charges, l_{pe} is the electrostatic persistence length, $\xi_M = \lambda_B / b$ is the so-called Manning parameter and $\lambda_B = e^2/(4\pi\varepsilon_0\varepsilon' k_B T)$, the *Bjerrum length*, is the distance at which the Coulombic interaction between two unscreened elementary charges is equal to the thermal energy. This shows how a polyelectrolyte chain becomes more flexible in high ionic strength solvents, where $l_{pe} \cong 0$. Conversely a very stiff chain polymer, i.e. one where l_{pi} is high, say >50 nm, shows relatively little 'polyelectrolyte effect', since the maximum value of l_{pe} at low ionic strength is typically <10 nm.

In practice, physical gels as we recognize them are never formed from dilute solution of flexible polymers such as poly(ethylene oxide), which entangle but do not form junction zones. Such a system cannot retain solvent (water) and has no 'solid' character, so the system flows even when the concentration is high. At one limit, the minimum concentration necessary for gelation ought to be related to the flexibility (persistence length) of the chain.

Both structure formation and solubility of polyelectrolytes are strongly influenced by the counterion species and the ionic strength. An increase in ionic strength changes the conformation of polyelectrolytes via the screening effect and also because of the thermodynamically controlled 'salting out' effect which makes the solvent quality poorer. Coulombic interactions play the dominant role, while other intermolecular forces (including van der Waals forces, hydrogen bonding and hydrophobic interactions) also modify structure formation. More particularly, hydration effects and the Coulombic interaction are important for monovalent cations, while divalent and multivalent cations can 'cross-link' polyelectrolytes by sharing charges between different chains.

Most gel-forming polymers are stiff, at least in their ordered (e.g. helical) form, and the effect of salt is different from that for flexible polymers. As will be described later, in most cases salts also promote the formation of these ordered structures, including multiple helices – double (κ- and ι-carrageenan, gellan, DNA) or triple (gelatin; see Chapter 7) – and lead to the formation of more complex structures. In this chapter we restrict ourselves to certain gelling polysaccharides including the carrageenans, pectins, alginates and gellans. A number of other systems, including the liquid crystal 'gels' which can be formed from such persistent ordered structures, are discussed in Chapters 3 and 9.

5.4 Gelation of carrageenans and gellans

Two of the most important classes of gelling polysaccharides are the carrageenans, from marine algae, and the gellans, which are microbial fermentation products. Both are of importance in food, cosmetic and personal products, although the use of gellans in food products has been restricted in some jurisdictions (only cleared for food use in Europe in 1994, for example). More recently, gellan has been suggested as an injectable formulation in tissue engineering (Chapter 11). A comprehensive review on gelation of gellan has been published recently (Morris *et al.*, 2012).

The behaviour of polyelectrolyte solutions depends strongly on temperature, shear rate, stiffness, branching and chemical structure. Certain ionic polysaccharides, such as some marine polysaccharides (carrageenans, alginates), plant polysaccharides (pectins) and microbial polysaccharides (gellans) form thermoreversible gels by cooperative Coulombic interactions with counterions, by hydrogen bonding and by hydrophobic interactions (see Chapter 6). As mentioned above, this usually involves some pre-ordering of chains into more complex secondary structures.

The gelation of both κ-carrageenan and deacylated gellan has been studied extensively, and so is described in detail here. Helix formation is known to be a prerequisite for gel formation for such anionic polysaccharides, including κ- and ι-carrageenan and gellans (and also for some essentially neutral polymers such as gelatin and agarose; see Chapter 7). These biopolymers are known to be in a different, so-called 'random coil conformation' at higher temperatures (although data for their persistence length in this form is very limited), and do not form gels. At lower temperatures they undergo a disorder–order (coil–helix) transition and form aggregates in which it is now largely accepted that shared helical portions form the cross-links of a three-dimensional network. Since these physical cross-links have a finite size and so differ from the chemical point cross-links in covalent gels, they are now all but universally referred to as *junction zones*.

In extensive studies of the optical rotatory power of polysaccharides in various conformational states, Rees (1969) established a semi-empirical relationship between torsion angles which specified the relative orientations of residues in these saccharide polymers, and the magnitude and sign of the optical rotation (Morris, 1994). Applying this relationship to data for cold-set gelled solutions of carrageenans, and in particular to data on optical rotation versus temperature, Rees and co-workers established an initial model for the cross-linking process; see Figure 5.1.

In this model, the cross-links were described as cooperatively formed junction zones arising when regions of the carrageenan polymers combined to form ordered double helices (and, in some cases, clusters of these helices). Cooperativity was inferred from the suddenness of the optical rotation change in relation to temperature, and the presence of a double-helical conformation was deduced by combining the lowest temperature optical rotation results, and confirmed with X-ray fibre diffraction data.

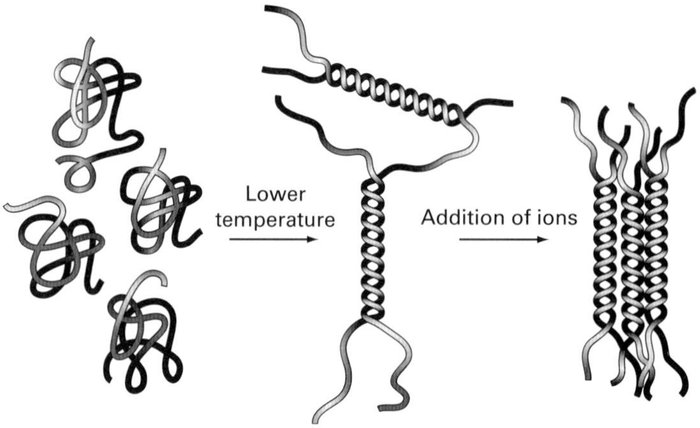

Figure 5.1 Schematic representation of the original gelation mechanism for ι- and κ-carrageenan. In the middle, the chains are cross-linked by isolated double helices, while, on the right, double helices are aggregated to an unknown extent. Reproduced from Rees (1969) by permission of the Royal Society of Chemistry.

Illustrating the ion specificity of the salt-induced conformational transition of κ-carrageenan, the Group I cations, potassium, caesium and rubidium were reported in early work to bind only to helical κ-carrageenan and not to coils (Rinaudo *et al.*, 1979; Smidsrød and Grasdalen, 1984; Belton *et al.*, 1985), while in other work, based on optical rotation and intrinsic viscosity measurements, it was reported that sodium and potassium ions also bind to disordered κ-carrageenans (Norton *et al.*, 1983). In a more recent study, Nilsson and Piculell estimated cation binding using a self-consistent treatment (Poisson–Boltzmann equation) and compared thermodynamic data obtained by optical rotation measurements with counterion NMR data on the binding of ions to the helix. They found good agreement, and concluded that cations bind to κ-carrageenan in the helical conformation and not to the same polymers in the disordered (coil) conformation (Nilsson and Piculell, 1991; Zhang *et al.*, 1994).

A number of studies have been carried out using the specialized but powerful techniques of specific nucleus NMR, especially illustrated by ^{133}Cs resonance. Such investigations showed no evidence of specific Cs^+ interaction with κ-carrageenan in the coil form (Grasdalen and Smidsrød, 1981), whereas, with the formation of helices, significant line broadening of the ^{133}Cs signals occurs. Under these conditions, the peak changes from a single Lorentzian to a multi-Lorentzian line shape, and changes in the chemical shift are observed, indicating a specific ion binding to the helix. Using this method, Grasdalen and Smidsrød also showed that iodide ions promote helix formation but prevent the aggregation of κ-carrageenan molecules.

Entangled flexible chains tend to disentangle simply by reptation (Doi and Edwards, 1999), while aggregated helices, because of their geometry, have a far longer lifetime. Using small-deformation dynamic viscoelastic measurements (Chapter 2), Grasdalen, and Smidsrød (1981) also showed both the storage and loss moduli to be almost frequency-independent at lower temperatures when these polymers are in a helical conformation, while both moduli showed strong frequency dependencies at higher temperatures.

Viebke *et al.* (1994) fitted the experimental data reported by Rochas *et al.* (1990) for helix fraction as a function of temperature, using the Zimm–Bragg theory on helix–coil transition (Figure 5.2), and found good agreement with a viscometric experiment performed under non-aggregating conditions (in 0.2 M LiI), and so showed that the relative viscosity of a κ-carrageenan solution was a monotonic function of the helix fraction in the system.

To separate helix formation and aggregation, they first prepared a non-gelling solution of all-helical κ-carrageenan in a LiI solution. This solution was subsequently dialyzed against KCl; i.e. it was exposed to salt conditions that induce aggregation. (With a normal cold-setting procedure, such a combination gives a strong gel.) They found that a gel, visibly indistinguishable from the cold-set gel, was also formed by the dialysis procedure.

It should be noted that the dialysis conditions were such that the coil content should have been insignificant at all stages. This conclusion is based on studies of the helix formation of κ-carrageenan in solutions of mixed salts of a gelling cation and a helix-inducing anion, which showed that the helix is, in fact, more stable in the mixed salts at a constant ionic strength (Zhang *et al.*, 1992; Viebke *et al.*, 1998). The salt-induced

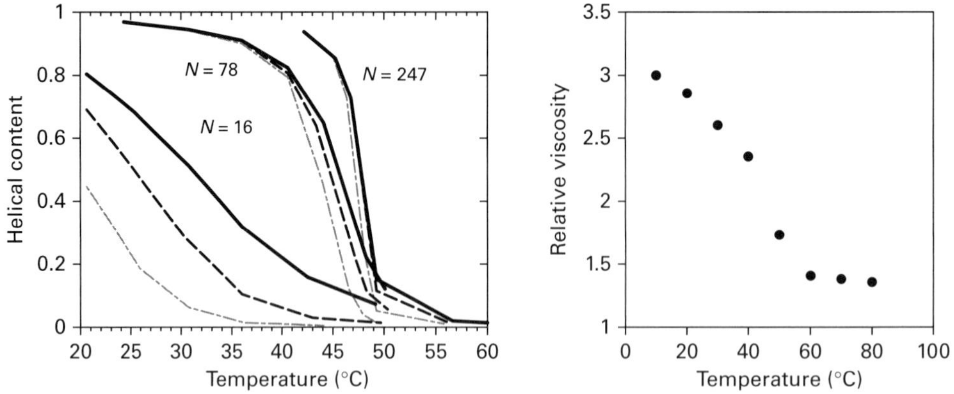

Figure 5.2 Left: helical content θ estimated for κ-carrageenan with various molar masses using the Zimm–Bragg treatment: κ-carrageenan fractions with, from left to right, 16, 78 and 247 repeating disaccharide units, at total concentrations of 13 (solid), 1.3 (dashed) and 0.13 (dot-dashed) mM. Right: relative viscosity of 0.1% κ-carrageenan in 0.2 M LiI as a function of temperature. Reprinted with permission from Viebke *et al.* (1994) © 1994 American Chemical Society.

gelation was also found to be reversible; dialysis of the gel against LiI again gave a clear solution. A similar dissolution in LiI was also found for a cold-set gel under the same salt conditions. Viebke *et al.*(1994) suggested that this was a strong indication that branching on the helical level is not required for the gelation of κ-carrageenan and that the network, whether cold-set from initially randomly coiled molecules or salt-set from helical molecules, is essentially formed on the superhelical level. The conclusion that the κ-carrageenan gel network is not created by each chain forming double-helical junctions with, on average, two or more partners is consistent with the so-called domain model proposed by Morris *et al.* (1980b).

These workers showed that ι-carrageenan formed a helix in a solution, but not a gel, when tetramethylammonium ions (Me_4N^+) are the sole counterions, but did form a gel when K^+ ions were added. Viebke *et al.* (1998) showed that a non-gelling solution of helical rods of κ-carrageenan could be reversibly transformed into a gel by dialysis against an appropriate salt solution under conditions such that the carrageenan molecules never pass through the coil state. This also implies that the branching of helical chain portions caused by the presence of 'kink' residues is not necessary for gel formation.

The model shown in Figure 5.3a is oversimplified, since the helical part should include aggregated double helices. The difference between the models shown in Figure 5.3a and Figure 5.3b seems to lie in the absence of flexible chains connecting junction zones in the latter. Model (b) resembles the 'fibrous model' proposed from atomic force microscopy (AFM) studies (Morris, V. J. *et al.*, 1999).

Cryo-TEM observation was performed for κ-carrageenan molecules without removing water, and rigid superhelical rods and aggregates of such rods were found (Borgström *et al.*, 1996), in good agreement with observation for κ-carrageenan molecules deposited on mica surface by electron microscopy (Hermansson *et al.*, 1991) and by AFM (Ikeda *et al.*, 2001). However, this representation does not reflect the presence of the so-called

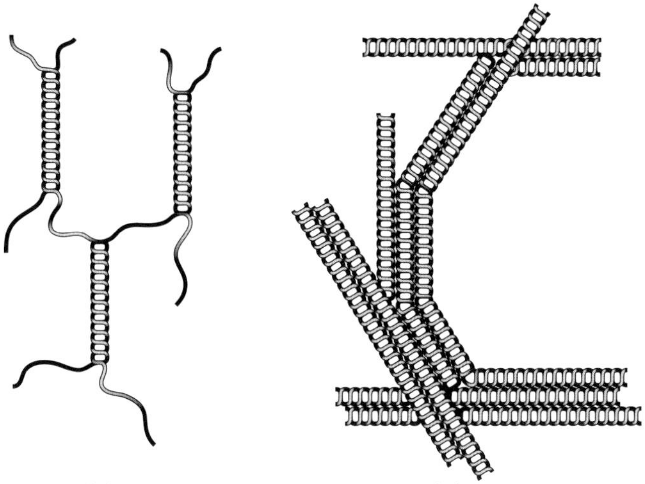

Figure 5.3 Schematic drawings of possible mechanisms of network formation in helical gels: gelation on the helical (a) and superhelical (b) levels. In (b), no branching from helical molecules occurs. Reprinted with permission from Viebke *et al.* (1994) © 1994 American Chemical Society.

chemical 'rogue residues' (Morris *et al.*, 1980a), which act as 'helix stoppers' and so do not allow 100% of the units to be involved in a double helix. Consequently, it seems unlikely that a perfect superhelix can be formed. Here, advances are urgently required to improve the technique so that a 'raw' gel can be examined without depositing a polymer solution on to the mica. Small-angle X-ray scattering (SAXS) data is useful for discussing the structure of gels, and will be described later.

Helix formation does not always lead to gel formation, but sometimes, because the polyelectrolyte in solution contains helices, the overall profile becomes stiffened (even rod-like, in the limit of stiffness of polymer chains) and so can form either an isotropic or anisotropic solution, or even precipitates. Rigid (or semi-flexible) polymer chains in solution behave differently than flexible chains in many respects. Because of their rod-like structure, there is a possibility of formation of liquid crystalline ordered phases and gels (Chapter 3). It should also be pointed out that a helix–coil transition can occur even in a high modulus gel, as discussed below.

Figure 5.4 shows the temperature dependence of the complex Young's modulus and of the mechanical loss tangent from observation of the longitudinal vibration of a cylindrically moulded gellan gel (Nitta *et al.*, 2001). This method is free from slippage which, for these systems, sometimes causes misleading interpretations (see Chapter 2). On heating, in the temperature range where the loss tangent showed a step-like change, the ellipticity showed a steep change and an endothermic peak appeared, originating from the helix–coil transition. All these changes were also observed on cooling, although with a slight thermal hysteresis. Since the cylindrical gel is of high modulus, this indicates that the helix–coil transition can still occur within such a gel, and suggests that such pertinent molecular motion is not inhibited.

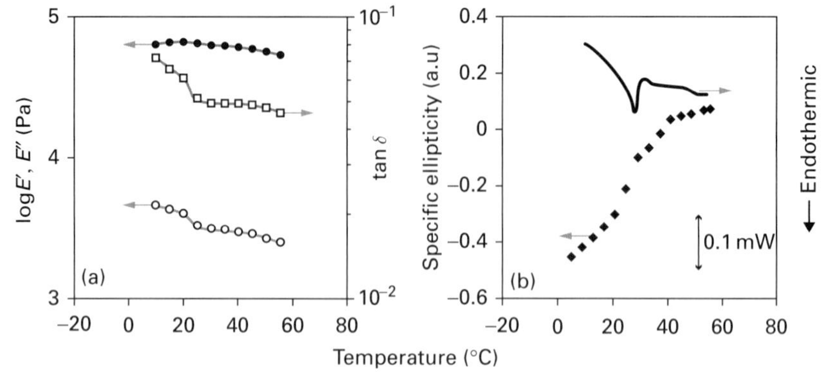

Figure 5.4 (a) Temperature dependence of storage Young's modulus E' (●), loss Young's modulus E'' (○) and mechanical loss tangent (□). (b) Heating differential scanning calorimetry (DSC) curve (solid curve) and temperature dependence of the specific ellipticity at 202 nm (♦) of a 1.6% w/w gellan gel on heating at 0.5°C min^{-1}. Reprinted from Nitta *et al.* (2001) with permission from the Materials Research Society of Japan.

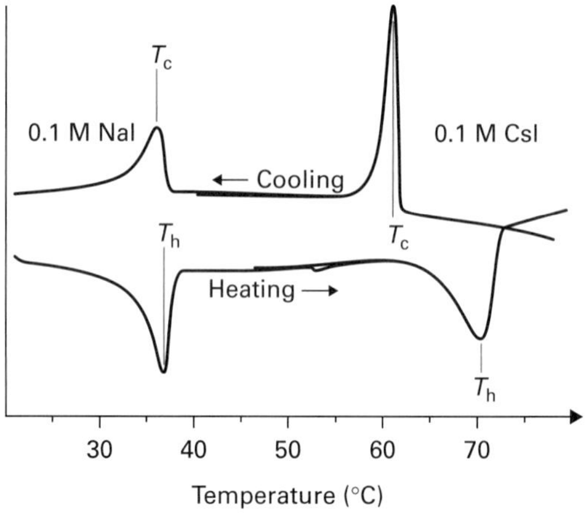

Figure 5.5 Two typical DSC thermograms for κ-carrageenan. The cooling and heating curves for both molar fractions $X_{Cs} = 0$ (0.1 M NaI) and $X_{Cs} = 1$ (0.1 M CsI) are displayed. Reprinted with permission from Viebke *et al.* (1998) © 1998 American Chemical Society.

5.4.1 Effects of ions on structure formation

Viebke *et al.* (1998) examined further the dependence of the coil–helix-aggregate transition of κ-carrageenan on salt composition. The transition enthalpy on cooling was measured by differential scanning calorimetry (DSC; see Chapter 2), and the salt composition was varied systematically using mixtures of the four salts NaCl, NaI, CsI and CsCl, which includes those generating both aggregating and non-aggregating conditions. In this work, either the anion or cation was exchanged for another, but with the total salt concentration kept constant (Figure 5.5).

Viebke *et al.* (1998) used an electrostatic model to describe the specific binding of caesium and iodide ions to the κ-carrageenan helix. This modelling resulted in a semi-quantitative description of the variation of the ion binding and the charge density in the various salt mixtures. It was found that the model reproduces the relevant transition enthalpies in the non-aggregated systems very well, while in the aggregated systems the model predictions deviate markedly from experimental results. This deviation, interpreted in terms of an aggregation enthalpy, varies with the extent of thermal hysteresis. The hysteresis occurs only when the charge density is lower than the charge density for the bare κ-carrageenan helix (without bound ions). Another interesting observation is that the width of the DSC peak obtained on heating increases drastically when the aggregation occurs. Finally, the 'gel strength', here defined as a maximum stress at a breaking point determined by uniaxial compression, was found to depend linearly on the extent of the hysteresis measured by DSC, and so reflects the degree of aggregation.

Gel formation of helix-forming anionic polysaccharides can be induced at a constant temperature by increasing ionic strength. Using the dialysis technique mentioned above, Piculell *et al.* placed a κ-carrageenan solution in LiI (all helical conformation), immersed it in KCl salt solution and found a gel was formed (Piculell *et al.*, 1993; Viebke *et al.*, 1994). From this evidence, they again concluded that the κ-carrageenan gel network is created essentially by helix aggregation.

Much of the structuring effects noted above also apply to the gellans, but, for these systems, it has been observed recently that the immersion of such gels in water or electrolyte solution induces chain release, and this release is more noticeable for shorter chains. Ultimately the gel becomes eroded and then disintegrates, and the rate of collapse depends on polymer concentration, original molecular mass and the initial salt content of the gels and the solvent. Salt diffusion from the gels into the solution is faster than chain release; chains which lose condensed or bound ions cannot retain a helical conformation, and so they then diffuse out into the solution. The storage Young's modulus E' of gellan gels immersed in various solvents was observed as a function of time (Figure 5.6). For the first 3 h, E' increased both in water and in Me_4N^+ Cl^- solution, a solution which inhibits the helix aggregation of gellan. Subsequently, E' for a gel immersed in water decreased because of the release of chains contributing to the network (Hossain and Nishinari, 2009). This process should be studied further, but is consistent with the release of some of the sol fraction, as found in lightly cross-linked chemical gels.

5.4.2 Effects of cations on the transition temperature

The gel–sol transition temperature T_m of polyelectrolyte gels depends on the salt concentration, and a linear relation between the inverse of the transition temperature and the salt concentration has been found for many polysaccharide gels. Snoeren and Payens (1976) reported an equation for the salt dependence of the sol–gel transition temperature for κ-carrageenan solutions from both optical rotation and light scattering measurements. This showed a linear dependence on the logarithm of the salt concentration. However, the slope of this linear relation is quite different from that expected from an earlier theoretical

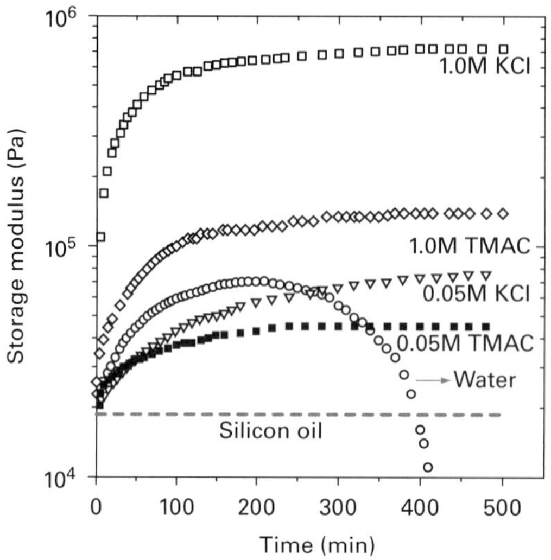

Figure 5.6 Plot of the storage Young's modulus E' as a function of time for 2.0 wt% gellan gel immersed in various salt solutions. The dotted line represents E' of gellan gels immersed in silicon oil. Here E' was measured by observing the longitudinal vibration of a cylindrical gel, a technique which helps to eliminate the usual slippage effects. Reprinted with permission from Hossain and Nishinari (2009) © 2009 Springer.

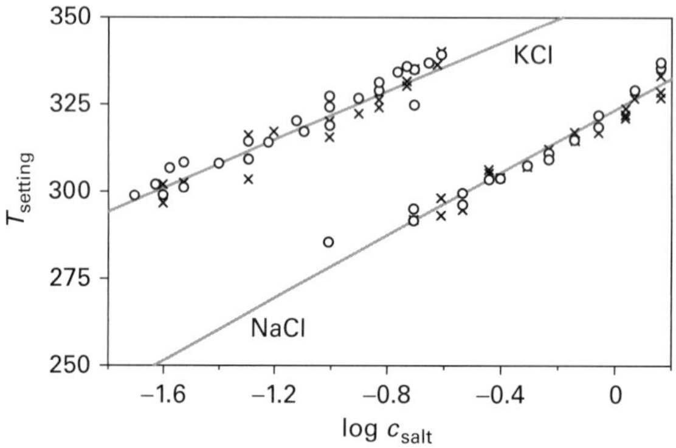

Figure 5.7 Dependence of the gelation temperature of κ-carrageenan gels on electrolyte concentration as measured by optical rotation and light scattering: optical rotation, 0.25% κ-carrageenan (○); light scattering, 1% κ-carrageenan (×). Reprinted with permission from Snoeren and Payens (1976) © 1976 Elsevier.

prediction by Record (1967). Since it is likely that counterions condense on to the carrageenan chains, Snoeren and Payens estimated the slope by Manning theory (Manning, 1988) but still found a large deviation. They attributed this deviation to the Debye–Hückel approach on which both the treatments of Record and Manning are based (Figure 5.7).

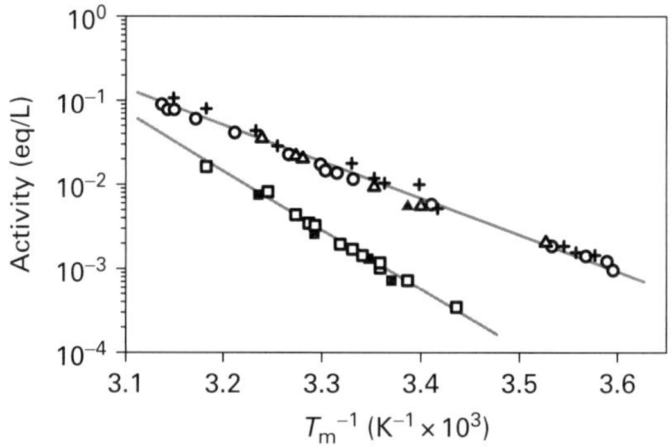

Figure 5.8 Dependence of the inverse of the temperature of conformational transition (T_m^{-1}) on counterion activity: Li-gellan, LiCl (▲); Na-gellan, NaCl (+); K-gellan, KCl (○); Me_4N-gellan, Me_4N^+ Cl^- (Δ); Ca-gellan, $CaCl_2$ (□); Mg-gellan, $MgCl_2$ (■). Reprinted with permission from Milas and Rinaudo (1996) © 1996 Elsevier.

Rochas and Rinaudo (1980) found a linear relation between $1/T_m$, the inverse of the melting temperature, and the logarithm of the total salt concentration c_T. They found a similar relation for gellan (Milas and Rinaudo, 1996), as shown in Figure 5.8.

For gellan, determination of the activity of counterions gives information on the conformation of the polymer in dilute solutions, in particular in determining whether the polymer adopts a single- or a double-helical structure. For κ-carrageenan, a similar relation between the transition temperature and the total concentration of free counterions was reported by Hugerth *et al.* (1999), based on fluorescence measurements, dynamic viscometry and optical rotation.

5.4.3 Role of monovalent cations

Since the pioneering work of Hofmeister almost a century ago, it has been recognized that simple salts may drastically alter the solubility and other solution properties of macromolecules. In general, the relative effectiveness of various ions follows a well-defined order (the lyotropic series), which extends to phenomena far removed from colloid chemistry (such as the surface tension of salt solutions, the miscibility of liquids and the kinetics of organic chemical reactions in solution). In the cation-induced gelation of negatively charged polysaccharides, however, the normal lyotropic order may be distorted or scrambled by selective interactions of the polymer and counterion.

In the carrageenan series, the cation requirements for gelation and adoption of the underlying, ordered junction-zone geometry are dependent on the degree of sulphation of the polymer. For example, K^+ ions are more effective than Ca^{2+} in promoting gelation of κ-carrageenan, whereas, for the more highly charged ι-carrageenan, the order is reversed (Payens and Snoeren, 1972; Rinaudo *et al.*, 1979; Morris and Belton, 1980; Morris and Norton, 1983).

The cations K^+, Cs^+, Rb^+ and Cs^+ have a stronger gelation-enhancing effect than Na^+ and Li^+, probably because of the difference in hydration. It has been reported that water

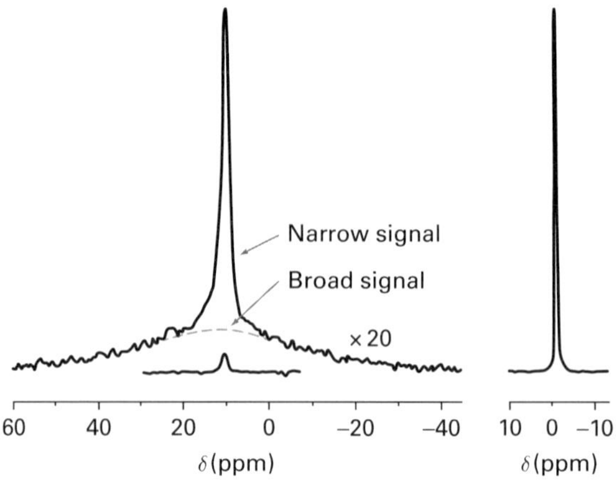

Figure 5.9 ^{133}Cs NMR spectra of 3% w/v caesium κ-carrageenate in D_20 at 80°C (right) and 25°C (left) obtained at 13 MHz (JEOL FX-100 NMR spectrometer). Chemical shift is positive downfield relative to ^{133}Cs ions in 10 mM CsCl in D_2O, contained in a central coaxial tube. Reprinted with permission from Grasdalen and Smidsrød (1981) © 1981 American Chemical Society.

molecules around K^+ move faster than those around Na^+, and that the bond length of M^+–H_2O ($M^+ = K^+, Cs^+, Rb^+, Cs^+, Na^+, Li^+$ etc.) decreases in the order $Cs^+ > K^+ > Na^+ > Li^+ > NH_4^+$, based on the dynamic hydration number of these cations determined from NMR, neutron scattering and simulations (Ohtaki and Radnai, 1993).

Using ^{133}Cs NMR, Grasdalen and Smidsrød (1981) concluded that caesium ions bind to κ-carrageenan by observing the restriction on mobility which appears as a broadening of the spectra for a 3% w/v solution of Cs^+ κ-carrageenate in D_2O as the temperature is lowered from 80°C to 25°C. This clear broadening was not observed in ι-carrageenan, where only a slight broadening, due to increased viscosity, was observed at low temperature; see Figures 5.9 and 5.10.

A motional restriction of at least three orders of magnitude for bound ions relative to free ions is indicative of strong and long-lived binding of the Cs^+ ions at well-defined sites on the carrageenan molecules. Grasdalen and Smidsrød (1981) argued that, in addition to the sulphate groups, these sites consist of one or more groups with hydroxyl and ring oxygens to stabilize the binding of the gel-promoting larger alkali metal ions, and arranged to create a cavity. The temperature dependence of the ^{133}Cs NMR line shifts for 3% w/v caesium κ-carrageenate in D_2O on cooling and heating show a similar behaviour to the sol–gel transition observed by rheology, including hysteresis, and was attributed to gel formation steps, i.e. the formation of junctions zones, accompanied by site binding of counterions.

The ^{133}Cs line shift as a function of concentration of Cs^+ κ-carrageenate showed a steep increase at 1% w/v, and the curve was sigmoidal, indicating the cooperative nature of the sol–gel transition. Smidsrød and co-workers also found that gelation of κ-carrageenan produces highly selective binding sites for alkali metal ions, in which Cs^+ and K^+ ions bind much more strongly than Li^+ or Na^+. The so-called binding selectivity coefficient value at low potassium contents was ~10 for κ-carrageenan,

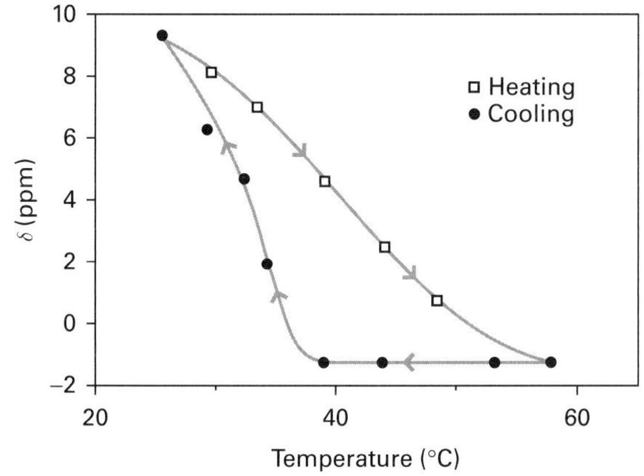

Figure 5.10 Temperature dependence of ^{133}Cs NMR line shift for the sample defined in Figure 5.9. The figure shows the effect of cooling and heating through the sol–gel transition, allowing 10 min for equilibration at each temperature. The gel formation temperature is just below 40°C. Reprinted with permission from Grasdalen and Smidsrød (1981) © 1981 American Chemical Society.

compared to ~2–3 for ι-carrageenan gels (Smidsrød *et al.*, 1980). This finding suggests the formation of ion-selective 'salt bridges', as proposed by Bayley (1955).

NMR peak intensity, peak width and chemical shift for κ-carrageenan solutions in the presence of Na^+, Rb^+, Cs^+ and K^+ ions suggest that gel-forming cations (Rb^+, Cs^+ and K^+) are bound to κ-carrageenan in the gel state, but that the Na^+ ion is not (Tokita *et al.*, 1989). In the domain model (Morris *et al.*, 1980a, 1980b), gelation of κ-carrageenan involves a coil–helix transition (non-gelling) followed by the formation of cation (K^+, Rb^+ and Cs^+) -mediated side-by-side aggregates of double helices (formation of junction zones = gelling).

Several workers, however, have argued for a single-helix model for κ-carrageenan gelation (Smidsrød *et al.*, 1980; Smidsrød and Grasdalen, 1984). In this model, it is also assumed that K^+, Rb^+ and Cs^+ are bound to the junction zone but that this is formed by cation-mediated side-by-side aggregation of single helices. In either model, the line shape of NMR spectra of gel-forming cations becomes broad when a gel is formed because of the restriction of the molecular motion of these cations in the formation of junction zones. These NMR results are direct evidence of the very highly selective interaction between the gel-forming group I cations and κ-carrageenan. However, for κ-carrageenan the ionic selectivity for divalent counterions is low.

The mid-point temperatures (T_m) of the disorder–order transitions were determined at a constant ionic concentration, and in the presence of monovalent counterions they varied according to the following sequence: $Rb^+ > K^+, Cs^+ > Na^+ > Li^+, NH_4^+ > R_4N^+$. No selectivity was observed for monovalent anions with the exception of I^-, which stabilizes the helix but prevents gelation (Rochas and Rinaudo, 1980; Borgström *et al.*, 1996; Rinaudo, 2008). Rochas and Rinaudo (1980) measured the activity coefficient $\gamma_M{}^+$ of sodium and potassium counterions as a function of the square root of κ-carrageenan concentration. Results showed that both $\gamma_{Na}{}^+$ and $\gamma_K{}^+$ extrapolated to a value of 0.72 at infinite dilution (Figure 5.11).

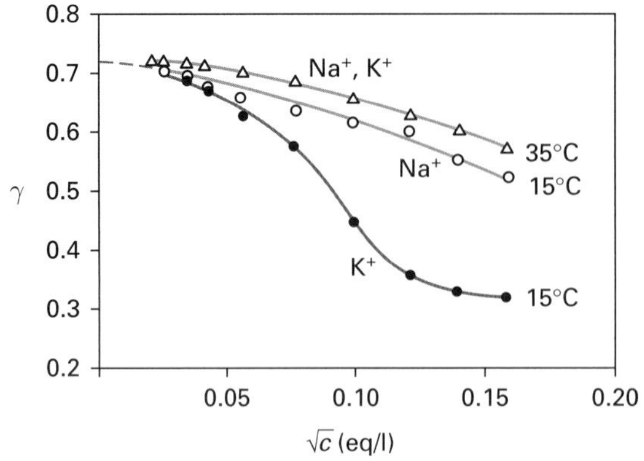

Figure 5.11 Activity coefficient of monovalent counterions (K^+ and Na^+) as a function of polymer concentration c at 15°C and 35°C. Reprinted with permission from Rochas and Rinaudo (1980) © 1980 John Wiley & Sons.

At higher polymer concentrations, γ_{Na^+} was not so dependent on temperature in the range 15°C to 35°C, while γ_{K^+} was 0.37. Since $\gamma_1 = (2\,\lambda_P)^{-1}$ for monovalent cations and $\gamma_2 = (4\,\lambda_P)^{-1}$ for divalent cations, where λ_P is the so-called charge parameter (or the inverse of the distance between two charged groups), the value of λ_P was doubled. Also, the melting enthalpy determined from the DSC endothermic peak in water is double the value observed in non-aqueous solvents (formamide and DMSO) in which a coil-single helix transition is thought to occur. With these results and the doubling of the molar mass under non-gelling conditions, Rochas and Rinaudo (1980) concluded that double- rather than single-helix formation occurs in water and electrolyte, and that now appears to be the consensus.

For gellan, Morris, E. R. *et al.* (1999) reported that the shear storage modulus G', breaking stress σ_b and gelation temperature T_{gel} at a fixed Na-gellan concentration (0.5–0.7%) as a function of NaCl showed a maximum. In other words, all parameters increased and then decreased with increasing concentration of NaCl, with the maximum value corresponding to a NaCl concentration of about 300 mM for G', and σ_b, and 750 mM for T_{gel}. The Young's modulus and the breaking force of gellan gels as a function of potassium chloride concentration showed a similar maximum (Milas and Rinaudo, 1996). This experimental finding, that the elastic modulus of these polyelectrolyte gels as a function of added salt shows a maximum, seems to be universal: the storage Young's modulus of κ-carrageenan gels as a function of added alkali metal salts was also reported to show a maximum (Watase and Nishinari, 1982). Work by Kasapis *et al.* (1999) showed a maximum of the storage shear modulus of gellan gels as a function of added $CaCl_2$ and NaCl.

The increase in elastic modulus or transition temperature may be understood by the shielding of electrostatic repulsion between, respectively, the sulphate groups in κ-carrageenan or the carboxyl groups in gellan, helping promote helix formation, but the reason for the subsequent decrease is not yet fully understood. On addition of excess salt, other phenomena, such as phase separation, may occur and so 'weaken' the overall gel structure. Morris *et al.* (1995) explained this behaviour in terms of the transition from a homogeneous to an inhomogeneous gel structure. In order to make definitive conclusions in this area, it is necessary to perform experiments using monodisperse samples with narrow molar mass distribution and have them converted into the sodium cation form. Few if any such studies appear in the literature.

5.4.4 Role of anions

In comparison with studies of the effects of cations on the gelation of anionic polysaccharides, the corresponding effects of anions have not so far been clarified. Rochas and Rinaudo (1980) and Piculell and Nilsson (1989) reported that the nature of the anions had no influence on the gelation of carrageenans, and thus it was suggested that anions do not take part in the gelling processes. As mentioned above, Grasdalen and Smidsrød (1981) showed the exceptional behaviour of iodide anions: from the observation of ^{127}I NMR resonance in the presence of κ-carrageenan below the helix–coil transition temperature, they found the increase of the excess linewidth was proportional to the concentration of κ-carrageenan, which was attributed to the binding of a fraction of iodide anions to the κ-carrageenan (Figure 5.12).

Since the temperature dependences of the excess linewidth and of the specific optical rotation for tetramethylammonium (Me_4N^+) κ-carrageenate in the presence of Me_4N iodide resemble each other, they concluded that iodide binding to κ-carrageenan gave rise to the stabilization of its helical conformation. They also concluded that the conformational transition is an intramolecular event, i.e. a single-helix formation, based on the observation that the transition temperature was independent of the concentration of κ-carrageenan, that the measured number average molecular mass of κ-carrageenan was independent of anions (for I^- and Cl^-) and that the intrinsic viscosity in $Me_4N^+ I^-$ solution is much greater than that in $Me_4N^+ Cl^-$ solution.

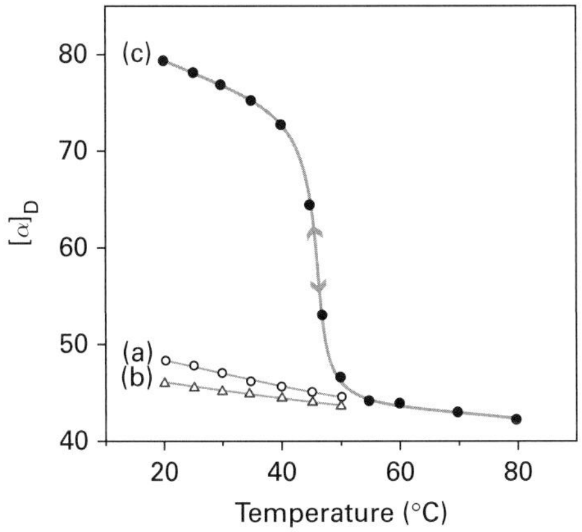

Figure 5.12 Specific optical rotation of 1% w/v aqueous tetramethylammonium κ-carrageenate (M_n = 150 000 g mol^{-1}) at various temperatures: (a) in the absence of salt; (b) in the presence of 0.15 M $Me_4N^+ Cl^-$; (c) in the presence of 0.15 M $Me_4N^+ I^-$ and 10^{-3} M $Na_2S_2O_3$. Reprinted with permission from Grasdalen and Smidsrød (1981) © 1981 American Chemical Society.

The absence of turbidity and thermal hysteresis in the temperature dependences of the specific optical rotation for Me_4N^+ κ-carrageenate (0.1–3% w/v) in the presence of Me_4N^+ I^- also led Grasdalen and Smidsrød (1981) to propose that I^- ions promote helix formation but prevent the aggregation of helices. Although they did not find a steep increase in the specific optical rotation in the presence of Me_4N^+ Cl^-, but only in Me_4N^+ I^-, Norton *et al.* (1984) found an analogous transition in the presence of Br^- and Cl^- co-anions. At all experimentally accessible concentrations of Me_4N^+ Cl^-, Br^-, I^- and NO_3^-, the results obtained on heating and on cooling were closely superimposable, no gel formation was observed and the solutions remained optically clear at low temperature. In the presence of Me_4N^+ sulphate (and to a lesser extent fluoride), by contrast, significant hysteresis was observed between heating and cooling scans, conformational ordering was accompanied by an increase in turbidity and a weak but cohesive gel structure was formed. These effects increased with increasing concentration of salt, and became undetectable at low levels of salt.

The mechanical spectra of κ-carrageenan helices dispersed in KCl and NaI aqueous media were compared (Ikeda and Nishinari, 2001). A dispersion of 1.5% w/w non-aggregated κ-carrageenan in 0.2 M NaI solution (which prevents aggregation of helices) exhibited a 'structured liquid' type dynamic mechanical spectrum at 20°C. In other words, the storage modulus G' predominated over the loss modulus G'' in the entire frequency range examined (0.5–100 rad s^{-1}), both moduli showed a slight frequency dependence and $\tan\delta$ was not so small (>0.1) (Figure 5.13). On the other hand, 0.15% w/w κ-carrageenan in an aggregating condition, i.e. in 0.2 M KCl solution, at 20°C showed elastic gel behaviour. However, under large deformation – large enough for conventional gels to rupture – the structured liquid systems flowed but never ruptured, suggesting that the material (sometimes called 'weak gel') rheological properties of the κ-carrageenan dispersions were the result of a sufficiently long

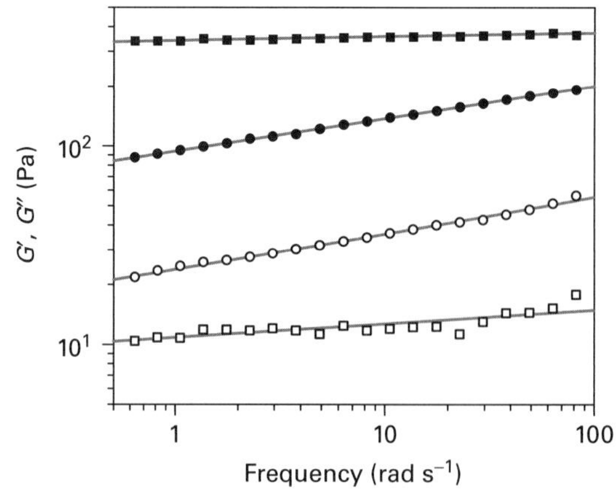

Figure 5.13 Frequency dependence of G' (solid symbols) and G'' (open symbols) of 1.5% w/w κ-carrageenan in 0.2 M NaI (circles) and 0.15% w/w κ-carrageenan in 0.2 M KCl (squares) at 20°C. Reprinted with permission from Ikeda and Nishinari (2001) © 2001 American Chemical Society.

relaxation time for topological entanglements among double-helical conformers but not for the formation of a three-dimensionally percolated permanent network.

Again, the mid-point temperature of the disorder–order transitions increases systematically with the Hofmeister number for the anion through the lyotropic series, so that T_m values were in the order $SO_4^{2-} < F^- < Cl^- < Br^- \leq NO_3^- < I^- < SCN^-$ when salt concentration and cation (Me_4N^+ or K^+) are held constant. A corresponding increase was observed in transition enthalpy (ΔH_{cal}) and entropy (ΔS_{cal}). Helix–helix aggregation (as indicated by turbidity, gel formation and hysteresis between heating and cooling scans) also shows a systematic dependence on the Hofmeister number for the anion, but in the opposite sense. Thus, with Me_4N^+ as the sole counterion present, clear solutions with no thermal hysteresis in the order–disorder transition are observed at all temperatures with I^-, Br^-, NO_3^- or Cl^- as co-anion, whereas lower modulus turbid gels, with a significant thermal hysteresis between melting and re-forming, are created in the presence of SO_4^{2-}, and to a lesser degree F^-. With K^+ as counterion, a similar regular progression is observed through the anion lyotropic series, from rapid formation of very turbid gels in the presence of F^- to very slow development of clear gels with I^- or SCN^- (Figure 5.14).

In agreement with previous studies, an increase in ^{127}I NMR linewidth was observed on conformational ordering of κ-carrageenan (Me_4N^+ salt form) in the presence of $Me_4N^+ I^-$. However, closely similar behaviour was observed for ^{35}Cl and ^{81}Br, indicating a simple charge-cloud interaction rather than the specific site-binding of I^- which has previously been suggested. The change in enthalpy during (cooperative) helix formation can be determined by ΔH_{app} from a van't Hoff plot (Chapter 3), and the corresponding value ΔH_f for disaccharides observed by direct calorimetric measurement. Norton *et al.* obtained the apparent number n_{app} of disaccharides involved in the cooperative transition of helix formation (given by the ratio $\Delta H_{app}/\Delta H_f$) (Norton *et al.*, 1984; Austen *et al.*,

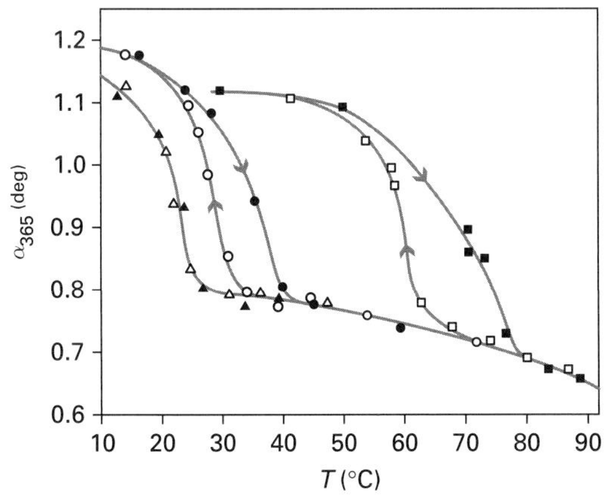

Figure 5.14 Conformational transition of κ-carrageenan (Me_4N^+ salt form) in the presence of 0.167 M (■,□), 0.25 M (•, ○) and 1 M (▲, Δ) $(Me_4N)_2SO_4$; OR conditions 365 nm; 10 cm path length; polymer concentration 0.5% w/v. Filled and open symbols are for heating and cooling, respectively. Reprinted with permission from Norton *et al.* (1984) © 1984 Elsevier.

1988; Uedaira, 1990). Uedaira (1990) pointed out that the Hofmeister number is empirical and has no clear physical meaning. Instead he introduced the 'dynamic hydration number' n_{DHN} to interpret the experimental data of Norton *et al.* (1984), and showed that both the DSC mid-point temperature (T_m) and the apparent number of effective (active) disaccharide units n_{app} are closely correlated with his new n_{DHN}. The iodide ion was most effective in increasing the temperature of helix formation and gave the smallest n_{app}, which is induced by the activated motion of macromolecular chains at higher temperatures.

To understand the overall effect of charge on carrageenan gelation, Piculell and Nilsson (1989) employed the neutral but structurally quite similar agarose (Chapter 7) to study the effect of anions. Uedaira (1990) also estimated the concentration dependence of the coil–helix transition temperature experimentally observed by Piculell and Nilsson, and displayed it as a function of the dynamic hydration number of anions. The gelation temperature shifts to lower temperatures with increasing concentration of the added salt. The thiocyanate ion shows the strongest negative hydration and thus most strongly depresses the gelation temperature.

What, then, is the overall conclusion regarding the structure of the carrageenan and gellan gels formed in electrolyte solution? The consensus is that a double rather than a single helix is formed, but that gelation is driven by the specific nature of the cation. The original Robinson domain model for carrageenan gelation (Figure 5.1, right) suggested that helix aggregation was induced by, say, (addition of extra) K^+ ions. Indeed, by chemical treatment the so-called 'kink' points (points of galactose-6-sulphate substitution) which terminated a helix could be cleaved, leaving isolated double-helical 'segments' (Rees, 1969). One current picture (Figure 5.3b) suggests that helix formation is more absolute and that a totally helical structure can be formed which then aggregates into the superhelix, although this seems to be in direct contradiction with the existence of these kink points. However, it is what has been suggested by, for example, AFM measurements (Ikeda *et al.*, 2001) and by DSC results (Piculell *et al.*, 1993), and it is fair to say that almost no-one now accepts the original model of Figure 5.3a.

Perhaps the best overall picture for the carrageenan gels is one which is somewhat less perfect than Figure 5.3b, in that some more flexible regions remain. The problem in resolving such issues is that perfect monodisperse chain length samples are not available, and so almost all published work is on different materials, and under slightly different conditions.

One exception to this, and a definitive approach, is to adopt a similar strategy for the carrageenans as was employed in the Japanese 'Round Robin' exercise on gellan. From this work, three special journal issues were published.[1] Although monodisperse

[1] (a) Nishinari, K., Kajiwara, K., Ogino, K. (eds), 1993. *Food Hydrocolloids* **7**, 361–456. Special issue: Gellan Gum.
(b) Nishinari, K. (ed.) 1996. *Carbohydr. Polym.* **30**, 75–207. Special issue: Gellan Gum: Structures, Properties and Functions.
(c) Nishinari, K. (ed.) 1999. *Progr. Colloid Polym. Sci.* **114**, 1–135. Special issue: Physical Chemistry and Industrial Application of Gellan Gum.

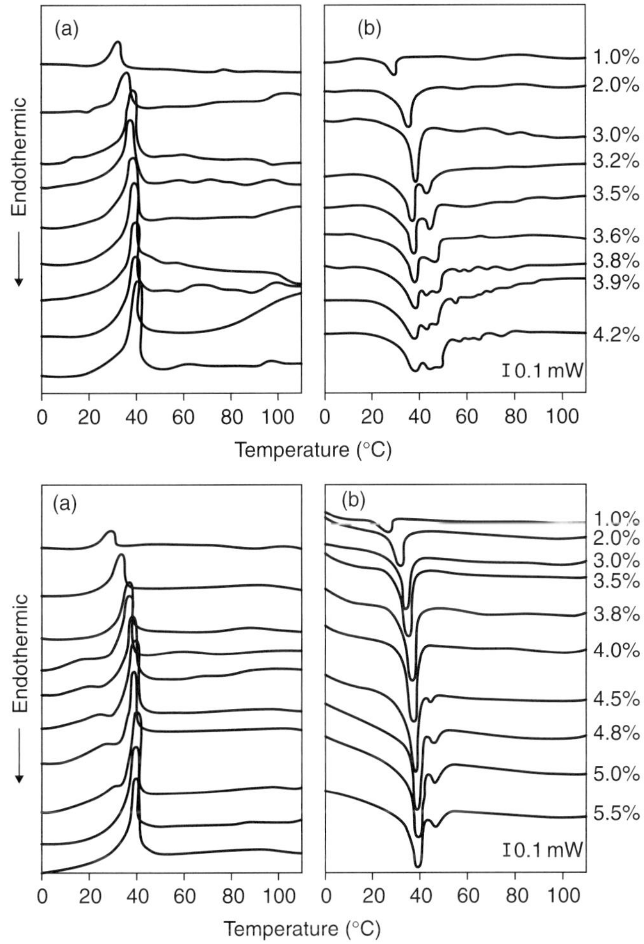

Figure 5.15 Cooling (a) and heating (b) μDSC curves for Na gellan solutions of various concentrations (indicated on the right-hand axis). Scan rate: 0.5°C min^{-1}. Upper: reprinted with permission from Miyoshi *et al.* (1996a) © 1996 Elsevier. Lower: reprinted with permission from Miyoshi *et al.* (1999) © 1999 Springer.

distributions could not be produced, common samples of a sodium type gellan were distributed among the group.

Figure 5.15 shows DSC heating and cooling curves for sodium type gellan samples used in the Japanese collaborative programme. An exothermic peak at 29°C in the cooling DSC curve and an endothermic peak at 30.5°C in the heating DSC curve for a 1% gellan gum solution shifted to higher temperatures with increasing gellan concentration. The effect of gel-promoting cations K^+, Ca^{2+} and Mg^{2+} was much reduced in the latter (1999) work compared to the effect reported in the earlier 1996 publication, as shown in these DSC curves.

Consequently, before definitive conclusions can be drawn, still further – ideally cooperative – work using pure sodium type gellan with a narrow molecular mass distribution is required. Ogawa *et al.* (2009) recently obtained three gellan samples converted to Na^+ salt form and with varying molar masses (7.1–12.7 × 10^3 g mol^{-1}). In this work they found that the temperatures for coil–helix transition and gel formation were almost independent of molar mass in Na^+, K^+ and Ca^{2+} solutions, but differed significantly in pure water, where the coil–helix temperature was up to 20°C higher.

5.5 Gelation of alginates and pectins

Alginate is a collective term for a family of exopolysaccharides produced from brown seaweeds and some bacteria, while pectins occur quite commonly in plant cell walls and most commercial pectins are extracted from apple pulp or citrus peel. Both have been extracted and used in a variety of products, alginates for example in foods, but also in advanced biomedical treatments. Suitable alginate solutions, usually prepared as the Na^+ salt form, gel very readily when mixed with a solution of Ca^{2+}. Pectins can be gelled by ions, but also by lowering water activity, for example by adding low molecular mass sugars – the traditional route in jam and preserve production. Unlike the carrageenans, the gels formed from alginates and pectins are not generally regarded as thermoreversible, although such gels do tend to weaken on heating. In this section we consider only pectin gelation with ions, and the structural parallels with alginate gelation.

5.5.1 Alginates

Chemically, alginates are linear copolymers of (1→4)-linked *β*-D-mannuronate (M) and α-L-guluronate (G) residues. Alginates have unique ion-exchange properties: most monovalent counterions (except Ag^+) form soluble alginate salts, whereas divalent and multivalent cations (except Mg^{2+}) form gels or precipitates. The affinity has been found to follow the order $Mn^{2+} < Zn^{2+}, Co^{2+}, Ni^{2+} < Ca^{2+} < Sr^{2+} < Ba^{2+} < Cd^{2+} < Cu^{2+} < Pb^{2+}$ (Rinaudo, 2008).

The residues are arranged in a blockwise pattern with G-blocks and M-blocks interspersed with MG alternating blocks. Based on X-ray fibre diffraction of polyguluronate and polymannuronate sodium salts, it has been shown that the former form a two-fold buckled helical shape, via diaxial 1, 4 linkages, which has large cavities between adjacent paired chains, while the latter adopt an extended ribbon-like conformation. Using a variety of methods including circular dichroism, electron micrography, dialysis experiments and model building, Grant *et al.* (1973) proposed a so-called 'egg-box' model for the gelation of alginates, where two or more chains are involved in cooperative binding, forming the structure. The buckled chain is shown as a two-dimensional analogue of a corrugated egg-box, with interstices in which the cations may pack and be coordinated like the eggs in a real egg-box (Figure 5.16).

For example, G-block sequences of alginate show a specific affinity for Sr^+, which is not shared by M-blocks. The origin of this behaviour appears to lie in the steric requirements for effective packing of cations within the egg-box junction zones which cross-link alginate gels.

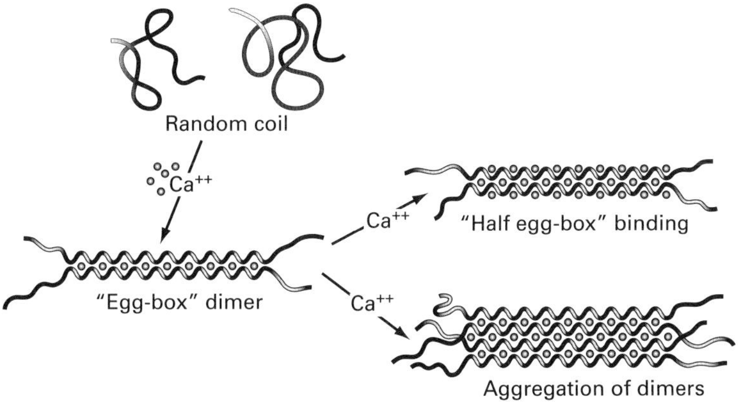

Figure 5.16 Schematic representation of Ca chelation to poly(L-guluronate) sequences. Open circles represent Ca^{2+} ions. Dimerization involves calcium chelation only to the interior faces of the participating chains (50% of the stoichiometric requirement). On more extensive aggregation, this ratio will tend towards 100% with increasing aggregate size, while site binding of calcium to the exterior faces of dimers ('half egg-box' binding), without further aggregation, raises the ratio to 150%. Reprinted with permission from Morris *et al.* (1978) © 1978 Elsevier.

The alginate model of Figure 5.16 was further examined by SAXS (Chapter 2) and the cross-sectional radius of gyration R_c was determined to be 0.31–0.56 nm (Stokke *et al.*, 2000). The SAXS data suggested that dimerization of chain segments was the principal association mode at low fractional Ca^{2+} saturation of G. Increasing the Ca^{2+} saturation of guluronic acid, either by concentration or selection of alginate source, yielded coexisting lateral association modes. These junction zone multiplicities occur because there is a delicate balance between the block length distribution of the α-L-GulA residues, polymer concentration and Ca^{2+}. The relationship between rheological properties and the structure determined by SAXS was studied (Yuguchi *et al.*, 2000). The storage modulus of gels was found to increase as the bundles composed of associated alginate chains grow (Figure 5.17). Gel elasticity was thought to be mainly sustained by single chains in the alginate sample with a low fraction of α-L-GulA. A good correlation was found between the storage modulus and the mean cross-sectional radius, and it was concluded that alginates with a high fraction of α-L-GulA associate into thicker bundles, which join to form a network. It was also suggested that gel elasticity is due to the flexible joints between bundles, since the fraction of single chains is extremely low.

The overall composition of M/G residues and their distribution patterns vary with seaweed species. However, very recently the availability of C-5 epimerases, specific enzymes which can convert MM- to MG-blocks in the polymer or can introduce G-blocks of various lengths, has been pioneered by the group led by Skjåk-Bræk (Campa *et al.*, 2004; Holtan *et al.*, 2006). This powerful approach has allowed many special 'tailor-made' alginate samples to be prepared, by tuning the fine structure and the composition and distribution patterns of M/G-blocks (Donati *et al.*, 2005). Monovalent sodium and smaller divalent magnesium ions cannot induce the gelation of alginate, but

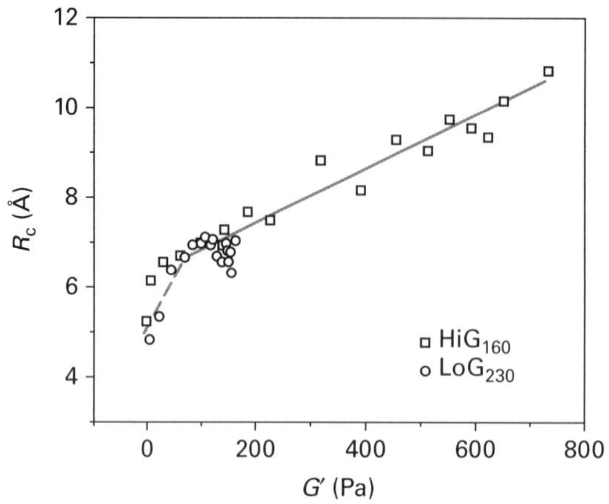

Figure 5.17 Correlation between mean cross-sectional radius and storage modulus for 1% alginate gels: (□) high guluronic acid content, $M_w = 16{\times}10^4$ g mol^{-1} (HiG_{160}), [Ca^{2+}] = 10 mM using CaEGTA; (○) low guluronic acid content, $M_w = 23 \times 10^4$ g mol^{-1} (LoG_{230}), [Ca^{2+}] = 10 mM using CaEGTA. Reprinted with permission from Yuguchi *et al.* (2000) © 2000 Elsevier.

calcium ions are very effective gel inducers. This ion selectivity can be explained by introducing a selectivity coefficient (Smidsrød and Haug, 1968).

It was confirmed that the affinity of alginate for divalent cations increased with increasing content of G unit, and that the physical properties of these polymers in aqueous media depend not only on the M/G ratio but on the distribution of the M and G units along the chain. Overall, then, the egg-box model is acceptable for describing the major features of the formation of alginate gels in the presence of the larger alkaline earth metals.

A recent study by Zhao and co-workers (Zhao *et al.*, 2010) has contrasted the behaviour of chemically cross-linked and Ca^{2+} cross-linked alginate gels using compression and subsequent stress relaxation (Ferry, 1980). The chemical gels show a predominantly elastic response with a finite equilibrium modulus, whereas the ionic gels show an almost monotonic decrease in log stress (or modulus, since strain is constant) when plotted against log (time).

At the same time, Zhao *et al.* monitored the mass of the samples. The chemical gel showed a decrease in mass as water was driven out, but this was recovered when the gel was re-swollen to equilibrium. By contrast, the ionic gel mass remained almost constant under compression. The authors explain this by assuming that the gel with covalent cross-links relaxes stress simply by migration of the water (compressive de-swelling), while in the ionic gel cross-links relax by breaking and then re-forming, as suggested by others, including Pines and Prins (1973) in their work on agarose gels (Chapter 7). However, the strain (15%) and strain rates used in the work are comparatively high for stress relaxation studies of physical gels, and it is not clear that all measurements are entirely within the linear viscoelastic region.

As a final comment, alginate gels can also be prepared by lowering pH below ~2.5. This produces an alginic acid gel, which has equilibrium swelling properties and seems to involve hydrogen bonding (Draget *et al.*, 1994).

5.5.2 Pectins

The gelling of pectins depends very significantly on the source and extraction procedures employed. The main structure is formed predominantly from linear sequences of poly(α (1→4) D galacturonic acid), with occasional L-rhamnose residues interrupting the sequence, although other residues do occur in the polymeric chain, including, for example, galactose and glucose up to ~10% level. Some of the galacturonic acid groups are methyl esterified and this tends to govern the gelation behaviour. For example, if the degree of esterification (DE) is relatively small, say <40% (so-called low methoxy pectins), then gelation is somewhat analogous to that for the alginates, both because it is governed by Ca^{2+} and because it forms a form of distorted egg-box quite similar to that for alginate gels (Powell *et al.*, 1982).

If DE > 60%, then gelation can be induced simply by lowering pH to 3.5 or below or reducing water activity by adding low molecular mass sugars – this combination is of course the basis of jam making, and most jams and preserves can be regarded as partially filled (by the fruit residue) high methoxy pectin gels. However, the precise gelation mechanism has not been elucidated; it is commonly suggested that both hydrogen bonding and hydrophobic interactions are involved, but without more detailed explanation.

5.5.3 Egg-box type binding models

Donati *et al.* (2006) recently devised a theoretical model to describe the binding of Ca^{2+} to alginate–pectin and chain associations from aspects of both the specific interactions and counterion condensation, and applied it successfully to experimental data for pectin. There is a difference, however, in that this model assumes that the egg-box dimers grow progressively rather than via an all-or-none process, or with an induction period. Fang *et al.* (2007) examined the details of the binding of Ca^{2+} to alginate to elucidate the pathway of chain–chain association (Figure 5.18).

With increasing concentration of Ca^{2+} ions, three distinct and successive binding steps were found. They were assigned to (i) interaction of Ca^{2+} with a single guluronate (G) unit, forming monocomplexes; (ii) propagation and formation of egg-box dimers via pairing of these monocomplexes; and (iii) lateral association of the egg-box dimers, generating multimers. The third step has different association modes depending on the molecular mass of alginate. The boundaries between these steps are reasonably critical, and they correlate closely with the Ca–guluronate stoichiometry expected for egg-box dimers and multimers with 2/1 helical chains. The formation of egg-box dimers and their subsequent association are thermodynamically equivalent processes and can be fitted by a model of independent binding sites.

A recent paper (Schuster *et al.*, 2011) has used both microrheology and SAXS to investigate the structure of Ca^{2+}–pectin gel junction zones. Conclusions from this work suggest that the final gel structure consists of a mixture of four or more chains, together with single chain flexible regions. Again the structure seems more complex than for the basic alginate egg-box.

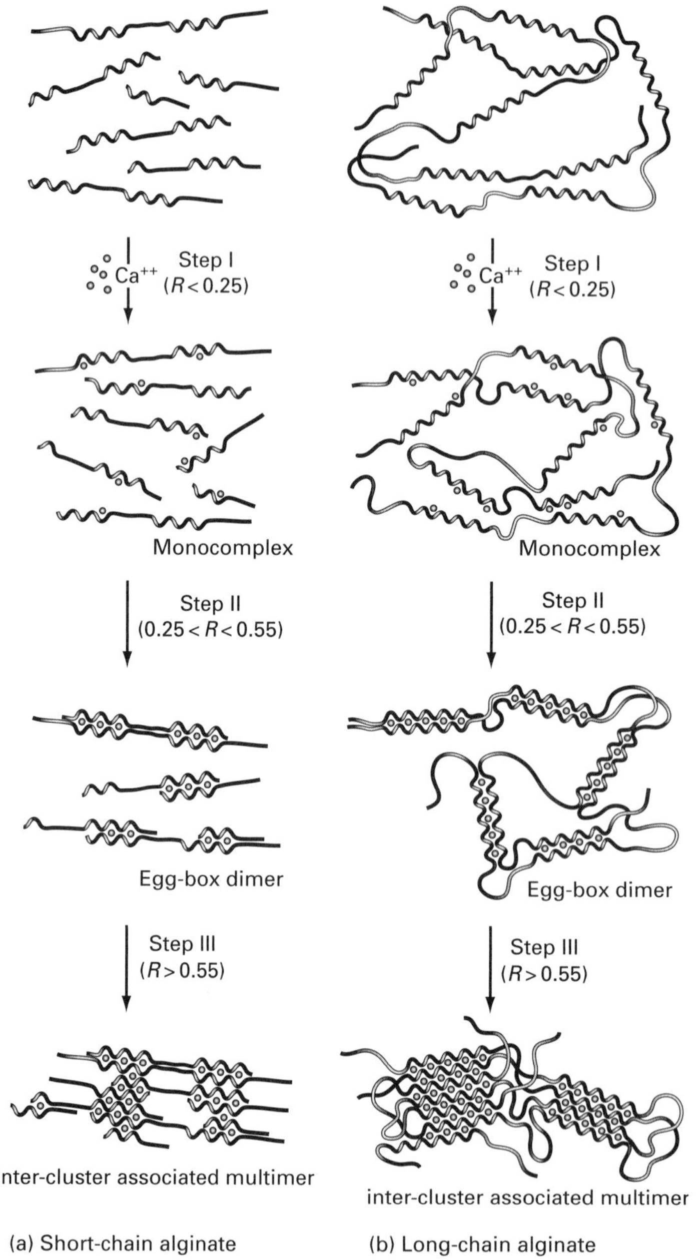

Figure 5.18 Schematic illustration of the multiple-step binding of Ca^{2+} to alginate: (a) short-chain alginate; (b) long-chain alginate. The zigzag lines, smooth lines and dots stand for G-blocks, M-blocks and calcium ions, respectively. R is the feed ratio of Ca^{2+} to G residue. Reprinted with permission from Fang *et al.* (2007) © 2007 American Chemical Society.

5.5.4 Gelation characteristics for alginates and pectins

One important aspect of these two systems is the very different kinetics of gelation. This already hints at the less than perfect egg-box structure formed for pectins. It has been known for many years that Ca^{2+} alginate gelation is extremely fast. Indeed this was the basis for a number of re-formed food products even in the 1960s. For example drop-wise addition of a Ca^{2+}-rich solution into a Na^{+} alginate solution all but instantaneously produces alginate gel beads, and the size of these can be controlled from, say, 1 mm down to microns by controlling the flow rate and/or orifice geometry. One current food application of this technology is an artificial salmon roe, *ikra*, produced in Japan (Nishinari, 1988). By contrast, low methoxy pectin gelation is a much slower process, and the resultant gels are generally of much lower modulus. Unfortunately there seem to have been surprisingly few rigorous rheological studies, particularly on pectin gels – exceptions being the work of Axelos, Gilsenan and their respective co-workers (Axelos and Thibault, 1991; Gilsenan *et al.*, 2000).

More recently, alginate gel beads have been applied in the biomedical area, with some considerable success. Some of this work is described in Chapter 11, but pioneering work by the Trondheim group has been concerned with encapsulation and subsequent implantation of alginate beads containing pancreatic cells into insulin-dependent diabetics; here the alginate gel surrounds serve to protect against immune reactive responses and so preserve cell viability (King *et al.*, 2003).

5.6 Xanthan

Xanthan, or xanthan gum, is an anionic microbial exopolysaccharide produced by *Xanthomonas* strains (particularly *X. campestris*) which is widely employed in applications as a structurant or stabilizer, mainly because of its rather unusual rheological properties. Chemically, it is very complex (Sutherland, 1998), as it consists of a β (1→4)-D-glucose (cellulose) backbone with trisaccharide side chains (1→3)-linked to alternate backbone residues. The side chains are β-D-mannose-(1→4) – β-D- glucuronic acid-(1→2)-α-mannose. An O-acetyl group is frequently present at the C-6 position of the internal mannose residue, while the terminal mannose may carry a (4→6)-linked pyruvic acid (strictly, pyruvate ketal) substituent. Varying the conditions under which the organism is cultured, or the microbial strain, alters the proportion of the side chains substituted (Sutherland, 1998), which also tends to modify rheological properties.

At low temperatures and/or in the presence of salt, xanthan adopts an ordered structure, generally now accepted to be a double helix, and makes it a very stiff polymer with a persistence length $l_p \sim 120$ nm (Sato *et al.*, 1984; Coviello *et al.*, 1986). (Such a l_p value is considerably higher than that for double-stranded DNA, for example, at *c.*50 nm.) Because of this intrinsic stiffness, l_p also hardly changes with ionic strength. This means it has been described as rod-like, although this is incorrect. This is because most commercial samples ($M_w \sim 1\text{–}2 \times 10^6$ g mole^{-1}) have contour lengths *c.*4–10 × the persistence length, and so are actually tending towards coil-like behaviour.

Xanthan does not form ionic gels of the type discussed in earlier sections, but the properties do vary with the associated counterion species (and also the ionic composition of the solvent). Unfortunately, the order–disorder transition temperature shifts both with ionic strength and with the amount of pyruvyl and acetyl substituents. The consequence of this is that in pure water it can pass into the disordered form or vice versa, at or below room temperature, depending on the polymer (or, more accurately, the associated counterion) concentration. That is mostly likely why studies of xanthan rheology in water, for example (Choppe *et al.*, 2010), while quite numerous, are often contradictory (though, of course, such a combination may be necessary in applications). For this chapter, we largely restrict subsequent discussion to solution properties measured fully in the ordered form, and in electrolyte solution.

As far as small-deformation oscillatory shear measurements for semi-dilute solutions – say ~ 1% w/w – are concerned, xanthan–electrolyte solutions give a reasonably flat gel-like mechanical spectrum with $G' > G''$ (Ross-Murphy *et al.*, 1983). On the other hand, this holds only for small deformations – shear strains, say, <5% – since when subjected to larger shear strains such a system does not fracture and fail like the systems above, but instead tends to flow rather like a 'normal' viscoelastic liquid (Ross-Murphy *et al.*, 1983). For this reason it has been called a 'weak gel' system, although we prefer the rheologically more accepted term 'structured liquid'.

Moreover, under these circumstances, the Cox–Merz 'rule' – the superposition of steady shear viscosity η at shear rate $\dot{\gamma}$ and dynamic viscosity η^* at frequency ω – is not obeyed as it would be for our normal polymer solution, and η^* seems always to be $> \eta$. It has been suggested that this response is due to more specific ion-mediated inter-chain couplings between xanthan molecules (Frangou *et al.*, 1982) in addition to the topological restraints, but an element of liquid crystalline behaviour (Chapter 3) has also been advocated. Here, there seems no doubt that there is evidence for some liquid crystallinity, but at significantly higher concentrations, where polymer molecular solubility itself can become compromised.

In steady shear, apparent viscosity–shear rate power-law exponents for xanthan solutions are high (~0.9), and the overall viscosity versus shear rate profile is more like that of a so-called Bingham viscoplastic material than a conventional 'shear thinning' polymer solution. Confirmatory evidence can be obtained from so-called stress overshoot and start-shear history experiments (Richardson and Ross-Murphy, 1987). Such work shows that the stress overshoot recovery time for xanthan extends out to times of at least 10^4 s, whereas at comparable concentrations for 'simple' polysaccharide solutions this tends to be complete after, say, 10–100 s.

What is important to point out is that such rheological behaviour is not unique to xanthan solutions, nor is it always seen with these. Indeed, Milas and co-workers have demonstrated that a xanthan solution prepared directly from the culture broth without pasteurization or freeze-drying shows only very small deviations from the Cox–Merz rule, even at quite high concentrations (Milas *et al.*, 1990). Correspondingly, other workers have published flow curves for commercial xanthan samples in water above the overlap concentration c^* which are quite as expected for polymers in solution. This implies that sample history is rather crucial; the 'typical' results above apply to solutions prepared from commercial samples usually, but not exclusively, ion-exchanged to a single or restricted cation form, and with added electrolyte. Under these conditions there is almost a power-law behaviour of η ($\dot{\gamma}$), and no indication of an η_0 plateau.

Among the most extensive and precise studies of xanthan rheology are those by Lee and Brant (2002a, 2002b, 2002c), who improved upon almost all earlier work by using molecular mass (or 'molecular weight', MW) fractionated samples. They demonstrated, with their fractionated lowest-MW sample ($M_w \sim 1.5 \times 10^5$ g mol^{-1}, contour length $\sim 2/3 \times l_p$), a definitive transition (maximum) in the viscosity at *c.*10% w/w (Lee and Brant, 2002a), as expected at, or close to, an isotropic–biphasic–nematic transition (Wissbrun, 1981). At lower concentrations, the behaviour was largely as described above, with no Cox–Merz superposition except for dilute solutions. For their 'medium'-MW sample, where $L \sim 2l_p$, the transition, which now occurred at *c.*7% w/w, was less pronounced (Lee and Brant, 2002b). They argued that this reflects the increased tendency of the higher-MW sample to form chain entanglements.

In the third paper of their series, Lee and Brant (2002c) investigated the temperature dependence of viscosity for their low- and medium-MW samples. First they mapped out the xanthan order–disorder transition temperature, but then were able to compare their results from viscosity measurements to a Flory chimney type phase diagram (Chapter 3). For the higher-MW sample, and at the lowest ionic strength (0.01 M, where the xanthan chain order–disorder transition occurs at *c.*60°C), they demonstrated how the extent of both isotropic and biphasic boundaries increased with concentration. However, it appears that the interesting liquid crystalline phenomena lie well above concentrations of practical interest for higher-MW commercial samples.

For these high-MW xanthans, preparation of a solution at above, say, 2% w/w becomes increasingly difficult. Other factors also begin to take over; liquid crystalline transitions tend to be 'smeared out' by polydispersity, and most such samples have $M_w/M_n \gtrsim 2$ (M_n is the number average molecular mass). Also, as pointed out by Lee and Brant, as chain length increases, entanglement effects begin to dominate. Overall, then, this suggests that liquid crystallinity will tend to be observed only under conditions rather far from those of practical interest.

Interestingly, the failure of the Cox–Merz rule, and the observation of power-law (log–log) viscosity versus shear rate curves, have also been reported for a number of other polysaccharides, including a chemically modified carboxymethylcellulose (Westra, 1989).

5.7 Chitin and chitosan

Chitin is one of the most commonly occurring materials on earth because it is the main component of the exoskeleton of crustaceans such as shrimps, and more particularly krill. The attached pendant N-acetyl groups of natural chitin are partially removed under alkaline conditions to yield a water-soluble and consequently more reactive derivative, called chitosan. Chitosan itself can be regarded formally as a copolymer of β (1→4)-linked 2-acetamido-2-deoxy-D-glucopyranose and 2-amino-2-deoxy-D-glucopyranose.

The effectiveness of the deacetylation process can be followed by infrared or NMR spectroscopy. The main parameters influencing the characteristics of chitosan are its molecular mass and its degree of deacetylation. The amino group has a pK of 6.6, which means it maintains a positive (cationic) charge in acidic to neutral solution, dependent on both pH and its degree of deacetylation.

However, as pointed out earlier, chitosan is not, on its own, a gelling polysaccharide, even though many biomedical papers still refer to chitosan gels, when these are simply high-viscosity entangled solutions. Chitosan receives a great deal of attention in medical and pharmaceutical applications, and one of the reasons is this charge effect. For example, every other polysaccharide described in this chapter is anionic, and indeed chitosan remains one of the few cationic natural polymers. However, allied to this, chitosan is known for its biocompatibility, allowing its use in medical applications such as topical ocular application, implantation or injection. Moreover, chitosan is metabolized by certain human enzymes, e.g. lysozyme, and can be considered as biodegradable. Owing to these positive charges at physiological pH, chitosan is also bioadhesive, which increases retention at the site of application. Many researchers and others claim that chitosan also promotes wound-healing and has bacteriostatic effects (Ueno *et al.*, 2001). When chitosan is dispersed as particles with sizes in the range 75–220 μm (Lima and Airoldi, 2004; Monteiro and Airoldi, 2005) in aqueous solution, it may also find industrial application, such as cation removal from wastewater or industrial effluents through biopolymer–cation interactions at the solid–liquid interface, with a preference for copper cations.

More detailed applications of chitosan are described in Chapter 11, in particular its use in biomedical applications and, together with a variety of anionic polyelectrolytes, in so-called polyelectrolyte complexes (PECs).

5.8 Conclusions

In spite of efforts by many research groups to clarify the precise mechanisms of gelation of ionic polysaccharide gels, few if any have been established unambiguously because of the absence of ideal samples, i.e. samples with a narrow distribution of molecular mass and completely converted into a single ionic salt form. Such improved samples, such as those used in the Japanese collaborative research group on gellan, could be distributed to laboratories in various disciplines, and the results compared and discussed. Non-perturbatory techniques such as SAXS combined with rheological studies should be performed in parallel to explain any unique gel properties. Despite these difficulties, this is an extremely active area with many innovative applications, some of which reappear in Chapter 11.

References

Austen, K. R. J., Goodall, D. M., Norton, I. T., 1988. *Biopolymers* **27**, 139–155.
Axelos, M. A. V., Thibault, J.-F., 1991. *Int. J. Biol. Macromol.* **13**, 77–82.
Bayley, S. T., 1955. *Biochim. Biophys. Acta* **17**, 194–205.

Belton, P. S., Morris, V. J., Tanner, S. F., 1985. *Int. J. Biol. Macromol.* **7**, 53–56.

Borgström, J., Quist, P.-O., Piculell, L., 1996. *Macromolecules* **29**, 5926–5933.

Campa, C., Holtan, S., Nilsen, N. *et al.*, G., 2004. *Biochem. J.* **381**, 155–164.

Choppe, E., Puaud, F., Nicolai, T., Benyahia, L., 2010. *Carbohydr. Polym.* **82**, 1228–1235.

Coviello, T., Kajiwara, K., Burchard, W., Dentini, M., Crescenzi, V., 1986. *Macromolecules* **19**, 2826–2831.

Dautzenberg, H., Jaeger, W., Kötz, J. *et al.*, 1994. *Polyelectrolytes: Formation, Characterization and Application*. Carl Hanser Verlag, Munich.

Dobrynin, A. V., Colby, R. H., Rubinstein, M., 2004. *J. Polym. Sci., Part B: Polym. Phys.* **42**, 3513–3538.

Doi, M., Edwards, S. F., 1999. *The Theory of Polymer Dynamics*, Reprint. Clarendon Press, Oxford.

Donati, I., Benegas, J. C., Paoletti, S., 2006. *Biomacromolecules* **7**, 3439–3447.

Donati, I., Holtan, S., Mørch, Y. A., Borgogna, M., Dentini, M., 2005. *Biomacromolecules* **6**, 1031–1040.

Draget, K. I., Skjåk Bræk, G., Smidsrød, O., 1994. *Carbohydr. Polym.* **25**, 31–38.

Fang, Y., Al Assaf, S., Phillips, G. O. *et al.*, 2007. *J. Phys. Chem. B* **111**, 2456–2462.

Ferry, J. D., 1980. *Viscoelastic Properties of Polymers*. Wiley Interscience, New York.

Frangou, S. A., Morris, E. R., Rees, D. A., Richardson, R. K., Ross-Murphy, S. B., 1982. *J. Polym. Sci., Part C: Polym. Lett.* **20**, 531–538.

Gilsenan, P. M., Richardson, R. K., Morris, E. R., 2000. *Carbohydr. Polym.* **41**, 339–349.

Grant, G. T., Morris, E. R., Rees, D. A., Smith, P. J. C., Thom, D., 1973. *FEBS Lett.* **32**, 195–198.

Grasdalen, H., Smidsrød, O., 1981. *Macromolecules* **14**, 229–231.

Hermansson, A.-M., Eriksson, E., Jordansson, E., 1991. *Carbohydr. Polym.* **16**, 297–320.

Holtan, S., Bruheim, P., Skjåk-Bræk, G., 2006. *Biochem. J.* **395**, 319–329.

Hossain, K. S., Nishinari, K., 2009. Chain release behavior of gellan gels. In *Gels: Structures, Properties, and Functions*. Springer, Berlin, pp. 177–186.

Hugerth, A., Nilsson, S., Sundelöf, L.-O., 1999. *Int. J. Biol. Macromol.* **26**, 69–76.

Ikeda, S., Morris, V. J., Nishinari, K., 2001. *Biomacromolecules* **2**, 1331–1337.

Ikeda, S., Nishinari, K., 2001. *J. Agric. Food Chem.* **49**, 4436–4441.

Kasapis, S., Giannouli, P., Hember, M. W. N. *et al.*, 1999. *Carbohydr. Polym.* **38**, 145–154.

King, A., Andersson, A., Strand, B. L. *et al.*, 2003. *Transplantation* **76**, 275–279.

Lee, H.-C., Brant, D. A., 2002a. *Macromolecules* **35**, 2212–2222.

Lee, H.-C., Brant, D. A., 2002b. *Macromolecules* **35**, 2223–2234.

Lee, H.-C., Brant, D. A., 2002c. *Biomacromolecules* **3**, 742–753.

Lima, I. S., Airoldi, C., 2004. *Thermochim. Acta* **421**, 133–139.

Manning, G. S., 1988. *J. Chem. Phys.* **89**, 3772–3777.

Milas, M., Rinaudo, M., 1996. *Carbohydr. Polym.* **30**, 177–184.

Milas, M., Rinaudo, M., Knipper, M., Schuppiser, J. L., 1990. *Macromolecules* **23**, 2506–2511.

Miyoshi, E., Nishinari, K., 1999. Rheological and thermal properties near the sol-gel transition of gellan gum aqueous solutions. In Nishinari, K. (ed.), *Physical Chemistry and Industrial Application of Gellan Gum*. Progress in Colloid and Polymer Science, vol. 114. Springer, Berlin, pp. 68–82.

Miyoshi, D., Takaya, T., Nishinari, K., 1996a. *Carbohydr. Polym.* **30**, 109–119.

Miyoshi, E., Takaya, T., Williams, P. A., Nishinari, K., 1996b. *J. Agric. Food Chem.* **44**, 2486–2495.

Monteiro, J., Airoldi, C., 2005. *J. Colloid Interface Sci.* **282**, 32–37.

Morris, E. R., 1994. Chiroptical methods. In Ross-Murphy, S. B. (ed), *Physical Techniques for the Study of Food Biopolymers*. Blackie Academic and Professional, Glasgow, pp. 15–64.

Morris, E. R., Nishinari, K., Rinaudo, M., 2012. *Food Hydrocolloids* **28**, 373–411.
Morris, E. R., Norton, I. T., 1983. Polysaccharide aggregation in solutions and gels. In Wyn-Jones, E., Gormally, J. (eds), *Aggregation Processes in Solution*. Elsevier, Amsterdam, pp. 549–593.
Morris, E. R., Rees, D. A., Thom, D., Boyd, J., 1978. *Carbohydr. Res.* **66**, 145–154.
Morris, E. R., Rees, D. A., Norton, I. T., Goodall, D. M., 1980a. *Carbohydr. Res.* **80**, 317–323.
Morris, E. R., Rees, D. A., Robinson, G., 1980b. *J. Mol. Biol.* **138**, 349–362.
Morris, E. R., Richardson, R. K., Whitaker, L. E., 1999. Rheology and gelation of deacylated gellan polysaccharide with Na+ as the sole counterion. In Nishinari, K. (ed.), *Physical Chemistry and Industrial Application of Gellan Gum*. Progress in Colloid and Polymer Science, vol. 114. Springer, Berlin, pp. 109–115.
Morris, V. J., Belton, P. S., 1980. *J. Chem. Soc., Chem. Commun.* 983–984.
Morris, V. J., Tsiami, A., Brownsey, G. J., 1995. *J. Carbohydr. Chem.* **14**, 667–675.
Morris, V. J., Kirby, A. R., Gunning, A. P., 1999. A fibrous model for gellan gels from atomic force microscopy studies. In Nishinari, K. (ed.), *Physical Chemistry and Industrial Application of Gellan Gum*. Progress in Colloid and Polymer Science, vol. **114**. Springer, Berlin, pp. 102–108.
Nilsson, S., Piculell, L., 1991. *Macromolecules* **24**, 3804–3811.
Nishinari, K., 1988. Food hydrocolloids in Japan. In Phillips, G. O., Wedlock, D. J., Williams, P. A. (eds), *Gums and Stabilisers for the Food Industry* **4**, IRL Press, Oxford, pp. 373–390.
Nitta, Y., Ikeda, S., Takaya, T., Nishinari, K., 2001. *Trans. Mater. Res. Soc. Jpn.* **26**, 621–624.
Norton, I. T., Goodall, D. M., Morris, E. R., Rees, D. A., 1983. *J. Chem. Soc., Faraday Trans. 1* **179**, 2745–2488.
Norton, I. T., Morris, E. R., Rees, D. A., 1984. *Carbohydr. Res.* **134**, 89–101.
Odijk, T., 1977. *J. Polym. Sci., Part B: Polym. Phys.* **15**, 477–483.
Ogawa, E., Yamazaki, K., Sugimoto, K. *et al.*, 2009. *Trans. Mater. Res. Soc. Jpn.* **34**, 481–484.
Ohtaki, H., Radnai, T., 1993. *Chem. Rev.* **93**, 1157–1204.
Oosawa, F., 1971. *Polyelectrolytes*. Marcel Dekker, New York.
Payens, T. A. J., Snoeren, T. H. M., 1972. *J. Electroanal. Chem.* **37**, 291–296.
Piculell, L., Nilsson, S., 1989. *J. Phys. Chem.* **93**, 5596–5601.
Piculell, L., Nilsson, S., Viebke, C., Zhang, W., 1993. Gelation of (some) seaweed polysaccharides. In Nishinari, K., Doi, E. (eds), *Food Hydrocolloids: Structures, Properties, and Functions*. Plenum Press, New York, pp. 35–44.
Pines, E., Prins, W., 1973. *Macromolecules* **6**, 888–895.
Powell, D. A., Morris, E. R., Gidley, M. J., Rees, D. A., 1982. *J. Mol. Biol.* **155**, 517–531.
Record Jr, M. T., 1967. *Biopolymers* **5**, 975–992.
Rees, D. A., 1969. *J. Chem. Soc. B*, 217–226.
Richardson, R. K., Ross-Murphy, S. B., 1987. *Int. J. Biol. Macromol.* **9**, 257–264.
Rinaudo, M., 2008. In Belgacem, M., Gandini, A. (eds), *Monomers, Polymers and Composites from Renewable Resources*. Elsevier, Amsterdam, pp. 495–516.
Rinaudo, M., Karimian, A., Milas, M., 1979. *Biopolymers* **18**, 1673–1683.
Rochas, C., Rinaudo, M., 1980. *Biopolymers* **19**, 1675–1687.
Rochas, C., Rinaudo, M., Landry, S., 1990. *Carbohydr. Polym.* **12**, 255–266.
Ross-Murphy, S. B., Morris, V. J., Morris, E. R., 1983. *Faraday Symp. Chem. Soc.* **18**, 115–129.
Sato, T., Norisuye, T., Fujita, H., 1984. *Macromolecules* **17**, 2696–2700.
Schuster, E., Cucheval, A., Lundin, L., Williams, M. A. K., 2011. *Biomacromolecules* **12**, 2583–2590.
Skolnick, J., Fixman, M., 1977. *Macromolecules* **10**, 944–948.
Smidsrød, O., Grasdalen, H., 1984. *Hydrobiologia* 116–117, 178–186.
Smidsrød, O., Haug, A., 1968. *Acta Chem. Scand.* **22**, 797–810.

Smidsrød, O., Andresen, I., Grasdalen, H., Larsen, B., Painter, T., 1980. *Carbohydr. Res.* **80**, C11–C16.
Snoeren, T. H. M., Payens, T. A. J., 1976. *Biochim. Biophys. Acta, Gen. Subj.* **437**, 264–272.
Stokke, B. T., Draget, K. I., Smidsrød, O. *et al.*, 2000. *Macromolecules* **33**, 1853–1863.
Stuart, M. C., de Vries, R., Lyklema, H., 2005. Polyelectrolytes. In Lyklema, H. (ed.), *Fundamentals of Interface and Colloid Science*. Elsevier, Amsterdam, pp. 1–84.
Sutherland, I. W., 1998. *Trends Biotechnol.* **16**, 41–46.
Tokita, M., Ikura, M., Hikichi, K., 1989. *Polymer* **30**, 693–697.
Uedaira, H., 1990. Hydrogels and water. In Nishinari, K., Yano, T. (eds.), *Science of Food Hydrocolloids*. Asakura Shoten, Tokyo, pp. 7–22.
Ueno, H., Mori, T., Fujinaga, T., 2001. *Adv. Drug Delivery Rev.* **52**, 105–115.
Viebke, C., Piculell, L., Nilsson, S., 1994. *Macromolecules* **27**, 4160–4166.
Viebke, C., Borgström, J., Carlsson, I., Piculell, L., Williams, P., 1998. *Macromolecules* **31**, 1833–1841.
Watase, M., Nishinari, K., 1982. *Rheol. Acta* **21**, 318–324.
Westra, J. G., 1989. *Macromolecules* **22**, 367–370.
Wissbrun, K. F., 1981. *J. Rheol.* **25**, 619–662.
Yuguchi, Y., Urakawa, H., Kajiwara, K., Draget, K. I., Stokke, B. T., 2000. *J. Mol. Struct.* **554**, 21–34.
Zhang, W., Piculell, L., Nilsson, S., 1992. *Macromolecules* **25**, 6165–6172.
Zhang, W., Piculell, L., Nilsson, S., Knutsen, S. H., 1994. *Carbohydr. Polym.* **23**, 105–110.
Zhao, X., Huebsch, N., Mooney, D. J., Suo, Z., 2010. *J. Appl. Phys.* **107**, 063509.

6 Hydrophobically associated networks

6.1 Introduction

In this chapter we are concerned with molecules which have a dual nature, with one part soluble in water (hydrophilic) and the other a non-polar part which is expelled from water and aqueous systems (hydrophobic); such molecules are sometimes known as *amphiphilic*. Molecules of this kind are forced to adopt unique orientations with respect to the aqueous medium and sometimes form organized structures. Such molecules were first recognized to play an important role in living matter, specifically in its organization, of which cell membranes are the best examples. The formation of these membranes is spontaneous and depends only on the fact that their constituent molecules are amphiphilic.

6.2 The hydrophobic effect

6.2.1 Origin of the hydrophobic effect

Hydrophobic substances dissolve readily in many non-polar solvents, but are hardly soluble in water. By contrast, hydrophilic molecules have a strong affinity for water, and tend to repel one another strongly; they contain electronegative atoms able to establish hydrogen bonds with water, and include alcohols, sugars, polar (non-ionic) groups, anionic (e.g. carboxylate or sulphonate) and cationic (trimethyl ammonium) groups and zwitterions (Israelachvili, 1992). The functional groups are believed to have a disordering effect and disrupt the local structure of liquid water, while hydrophobic molecules tend to increase the ordering of water molecules. Amphiphilic molecules such as *surfactants* (surface-active agents) have a particular architecture: one end (called the head) contains a hydrophilic group, while the rest of the molecule (the tail) is hydrophobic, usually with a long hydrocarbon chain.

Surfactants are known mainly for adsorbing on to surfaces (e.g. between liquid and gas phases) or interfaces (e.g. between two immiscible liquids). Soaps are a good example of surfactants; these are usually sodium or potassium salts of straight-chain fatty acids containing between 8 and 19 carbon atoms. Other characteristics of surfactants include the capacity, in solution, to form clusters of colloidal size known as *micelles*.

It was recognized in the early 1920s (McBain and Salmon, 1920) that the reversible formation of micellar aggregates in soap solutions was due to association between hydrocarbon chains driven by their mutual attraction. The free energy gained in association was equal to the energy difference between liquid and gaseous hydrocarbons, but this mutual attraction of non-polar groups plays only a minor role in the overall hydrophobic effect.

Understanding of the hydrophobic effect starts with a more precise description of aspects of water structure related to the formation of hydrogen bonds. Hydrogen bonds are strong and orientation-dependent bonds involving a particularly strong type of directional dipole–dipole interaction. A hydrogen bond between two groups XH and Y is usually denoted XH $\cdots$ Y, and its strength lies between 10 and 40 kJ mol^{-1}, larger than that of a van der Waals bond (1 kJ mol^{-1}) but still much weaker than a typical covalent or ionic bond (500 kJ mol^{-1}).

Hydrogen bonds can form weak three-dimensional structures in liquids, where they create a short-range order that is still significantly longer lived than in simple liquids; consequently the term 'associated liquids' is used. Hydrogen bonds play a crucial role in liquid water structure, since each oxygen and its two hydrogen atoms can participate in four linkages with other water molecules: two bonds involving its own H atoms and two involving its unshared lone-pair electrons with other H atoms. For this reason, liquid water has a tendency to retain an ice-like tetrahedral network structure, but the long-distance arrangement in the liquid is disordered and labile. This tetrahedral coordination of a water molecule is the origin of its unusual properties, including a maximum density at 4°C with solid ice less dense than liquid water, a low compressibility and a high dielectric constant.

Non-polar molecules such as alkanes cannot form hydrogen bonds with water (Israelachvili, 1992). If a non polar molecule is not too large, it is possible for water molecules to organize around the dissolved non-polar solute in so-called 'clathrate cages', forming labile structures in which water molecules are more ordered than in the bulk liquid. Thus, when water molecules come in contact with a non-polar molecule the main effect is to induce reorientation of the water molecules in such a way that they can participate in hydrogen-bond formation with the bulk water, but without breaking any hydrogen bonds, an effect called *hydrophobic solvation*. The reorientation of water molecules around non-polar (hydrophobic) solutes is entropically unfavourable and, surprisingly, the new structure of the surrounding water molecules is more ordered than in the bulk liquid.

Closely related to hydrophobic solvation is the *hydrophobic interaction* or *hydrophobic effect*, which describes an unusually strong attraction between hydrophobic molecules and surfaces in water, much stronger than their attraction in free space. Such hydrophobic solutes then assemble spontaneously into larger structures, and as two hydrophobic species come together there is a rearrangement of hydrogen-bond configurations. The net result for the free energy change of all the molecules involved in the spontaneous aggregation is always negative. Hydrophobic interactions play a central role in molecular self-assembly in micelle formation of amphiphilic molecules and in biological membrane formation. The thermodynamic principles of the self-assembly of amphiphilic molecules are well understood, and so will only be summarized here.

6.2.2 Self-assembly of amphiphilic molecules

The fundamental thermodynamic equations of self-assembly deal with the equilibrium between aggregated species and single molecules. The association of N amphiphilic molecules into an aggregate generates a variety of structures in aqueous solution. In most cases, the micelle structure is a sphere with a hydrocarbon core, the polar or ionic groups at the surface serving to maintain the solubility in water. The hydrocarbons in micelles are generally regarded to be in the bulk liquid medium. The core is a small volume of liquid hydrocarbon. However, parts of the hydrocarbon chains close to the polar surface are constrained and lie more or less perpendicular to the surface. The effects of such constraints are nevertheless quite small for simple amphiphilic molecules, as confirmed by spectroscopic relaxation methods (Tanford, 1991).

Equilibrium thermodynamics requires that, in a system of molecules that form aggregated structures, the chemical potential of all identical molecules in different aggregates is the same.

If N is the number of molecules in an aggregate, this condition gives

$$\mu = \mu_N = \mu_N^0 + \frac{kT}{N}\log\left(\frac{X_N}{N}\right) = const. \quad N = 1, 2, 3, \ldots \tag{6.1}$$

Here μ_N is the chemical potential of a molecule in an aggregate of aggregation number N, μ_N^0 is the standard chemical potential (the mean interaction free energy per molecule) in a micelle of size N, μ_1^0 is the standard chemical potential of a single molecule and X_N is the molar fraction or the activity of amphiphilic molecules in micelles of size N (Israelachvili, 1992). Following the law of mass action, one can also write this equilibrium as a function of the rate of association of molecules, A:

$$N A_1 \leftrightarrow A_N.$$

For ideal (dilute) solutions without interactions between aggregates, the equilibrium constant of this reaction is

$$K_N = \frac{[A_N]}{[A_1]^N}. \tag{6.2}$$

With

$$[A_N] = \frac{X_N}{N} \quad \text{and} \quad [A_1] = X_1,$$

we have

$$K_N = \frac{X_N}{NX_1^N} = \exp\left[-\frac{N(\mu_N^0 - \mu_1^0)}{kT}\right]. \tag{6.3}$$

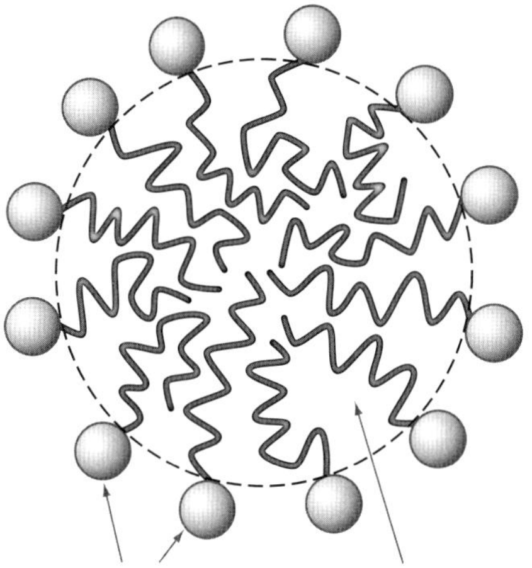

Figure 6.1 A spherical micelle of amphiphilic molecules.

The fraction of molecules in micelles of size N is

$$X_N = N\left[X_1 \exp\left(-\frac{\mu_N^0 - \mu_1^0}{kT}\right)\right]^N, \quad \sum_{N=1}^{\infty} X_N = C. \tag{6.4}$$

Here C is the molar fraction of amphiphilic molecules in solution. When micelles form, the difference in the standard free energies has to be negative:

$$\mu_N^0 - \mu_1^0 < 0.$$

The structure of a micelle is shown in Figure 6.1. In most currently used surfactants there is an optimum aggregate size or aggregation number $N = N_{agg}$ for a spherical micelle made of N_{agg} molecules. The value of N_{agg} is obtained by minimizing the free energy of the micelle versus the surface area of a molecule, taking into account two opposing effects:

- the attractive energy arising from the interfacial tension between the liquid hydrocarbon and water
- the repulsion between the hydrophilic heads due to steric hindrance or to electrostatic repulsion.

An equilibrium is established between single molecules and aggregates of size N_{agg}; when the concentration c increases, single molecules dissolve until the so-called critical micellar concentration (CMC) is reached. At this concentration, single molecules coexist with the first micelle:

$$(X_1)_{crit} = [\mathrm{CMC}] \approx \exp\left(-\frac{\left(\mu_1^0 - \mu_{N_{agg}}^0\right)}{k_B T}\right), \quad X_{N_{agg}} = N_{agg}. \tag{6.5}$$

The standard free energy of micellization $\Delta G_{\text{micelle}}$ for a micelle of N_{agg} molecules is given by

$$\Delta G_{\text{micelle}} = N_{\text{agg}}(\mu^0_{N_{\text{agg}}} - \mu^0_1) = -k_B T \ln K_{N_{\text{agg}}} = k_B T \ln[\text{CMC}]. \tag{6.6}$$

The enthalpy of micellization at temperature T and pressure P can be calculated from the Gibbs–Helmholtz relationship,

$$\Delta H_{\text{micelle}} = -RT^2 \left(\frac{\partial(\ln[\text{CMC}])}{\partial T} \right)_P, \tag{6.7}$$

and the entropy of micellization is, as usual,

$$\Delta S_{\text{micelle}} = \frac{\Delta H_{\text{micelle}} - \Delta G_{\text{micelle}}}{T}. \tag{6.8}$$

When the concentration of amphiphilic molecules is increased above the CMC, further addition of solute molecules results in the formation of additional aggregates, while the concentration of single molecules is more or less unchanged, remaining at the CMC value.

Temperature plays an important role in the behaviour of amphiphilic molecules. At low temperatures, the molecules have a very low solubility and most precipitate as hydrated crystals, in equilibrium with single molecules in solution. As the temperature is increased, the amphiphiles become soluble, while maintaining the equilibrium between micelles and single molecules. At the temperature called the Krafft temperature, the solubility equals the CMC, as shown in Figure 6.2. It is assumed that the Krafft temperature is related to the melting temperature of the crystalline hydrocarbon phase, the interior of the micelles being a liquid hydrocarbon.

CMCs for some common surfactants are listed in Table 6.1, determined for each surfactant above the Krafft temperature. There are several experimental techniques to determine this, the most usual being measurement of the surface tension at the interface

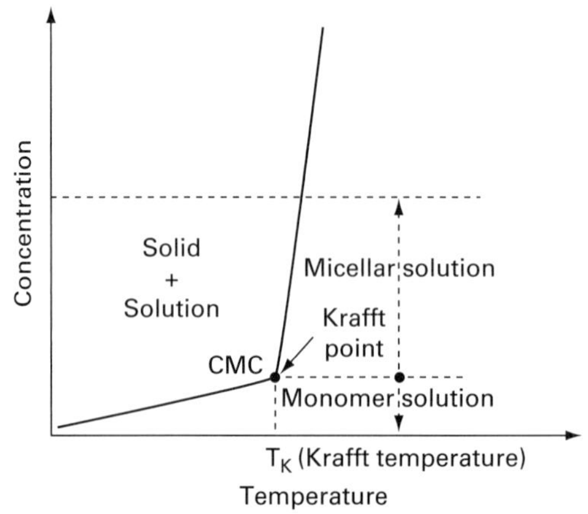

Figure 6.2 Solubility curve for amphiphilic molecules, showing the Krafft temperature.

Table 6.1 Critical micellar concentrations of some common surfactants.

Surfactant	CMC (mM)
Anionic (Sodium alkyl sulphates)	
$C_8H_{17} - SO_4^- \ Na^+$	130
$C_{10}H_{21} - SO_4^- \ Na^+$	33.2
$C_{12}H_{25} - SO_4^- \ Na^+$ (SDS)	8.1
$C_{14}H_{29} - SO_4^- \ Na^+$	2.1
Cationic (Alkyl trimethylammonium bromides)	
$C_{12}H_{25} - N(CH_3)_3^+ \ Br^-$	15
$C_{16}H_{33} - N(CH_3)_3^+ \ Br^-$ (CTAB or HTAB)	0.9
Non-ionic (alkyl polyoxyethylene monoethers)	
$C_{12}H_{25} - (OCH_2CH_2)_6OH$ ($C_{12}E_6$)	0.087

between air and the aqueous solution. Surface tension decreases with increasing amphiphile concentrations, then shows a break at the CMC and remains constant above the CMC. At the CMC an oriented monolayer of amphiphile molecules forms at the surface of the water: the polar or ionic head of the molecules is incorporated into the aqueous phase and the hydrocarbon tail is oriented away from the water.

Table 6.1 shows that, for a homologous series of surfactants bearing the same head group, the CMC decreases with the length of the hydrocarbon chain, because of the poorer solubility of the long hydrocarbon chain. For amphiphiles bearing the same length of hydrocarbon chain, the CMC is much lower (by a factor of 100) for a polar head than for an ionic head. The repulsive interactions due to charged head groups control micelle formation; in other words, ionic surfactants are more soluble as single molecules. Above the Krafft temperature the CMC varies very little with temperature (as shown by the dashed line in Figure 6.2). That said, it does vary strongly with the ionic strength of the aqueous solutions: for instance, adding 0.3 M of sodium chloride decreases the CMC of SDS from 8.1 mM to 0.7 mM.

Associating polymers behave similarly to surfactants with respect to the formation of micellar structures. The consequences of micelle formation for the rheological properties of the aqueous solutions are, however, very important and lie at the origin of the use of these polymers in numerous applications.

6.3 Hydrophobically modified water-soluble polymers

The notion of associative polymers was introduced in the early 1980s. These consist of a water-soluble hydrophilic backbone and a small amount (<2–5 mol%) of covalently linked hydrophobic chains (Glass, 1986, 1989; Shedge *et al.*, 2005). The molecular architecture of the grafted chains and the characteristics of the hydrophilic backbone allow a great variety of hydrophobically modified (HM) water-soluble polymers to be synthesized, including random, block and end-capped molecules.

In aqueous solution these associating polymers organize in a similar way to surfactant molecules, forming micelle-like structures with reversible intermolecular bonds. Such macromolecules have remarkable rheological properties and are widely used as thickeners in cosmetics, pharmaceuticals, detergents, textiles and many other areas.

A minimum concentration is required for intermolecular associations to take place, equivalent to the CMC of the polymer, but this limit is more difficult to determine because intramolecular associations of the hydrophobic moieties can take place in very dilute solution. At these very low concentrations, intramolecular association is the dominant effect and molecules adopt a compact conformation, while at higher concentrations the molecules form reversible networks.

The replacement of hydrocarbons with perfluoroalkyl groups in HM associating polymers is also of potential interest since the hydrophobic character of perfluoroalkyl groups is more pronounced than with their hydrocarbon analogues. The associative ability of CF_2 groups is 1.7 times higher than that of the CH_2 groups, as shown, for instance, in viscosity measurements for HMPAA by Zhou *et al.* (2001) and Shedge *et al.* (2005).

We now undertake a brief survey of the various types of HM water-soluble polymers and their synthesis methods.

6.3.1 HM polyelectrolytes

Free radical copolymerization of a hydrophilic monomer, acrylic acid (AA), with *n*-alkyl acrylates ($n = 8, 12, 18$), with the fraction of *n*-alkyl acrylate in the monomer mixture not exceeding 20 mol%, forms an interesting class of associating polymers (Philippova *et al.*, 1997). As poly(acrylic acid) (PAA) is a weak acid, its degree of ionization can be varied over a wide range by changing the pH of the medium. Two extreme cases exist: uncharged acrylic acid and completely ionized acrylic units, shown in Figure 6.3.

In HM polyelectrolytes, the electrostatic repulsion between polymer backbones tends to counteract the hydrophobic attractive effect. As a result, the 'viscosifying' behaviour depends on both pH and ionic strength. These polymers were also prepared as cross-linked hydrogels using N,N′-methylene bis(acrylamide) as a cross-linker. At low degrees of ionization (low pH), HMPAA gels are in a collapsed state, but their swelling increases with the degree of ionization, since swelling is controlled by the osmotic pressure exerted by the mobile counterions (Chapter 4). The fraction of hydrophobic units also controls the swelling transition, the pH of the transition

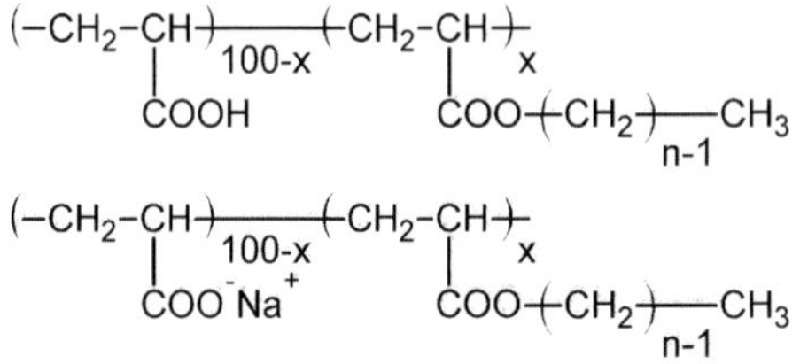

Figure 6.3 An example of MH polyelectrolyte (x is the percentage of grafted monomers): (a) uncharged and (b) completely ionized acrylic units.

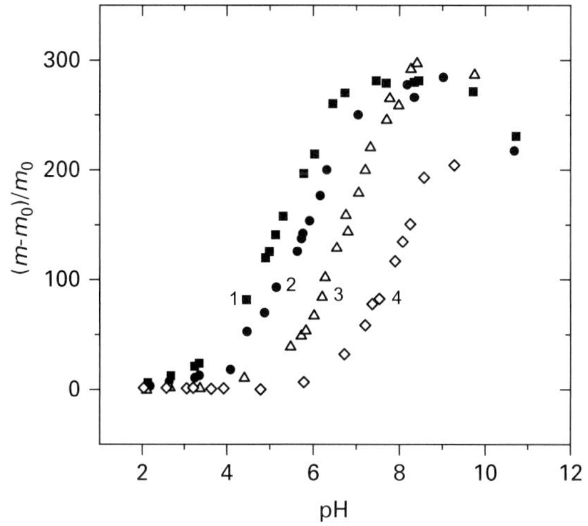

Figure 6.4 Dependence of the degree of swelling of PAA gels (1) and HMPAA gels: C_8–2.5% (2), C_8–10% (3) and C_8–20% (4) on the pH of the solution surrounding the gel sample. Equilibrium swelling is defined by the relative variation of the gel mass m, compared with that in the dry state, m_0. Adapted with permission from Philippova *et al.* (1997) © 1997 American Chemical Society.

becoming higher with increasing fraction and chain length of the hydrophobic units. NMR measurements (Philippova *et al.*, 1997) have demonstrated that the hydrophobic microdomains may be reversibly destroyed or created by changes in the pH of the medium. The swelling behaviour of these pH-responsive HMPAA gels is shown in Figure 6.4 over a range of pH which can, in turn, be varied between 5 and 8 as the mole fraction of hydrophobes ($n = 8$) increases from 0% to 20%.

6.3.2 Polysaccharide derivatives with pendant alkyl side chains

Polysaccharide derivatives are a class of products of interest in many industrial applications including textiles, printing, food, paper and pharmaceuticals. Typically the backbone is a polysaccharide to which lateral alkyl chains can be grafted. Examples include hydrophobically modified hydroxyethyl cellulose (HMHEC), ethyl(hydroxyethyl) cellulose (HMEHEC) and modified alginates (HMalginate) and dextrans (HMdextran). Figure 6.5 shows some characteristic HM polysaccharides.

In one study of interest (Nyström *et al.*, 2009), the HMHEC polymer contained 2 mol% of n-$C_{16}H_{33}$. The unmodified EHEC is a non-ionic amphiphilic polymer with hydrophilic and hydrophobic microdomains distributed randomly along the polymer backbone and exhibits demixing upon heating. In the HMEHEC sample, a small proportion of hydrophobic side chains (0.70%) of C_{12} or C_{14} alkyl chains was grafted on to the backbone.

As described in Chapter 5, alginates are marine anionic polysaccharides consisting of β-D-mannuronic acid (so-called M-block) and (1→4)-linked α–L-guluronic acid (G-block) units, in which GG-, MG- and MM-blocks of various lengths are arranged along the chain. A typical HMalginate contains approximately 5 mol% of *n*-octylamine (30% of C_8) groups

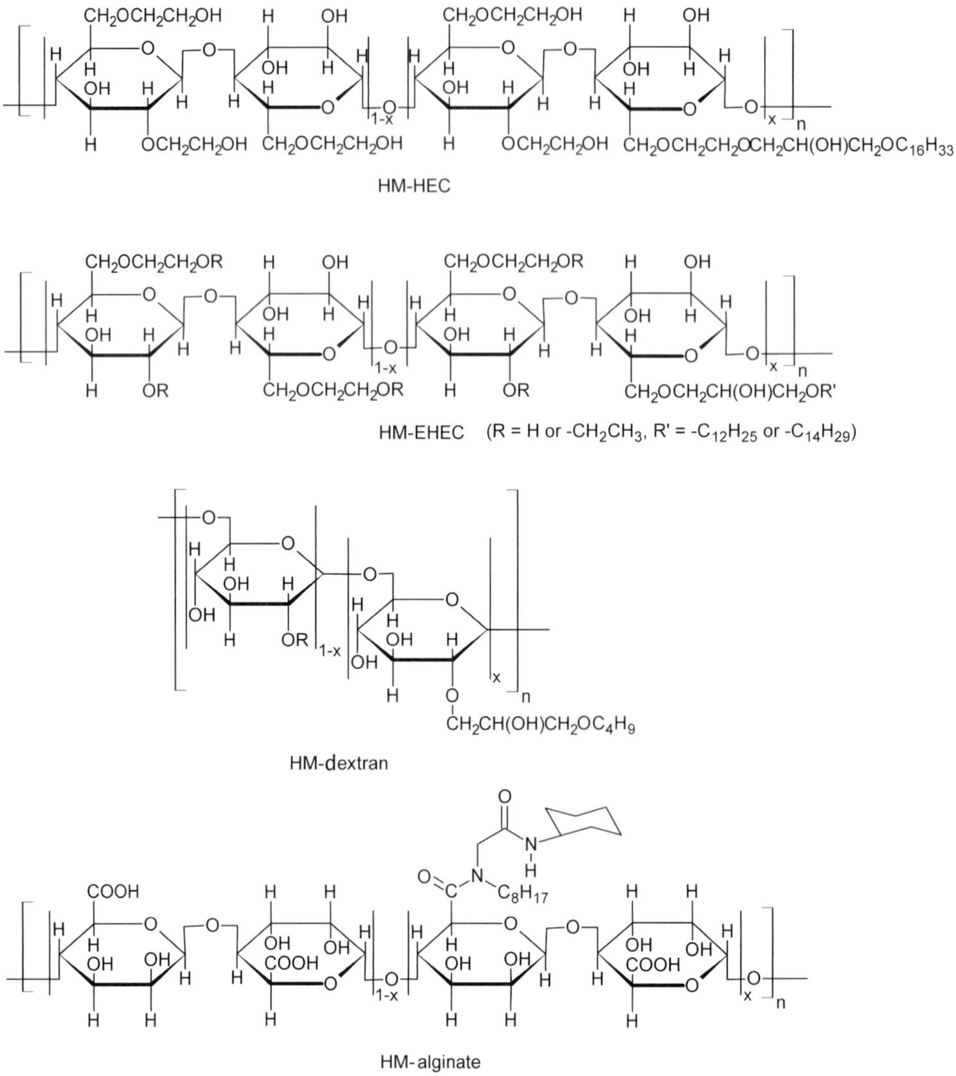

Figure 6.5 Schematic structures of HMHEC (degree of substitution $x = 0.02$), HMEHEC ($x = 0.0073$), HMdextran ($x = 0.016$) and HMalginate ($x = 0.054$). Reproduced from Nyström *et al.* (2009) with permission of the Royal Society of Chemistry.

grafted on to the backbone of the alginate. Dextran (Chapter 10) is a non-ionic microbial polymer composed of α-D-(1→6)-linked glucan chains with side chains that are mostly 1–2 glucose units long, attached to the O-3 of the backbone units. In the present work, the HMdextran typically contains 1.6 mol% of glycidyl butyl ether groups, as shown in Figure 6.5. In each case, the level of incorporation of the hydrophobic groups, and their chemical nature, is limited in order to retain the water solubility of the polysaccharide. Then the hydrophobic grafted groups tend to aggregate into micellar domains which behave as labile junctions.

Carboxymethylcelluloses (CMCell) with dodecylamide groups were synthesized by Cohen Stuart and co-workers by converting a small fraction of the carboxyl groups into dodecylamide groups with the degree of dodecyl substitution varying between 0.003 and 0.05 per residue (or glucose unit). The molecular mass of these samples was about 450 000 g mol^{-1}, so, although there are difficulties in the synthesis, one chain carries, on average, between 10 and 200 docecyl groups (Cohen Stuart *et al.*, 1998). Information about the distribution of the dodecyl groups along the chains should also be important, although such information is usually difficult to obtain. Alkyl chain attachment takes places in hot ethanol, where dodecylamine is quite soluble but CMCell is not, a difference which may lead to non-randomness in the substitution. From simple statistical considerations, it can be inferred that, at low degrees of substitution, the sample should be heterogeneous in the sense that different macromolecules may differ considerably in their degrees of substitution. A tendency for the alkyl groups to cluster adds to this heterogeneity.

6.3.3 Random-block copolymers obtained by microemulsion polymerization

Hydrophobically modified poly(acrylamide)s (HMPAm) make up an important class of associating polymers. One method used to copolymerize acrylamide with a hydrophobic co-monomer is a micellar polymerization technique in which the hydrophobe is solubilized into micelles dispersed in a water continuous medium (Volpert *et al.*, 1996). A schematic view of this micellar copolymerization is given in Figure 6.6. This process was shown to be well suited to the preparation of polymers with interesting viscosifying properties.

Because of the high local concentration within the micelles, hydrophobic groups tend to be distributed randomly along the poly(acrylamide) backbone in blocks. However,

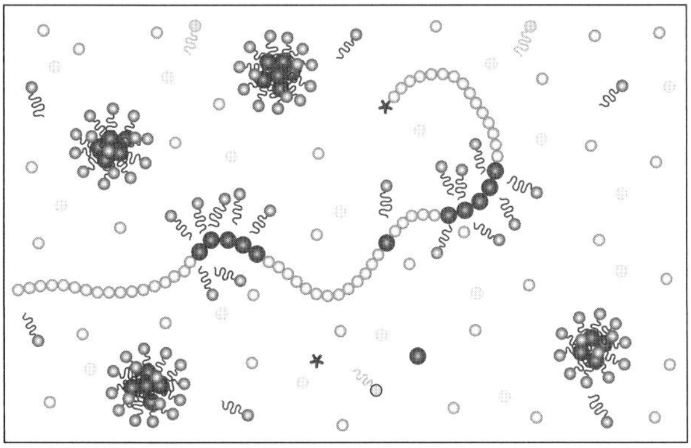

Figure 6.6 Schematic representation of the micellar copolymerization medium containing water-soluble monomers, hydrophobic monomers and surfactant molecules. Adapted with permission from Candau *et al.* (1998) © 1999 American Chemical Society.

precise information about the microstructure of these copolymers, such as the number and length of hydrophobic blocks, tends to be uncertain. In some cases, a drift in composition is observed which results in heterogeneous copolymer samples, making it impossible to draw any precise conclusion about the polymer's structure. The drift in copolymer composition was found to increase strongly upon increasing the number of hydrophobes per micelle, and the hydrophobe monomer is indeed consumed more rapidly than the acrylamide.

The authors showed that the replacement of N–H bonds by N–alkyl bonds avoids this drift in composition, so that, in N-methyl-N-hexyl acrylamide (MeHexAm) or N,N-dihexyl acrylamide (DiHexAm) copolymers, no corresponding drift was observed. In the case of DiHexAm the number of hydrophobes per chain is controlled by adjusting either the molecular mass of the polymer or the total hydrophobe content. Here again, in order to maintain solubility in water, the hydrophobe content of the modified poly(acrylamide)s is limited. These copolymers may be considered as model systems for random-block associating polymers. The rheological properties of these multi-block copolymers are strongly dependent on the hydrophobe/surfactant ratio, i.e. the initial number of hydrophobes per micelle, during the synthesis. This number also determines the hydrophobic block length, which was varied from 3 to 7. HMPAm polymers with the same hydrophobe content (1 mol%) and blockiness were synthesized with various molecular masses between 1.15×10^5 and 2.2×10^6 g mol^{-1}.

6.3.4 Telechelic copolymers

As mentioned in Chapter 4, linear polymers with functional groups at the two chain ends are known as telechelic, and one major class consists of the hydrophobic ethoxylated urethanes (HEURs), associative thickeners made of poly(ethylene glycol) (PEG) chains extended by di-isocyantes and end-capped by long-chain alkanols. Their description is found in papers by Jenkins *et al.* (1991), Lundberg *et al.* (1991) and Annable *et al.* (1993). Interest in these molecules results in part from their use as thickening agents in coating formulations but, because the positioning of the associating groups is known, they are also of great importance in more fundamental studies.

The positioning of the hydrophobe at the respective chain ends gives a particular type of rheology, discussed below, with Newtonian behaviour up to a relatively high shear rate. Molecular masses of the HEUR molecules are rather low, of the order of 10 000 to 30 000 g mol^{-1}, and with a relatively narrow distribution ($M_w/M_n \sim 1.4$). The end-cap length can be varied from C_{12} to C_{22}. The increase of the zero-shear viscosity relative to end-cap chain length is shown in Figure 6.7. A logarithmic increase of viscosity with hydrophobe length implies that the activation energy for disengagement (or, for a hydrophobe in the micelle(s), the absolute value of bonding susceptibility) increases with cap length. Extrapolation back to PEG solutions of similar molecular mass allows the minimum end-cap length required for viscosity enhancement to be determined: for these polymers it appears to be approximately 6 carbons. The activation energy derived from Figure 6.7 corresponds to 0.9 k_BT per methylene group.

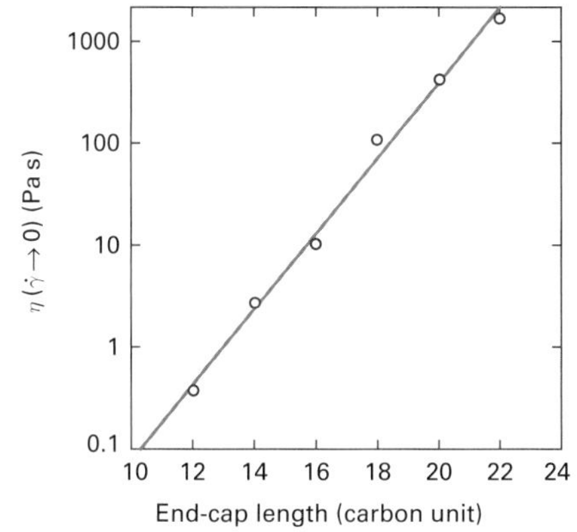

Figure 6.7 Effect of end-cap length on the Newtonian viscosity of an aqueous solution, $c = 2\%$ w/v of HEUR ($M_w = 3.5 \times 10^4$ g mol^{-1}) at room temperature. Reproduced from Annable *et al.* (1993) with permission of the American Institute of Physics for The Society of Rheology.

6.3.5 Triblock copolymers with a hydrophobe core

Poly(ethylene oxide)$_n$-poly(propylene oxide)$_m$-poly(ethylene oxide)$_n$ (PEO_n–PPO_m–PEO_n) triblocks, available commercially as Pluronic® (also called poloxamers), are non-ionic macromolecular surface-active agents. They are an important class of surfactants (Alexandridis and Hatton, 1995) with widespread industrial applications not only in detergents, emulsification, lubrication, cosmetics and inks, but also in pharmaceutical applications for drug solubilization and controlled release.

The PEO_n–PPO_m–PEO_n block copolymers are available in a range of molecular masses and PPO/PEO ratios. The mass of the PPO group can vary between 950 and 4000 g mol^{-1}, the PPO/PEO percentage varies between 10% and 80% and water solubility increases with PEO content. The cloud point, the temperature at the beginning of phase separation, for a polymer concentration of 1%, varies between 10°C and 100°C from low to high PEO content. Besides surface activity, these copolymers have a 'thickening power' (viscosity or minimum concentration required to obtain a gel at room temperature) which increases when the PPO block molecular mass increases and the PPO/PEO ratio decreases. In practice, the rheology of concentrated solutions needs to be carefully examined (see the next section). Micelle formation is observed with increasing temperature, and is associated with an endothermic enthalpy change.

6.4 Rheology of associating polymers

As already explained in Chapter 4, associating polymers have non-permanent (labile) junctions, giving rise to some interesting solution rheology. At low shear rates, the solutions exhibit liquid-like behaviour with a Newtonian plateau, but the main interest in applications of associating HM water-soluble polymers is the very important increase of the zero-shear viscosity compared to that of the unmodified polymer.

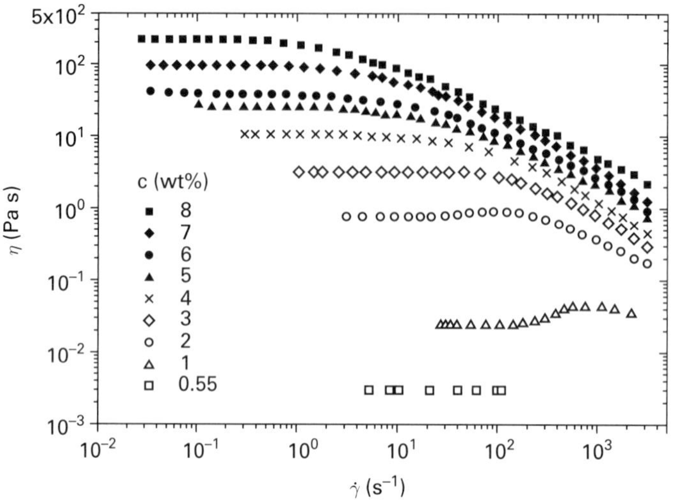

Figure 6.8 Steady-state viscosity versus shear rate at various concentrations of HM poly(acrylamide) (DiHexAm) samples in water: $M_w = 4.2 \times 10^5$ g mol^{-1}, hydrophobic block length 3.2 units, 9 hydrophobic blocks per chain. Adapted with permission from Regalado *et al.* (1999) © 1999 American Chemical Society.

Viscosity increases significantly:

- with the amount of substitution: for instance, for methyl cellulose, with the percentage of dodecyl groups to glucose units (Cohen Stuart *et al.*, 1998). When this percentage is varied from 3×10^{-3} to 5×10^{-2}, the zero-shear viscosity of a 1% solution increases by three orders of magnitude.
- even though it is not entirely desirable, with increasing degree of heterogeneity, as in the composition drift of HM poly(acrylamide)s prepared by micellar copolymerization (Volpert *et al.*, 1996). Here there is an order of magnitude difference between homogeneous and non-homogeneous samples.
- with polymer concentration, as shown in Figure 6.8.
- with the blockiness in random-block copolymerization (Candau *et al.*, 1998). The increase of viscosity with the length of the hydrophobic blocks is related to the lifetime of the (possibly binary) junctions between hydrophobes. As the lifetime of junctions is an exponential function of the activation energy for disentanglement (see Chapter 4), it is very sensitive to the length of the blocks. It was also observed that the viscosity of samples having the same hydrophobic block length, but increasing hydrophobic block density or increasing molecular mass, follows a scaling law with an exponent close to 4. This exponent suggests an analogy with the entangled regime of the homopolymer, where chain reptation (Doi and Edwards, 1986) is hindered by the presence of intermolecular junctions.
- with the length of the end cap in solutions of telechelic molecules (as shown in Figure 6.7).

In the semi-dilute range of concentrations, HM polymer solutions generally exhibit shear thinning behaviour consistent with predictions of the Tanaka–Edwards model. This can be an advantage, for example in applications where solutions are required to spread easily on a surface. Shear thinning is ascribed to the disruption of the network junctions: the rate of network breakdown under shear exceeds the rate at which the intermolecular associations take place. However, the Cox–Merz rule (Chapter 4; Cox and Merz, 1958; Graessley, 1974) is no longer obeyed. This rule states that the dynamic viscosity $\eta^*(\omega)$ can be superposed on data of the steady shear viscosity $\eta(\dot{\gamma})$, at equivalent low angular frequencies or shear rates. The rule is observed for almost all simple polymer solutions of high molecular mass or melts, and here the superposition extends further into the shear thinning range as well (Ferry, 1980). In associating polymer solutions, the shear viscosity is larger than the dynamic viscosity, the Newtonian plateau of the steady shear viscosity extending further than the dynamic viscosity plateau, as shown in Figure 6.9b. The opposite effect to this was seen in other structured liquid systems such as the xanthan polysaccharides (Chapter 5).

In some cases shear thickening (dilatancy) also occurs, as shown in Figure 6.9a. Shear thickening is also observed in HMPAm polymers with a statistical distribution of hydrophobes along the chains in the presence of 0.1 M NaCl. Telechelic HEUR solutions also exhibit shear thickening effects (Lundberg *et al.*, 1991). With increasing shear rates, viscosity increases and attains a maximum, and then falls again at higher shear rates.

In the case when the whole sample is uniformly sheared in volume, i.e. the shear is not localized into shear bands, and there are no wall slippage effects, Witten and Cohen (1985) outlined a mechanism of shear thinning in which flow produced an increase in inter-chain associations at the expense of those within the chain. They then calculated how chain elongation produced by flow altered the balance between these intra- and

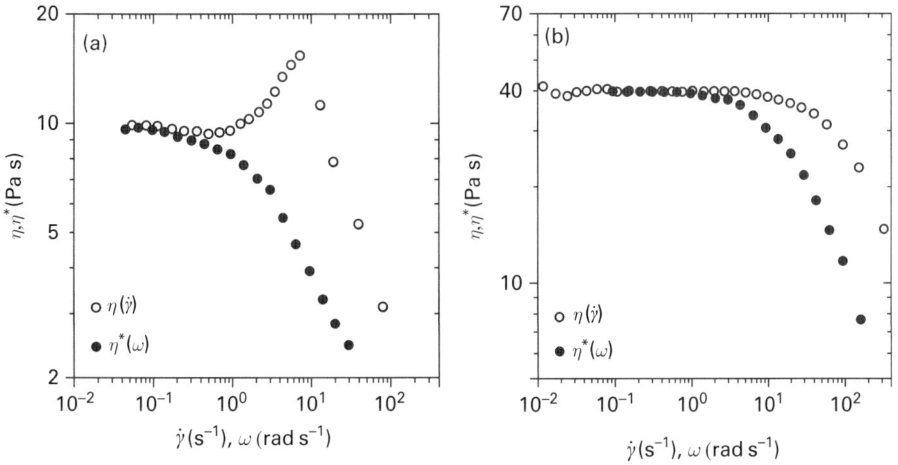

Figure 6.9 Steady-state viscosity $\eta(\dot{\gamma})$ and dynamic complex viscosity $\eta^*(\omega)$ as a function of the shear rate or angular frequency for two samples of the HM poly(acrylamide) DiHexAm: (a) $M_w = 4.2 \times 10^5$ g mol^{-1}, hydrophobic block length 3.2 units, 18 hydrophobic blocks per chain at $c = 2\%$ w/w; (b) $M_w = 1.4 \times 10^5$ g mol^{-1}, 6 hydrophobic blocks per chain at $c = 9\%$ w/w. Adapted with permission from Regalado *et al.* (1999) © 1999 American Chemical Society.

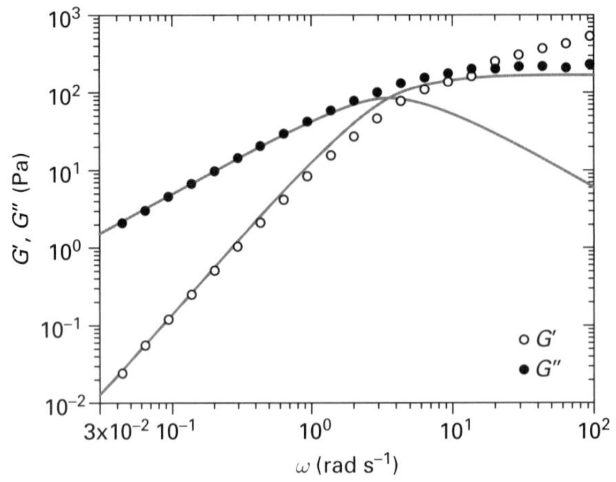

Figure 6.10 Storage and loss moduli versus angular frequency for HM poly(acrylamide): $M_w = 1.6 \times 10^5$ g mol^{-1} at $c = 5\%$ w/w with a hydrophobic block length of 7 units and 3.2 hydrophobic blocks per chain. Adapted with permission from Regalado *et al.* (1999) © 1999 American Chemical Society.

inter-chain associations, a cross-linking effect which will depend on molecular composition, and which was a maximum with only a few associating groups per chain. The model applied to solutions with concentrations slightly above c^*, and maximum viscosity was achieved when the time scale of deformation (or inverse of the shear rate) became shorter than the molecular relaxation time. At even higher shear rates, the stress was large enough for the rate of rupture to prevent inter-chain junctions re-forming (Lundberg *et al.*, 1991), and shear thinning was again observed.

Dynamic oscillatory measurements of HM water-soluble polymers exhibit a frequency dependence which corresponds to that of a viscoelastic liquid. A typical spectrum is shown in Figure 6.10. At low frequencies, the behaviour of the shear moduli is Maxwellian: $G'(\omega)$ and $G''(\omega)$ scale like ω^2 and ω, respectively. Then, at a certain angular frequency, the curves cross over each other and the inverse is taken as a characteristic time of the polymer solution. Deviations from this Maxwellian behaviour appear before the crossing point and especially at the high-frequency limit, and the shape of the curves at high frequency indicates higher relaxation modes. For telechelic HEUR associative polymers, by contrast, the dynamic spectrum shows almost perfect single Maxwell relaxation, in agreement with the Tanaka–Edwards model. At room temperature and moderate concentrations, this is also the case for PEO capped at one or both ends with hexadecyl groups (Renou *et al.*, 2009). In many other systems, the spectrum contains more than one Maxwell element, and appears more like the Rouse–Zimm type spectrum seen for 'simple' polymer solutions.

6.5 Interaction with surfactants

One of the characteristic features of HM water-soluble polymers is the very specific interactions that they have with surfactants, whereby hydrophobic interactions occur between the surfactant alkyl tails and the hydrophobic moieties of the polymer. The viscosity of the solutions goes through a pronounced maximum as the surfactant concentration is increased at a constant polymer concentration. This effect is important as it enhances the viscosity under conditions where the HM polymer itself would not show very effective thickening properties.

The effect is observed in many cases, for example with HM modified neutral polysaccharides such as ethyl(hydroxyethyl) cellulose (HMEHEC) (Nyström *et al.*, 1995), and anionic and cationic HMHECs, in the presence of anionic surfactant (SDS) (Beheshti *et al.*, 2008) or polyelectrolytes containing a low fraction of hydrophobic groups. HM polyelectrolytes can even associate with surfactants of the same charge; in this case, the attractive hydrophobic interactions overcome electrostatic repulsions between the polymer backbone and the surfactant ionic heads (Iliopoulos *et al.*, 1991). The most comprehensive studies are known for mixtures of oppositely charged surfactants and polyelectrolytes.

Magny *et al.* (1994) made a detailed analysis of the case of mixtures of cationic surfactants CTAC (dodecyltrimethylammonium chloride) and anionic HM polyacrylic acid (HMPAA). A steady-state fluorescence method, with pyrene as a fluorescence probe, was used to detect the formation of micelles or hydrophobic aggregates. The fluorescence emission spectrum of pyrene shows several so-called vibronic peaks, the intensity of which is a sensitive indicator of the polarity of the pyrene microenvironment. The ratio of the first and third vibronic peaks (I_1/I_3) is a sensitive indicator of the formation of micelles or hydrophobic aggregates because the pyrene probe preferentially lies close to, or inside, these microdomains. For instance, the ratio I_1/I_3 varies from 2 in pure water to 1.5 in a micellar solution CTAC. The end of the sharp transition of I_1/I_3 with increasing surfactant concentrations is taken as the CMC and is, in general, in good agreement with other determinations of micelle formation.

Owing to binding interactions, the critical aggregation concentration (CAC) of the surfactant differs from the CMC value (Magny *et al.*, 1994). In the presence of the unmodified polyelectrolyte, the transition of I_1/I_3 shifts to much lower surfactant concentrations (lower CAC). The introduction of hydrophobic groups influences the shape of the fluorescence curve and further decreases the CAC. The pyrene probe monitors the gradual increase in the hydrophobicity of the microdomains containing mixed micelles: the transition is stretched over one order of magnitude of surfactant concentrations and the amplitude of the transition decreases. Here, presumably, initially small aggregates of surfactants are formed, involving alkyl chains attached to the polymers, and the structure of the aggregates gradually changes with increasing surfactant concentration. Polymer hydrophobicity (degree of modification or length of the alkyl group) decreases the CAC of the surfactant by two orders of magnitude. As surfactant concentration approaches the CAC, a large increase of viscosity is seen (Figure 6.11). Upon further addition of surfactant, viscosity decreases again and the system suddenly approaches phase separation.

One possible mechanism which explains how such large changes in the viscosity occur is shown in Figure 6.12. This is the transition from inter-chain mixed aggregates to intra-chain binding of the surfactant into mixed micelles containing the polymer hydrophobes. At low surfactant concentrations, close to the CAC, mixed micelles containing

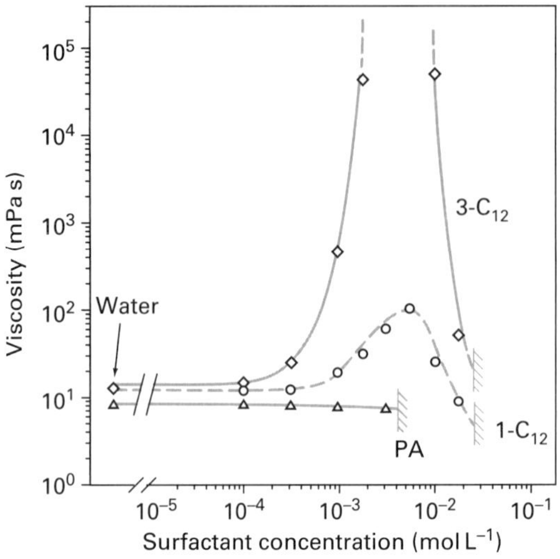

Figure 6.11 The change in viscosity of a 1% aqueous solution of precursor (PAA) and modified polymers (1% and 3% C_{12}) with CTAB surfactant concentration. The hatch symbol shows the limit of the phase separation. The viscosity measured at shear rates between 0.1 and 1 s^{-1} was reported to be Newtonian. Adapted with permission from Magny *et al.* (1994) © 1994 American Chemical Society.

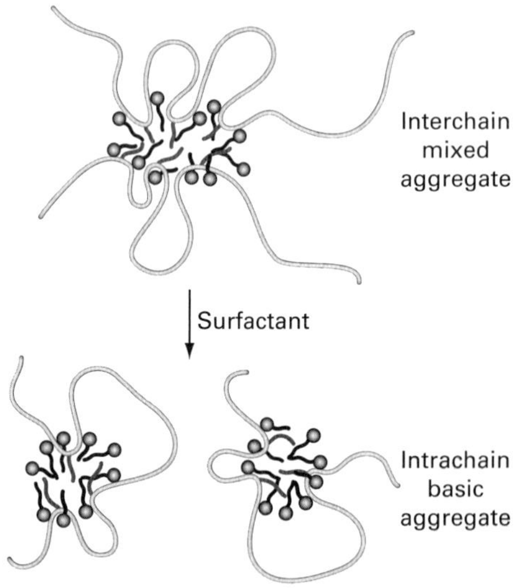

Figure 6.12 Schematic representation of surfactant–HM water-soluble polymer interactions: inter-chain mixed aggregates at low surfactant concentration are progressively converted into intra-chain mixed aggregates at high surfactant concentration. Adapted with permission from Magny *et al.* (1994) © 1994 American Chemical Society.

surfactants, and alkyl groups belonging to two distinct polymer chains, induce 'cross-linking' between the chains, which increases viscosity well above the contribution shown by the alkyl modification itself (although there is a small increase in the viscosity of these modified polymer as compared to the unmodified ones). At higher surfactant concentrations, the breakdown of the intermolecular aggregates decreases connectivity between the polymer chains and, consequently, the viscosity. The pre-phase separation surfactant concentration indicates a limit when all the alkyl groups in the mixed micelles belong to one polymer chain. Further addition of surfactant molecules results in the formation of more and more micelles free of any polymer alkyl groups, electrostatically bound to the polymer backbone, and the system approaches phase separation.

In their work, Magny *et al.* (1994) estimated the number of alkyl groups per micelle as the surfactant concentration increased: the micelle aggregation number for pure surfactant, close to 40, increases to 50 upon addition of the modified polymer. There is, however, no fixed stoichiometry between the polymer and surfactant contribution; rather, it depends on the surfactant concentration. A minimum limit was estimated to be around 6 alkyl groups per micelle for a PAA solution containing 1% to 3% HM monomers, with C_{12} alkyl groups, while the limit is lower with C_{18}, for the same surfactant. As shown in Figure 6.11, at the maximum viscosity the numbers of alkyl groups participating in the formation of a mixed micelle is approximately twice this value.

Similar mechanisms of association of surfactant (SDS) to the hydrophobic groups of EHEC and HMEHEC solutions were observed by Thuresson *et al.* (1995) by light scattering and titration microcalorimetry techniques, and by Nyström *et al.* (1995) using rheological methods. Structurally significant parameters such as the aggregation number, microstructure of the micelles and interconnectivity of the chains were not determined in these complex solutions because the heterogeneity of the local composition makes their determination problematic. All investigations of the polymer synthesis mentioned above highlight this difficulty, but also stress the fact that heterogeneous or block copolymers are more efficient viscosity enhancers than homogeneous samples.

6.6 Thermogelation or phase separation?

In most cases presented above, hydrophobic groups (mainly alkyl) spontaneously associate into micellar aggregates at room temperature. When the polymer is well solubilized at room temperature, the thickening effect appears on heating the solutions. There is a very important class of associating polymers which exhibit such large thermal effects, resulting from the hydrophobic interactions between long sequences of copolymers, that the phenomenon is called 'gelation'. Thermogelling systems, based on neutral block copolymers with hydrophobic cores such as PEO_n–PPO_m–PEO_n, and modified natural polymers such as methyl cellulose (MC) are also well known for exhibiting such an increase of viscosity upon heating. We discuss in the following section the mechanisms of hydrophobic association for these polymers.

We first examine solutions with hydrophobic core copolymers, known by their trade names as Pluronics® or Poloxamers. Temperature-dependent micellization and gel formation are two of the most characteristic properties of these block copolymers in aqueous solution. Because of the marked change in water solubility of the central PPO block, these copolymers form various aggregates depending on the degree of polymerization of each block. A number of triblock copolymers of this family have been shown to aggregate in the form of micelles with a core dominated by PPO and a corona dominated by hydrated PEO blocks (Alexandridis *et al.*, 1994).

Numerous publications deal with the critical micellization temperature (CMT) for Pluronics as a function of solution concentration. Pluronics with a high PPO content form micelles at lower concentrations and lower temperatures. Pluronics with the same hydrophobic segment and increasing hydrophilic segments have a small increase in their CMC and CMT, indicating that micelle formation becomes more difficult for the more hydrophilic molecules, but the effect of PEO is less pronounced than that of PPO: the primary factor in the micellization process of Pluronics is PPO. Such studies deal with rather dilute solutions ($c \sim 1\%$). Being concerned here with the gelation phenomenon, we must consider solutions with larger polymer concentrations.

Gelation of this type of polymer is basically different from those described in Chapter 3. Pham Trong *et al.* (2008) analysed the correlation between micelle formation and rheological changes of solutions of the material known as F127, where $n = 98$ and $m = 67$, with increasing temperature. Microcalorimetry is a very sensitive method for detecting micellization; the signal is endothermic during heating and perfectly reversible on cooling. The peak width, between the onset and peak end temperatures, is about 10°C. The peak areas of the signals were found to be directly proportional to the polymer concentration (between 0.025% w/w and 20% w/w), suggesting that, at the end of the enthalpic transition, almost all single polymers are included in micelles, so that the enthalpy of micellization per chain can be derived directly. The micellization temperatures vary with the concentration, between 30°C and 15°C (peak maximum).

As shown in Figure 6.13, at the end of the enthalpic peak and for the most concentrated solution a small endothermic peak is observed which corresponds to the crystallization temperature of the micelles. Figure 6.13 also shows the superposition of the enthalpy signal and the rheological measurements. Three domains appear, which correspond to three microscopic structures. Initially, in region (1) the solution is Newtonian; the polymers are well dispersed and overall viscosity and relative viscosity versus water both decrease as temperature increases, meaning that the polymer is less swollen; and its hydrodynamic volume decreases, so the solvent becomes less good. In region (2) micelles start to assemble and the viscosity increases, with no measurable storage modulus. The micelles repel each other, as grafted brushes on colloidal particles ('hairy grains') with a large number of arms, and the solution is expected to behave as a suspension of hard spheres. Assuming a constant aggregation number and a micelle radius from light scattering experiments (~11 nm), the variation of relative viscosity with micelle volume fraction should follow the phenomenological models of Krieger and Dougherty (1959). The solvent being a single-polymer solution with a decreasing concentration, interpretation of viscosity measurements suggests that the radius of these micelles steadily decreases with solvent quality, consistent with this poor solvent quality as PEO becomes less hydrophilic. Finally, region (3) corresponds to a colloidal crystal of spherical micelles (Artzner *et al.*, 2007).

For these systems, then, gelation is the transition from a colloidal suspension of micelles which behave as hard spheres to a colloidal crystal formed of these micelles.

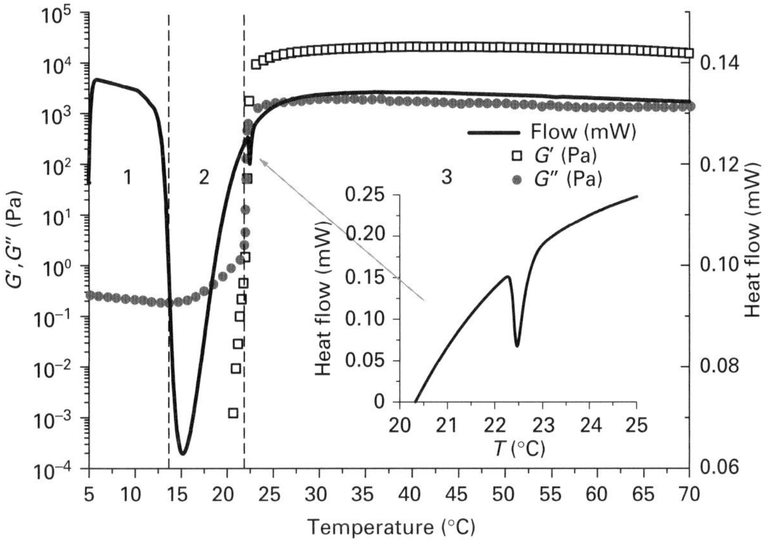

Figure 6.13 Data for micellization of F127 versus temperature, as determined by two different experimental methods: microcalorimetry (solid curve, showing endothermic peak) and rheological measurements (G' (□) and G'' (●) measured at 1 Hz) for $c = 20\%$ w/w. Adapted with permission from Pham Trong *et al.* (2008) © 2008 Elsevier.

For lower polymer concentrations, the crystallization peak is much smaller; possibly the solution is partially structured into ordered domains and the rheological measurements show a less steep temperature variation. In the latter case the transition corresponds to that of a viscoelastic liquid. The shear moduli of the crystalline phase in Figure 6.13 show little frequency dependence, but the loss modulus is high and close to the value of the storage modulus. The origin of the elasticity in the colloidal crystal phase has been discussed in the literature, and can be attributed either to the interfacial tension between water and PPO, by analogy with an emulsion system (Hvidt *et al.*, 1994), or to the entanglement of the PEO chain (Lau *et al.*, 2004), while the value of the loss modulus could be related to friction between micelles. There is no quantitative estimate of these contributions, and this is still an open field of investigation. During the reverse cooling cycle, while the calorimetric signal is reflected perfectly there is a large hysteresis in solution rheology; much higher temperatures are needed in order to erase the thermal history of the sample.

Another example of thermogelation is observed in solutions of poly[F127] such as EG56® from PolymerExpert. These are copolymers composed of three linear molecules with urethane and allophanate links (Pham Trong *et al.*, 2008). Their precise architecture is unknown; they may be branched or linear molecules. Formation of micelles during heating is detected by microcalorimetry measurements in parallel with the increase of the dynamic viscosity. For these associating polymers, the structure of the micelles formed by the PPO blocks is probably more disordered (aggregation number, functionality), but the enthalpy is approximately the same as for F127. The marked difference appearing in

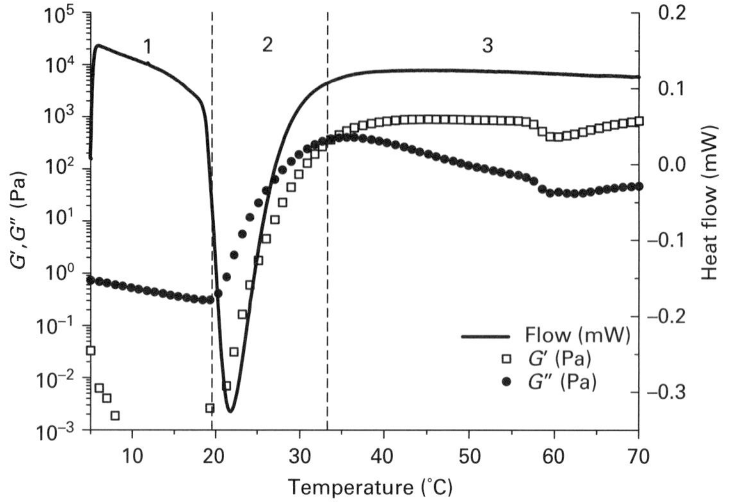

Figure 6.14 Data for micellization of EG56® versus temperature, as determined by microcalorimetry (solid curve) and by rheological measurements (G′(□) and G″ (•) at 1 Hz) for $c = 10\%$ w/w. Adapted with permission from Pham Trong *et al.* (2008) © 2008 Elsevier.

Figure 6.14 is that, in the temperature range where the main enthalpic transition occurs, when micelles form there is an increase in G' accompanying the increase of G''.

The frequency dependence of the shear moduli follows a frequency–temperature superposition master curve. The coefficient $a(T)$ for the frequency horizontal shift factor shows a very large variation, from 1 to 500, within a narrow temperature range, from 30° C to 42°C, while the vertical shift $b(T)$ factor which controls the amplitude of the shear moduli varies only by a factor of 2. The crossing point characterizing the lifetime of the junctions increases from 0.15 s to nearly 80 s and follows an Arrhenius plot as a function of temperature between 30°C and 42°C (measurements were not performed between 20°C and 30°C). In contrast with the telechelic HM polymers, the spectrum corresponds to a superposition of a number of Maxwell elements with different characteristic times. The frequency dependence of the complex viscosity and the master curve are shown in Figures 6.15a and 6.15b.

In these heat-set solutions the molecules are branched and, strictly speaking, existing theories cannot apply. However, in the LRC model (Chapter 4; Leibler *et al.*, 1991) for entangled solutions of polymers containing a fixed number of sticky groups, the terminal relaxation time T_d, or the inverse of the crossing frequency, is related to the average fraction of stickers engaged in an associated state; see Equation (4.26). The assumption for the mechanisms of stress relaxation in this model is that chain motion is controlled by topological constraints, which should not be too different from the polymers without stickers. For times longer than T_d, the structure of the network changes as the stickers detach from one tie point and reassociate at another point. For these heat-set networks, as temperature increases the terminal time is significantly lengthened relative to that of the non-aggregated polymer, as the average fraction of stickers in the associated state increases from 0 to a maximum of 1, because micelle

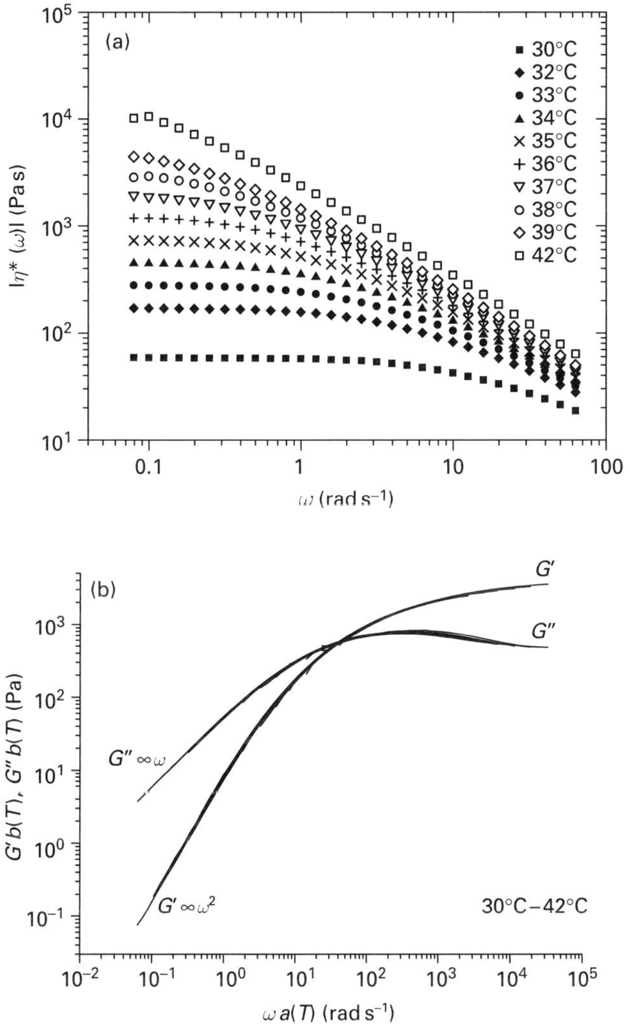

Figure 6.15 (a) Dynamic complex viscosity $|\eta^*(\omega)|$ and (b) master curve for the shear moduli with frequency–temperature superposition for poly[F127] (EG56®). Adapted with permission from Pham Trong *et al.* (2008) © 2008 Elsevier.

formation retards the diffusion process. For lifetimes shorter than T_d, the reversible network resembles a permanent network; in the data shown in Figure 6.15 the number of elastically active chains derived from the plateau value of the storage modulus increases slightly with temperature (vertical shift factor). According to the Tanaka and Edwards classification, the physical gel formed by the hydrophobic associations is definitely a non-permanent network.

Another well-known example of a thermogelling solution is methyl cellulose (MC) (Haque and Morris, 1993; Desbrieres *et al.*, 1998; Li, 2002). In contrast to Pluronics copolymers, where solutions and gels are completely transparent, indicating homogeneous phases, thermal gelation of MC is associated with phase separation in solution, and with a lower critical solution temperature (LCST). The hydrophobic association and entanglement of these high molecular mass polymers contributes to the formation of junctions which have a long lifetime. In the work of Li (MC sample with a degree of substitution of 1.8 and an average molecular mass $M_w = 310\,000$ g mol^{-1} (SM 4000)), solutions with concentrations of 3–4% w/w exhibit a sol–gel transition versus temperature, with a finite low-frequency storage modulus which increases with temperature. Although approaching the solubility limit of MC at high temperature, the system shows no indication of a significant fall in the shear moduli, even at temperatures as high as 80°C. According to the author, the Winter–Chambon criteria do not apply to this transition. The spectra illustrated by Li resemble those of a chemical gelation with 'permanent' cross-links, and the transition was analysed as if it were a percolation transition versus temperature. The solution is viscoelastic at room temperature, but exhibits a large increase in G' and G'' as temperature approaches 60°C, following the enthalpic transition.

It is known that all commercial MC samples are heterogeneous (Desbrieres *et al.*, 1998) and often poorly characterized. Li observed a structuring of the solution before the enthalpic transition (60°C), where G' starts increasing, while G'' decreased, possibly related to local association of the hydrophobic groups. The incipient phase separation induces a much larger organization of the entangled solution, whereby phase separation creates large topological reorganizations which are very different from the previous examples in this chapter.

A more detailed analysis was reported by Desbrieres *et al.* (2000), who investigated their specially synthesized and well-characterized MC samples. They used a similar degree of substitution (1.7) and a viscosity average molecular mass of 140 000 g mol^{-1} (Methocel A4). Rheological measurements were performed over a very large frequency range and clearly showed two regimes: at low temperature (20–40°C) the viscoelastic behaviour of an entangled solution was dominated by the non-associative part of the macromolecular chains, with a distribution of relaxation times and a Newtonian viscosity which decreases with temperature. During heating a discontinuous change of the terminal relaxation time was seen in a turbid solution, corresponding to the incipient development of strong hydrophobic interactions: the angular frequency at the crossing point ($G' = G''$) varied from 1000 to 1 rad s^{-1} around 60°C. This very important discontinuity was not reported in other types of systems; it may be the signature of the phase separation of the entangled solutions. A master curve at temperatures between 60°C and 68°C was established, as shown in Figure 6.16.

The master curve at the high-temperature limit clearly shows some relaxation at low frequencies, as expected in non-permanent networks, whereas the experiments by Li do not show this, which may in this case correspond either to unexplored lower frequencies or to the nature of the commercial sample employed. According to Desbrieres *et al.* (2000), the gelled solution of MC is a non-permanent network, in agreement with the hydrophobic association mechanisms reported in this chapter. The other interesting and surprising feature of the viscoelastic spectrum of MC at high temperatures is that a single relaxation mechanism can fit the low-frequency relaxation, as with telechelic polymers.

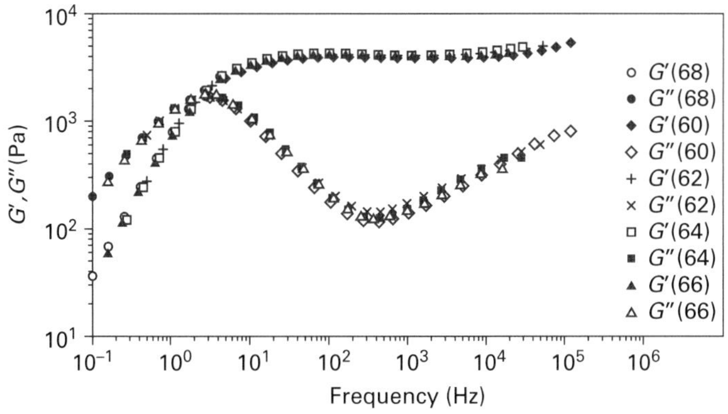

Figure 6.16 Master curve of the viscoelastic spectrum of MC solution ($c = 4\%$ w/w) at high temperature (60°C < T < 68°C). Adapted with permission from Desbrieres *et al.* (2000) © 2000 Elsevier.

The plateau value of G' corresponds to a small number of junctions per chain (3 or 4) if the rubber-like elasticity model applies. Evidently thermogelation for MC corresponds mainly to a step-like shift (by three orders of magnitude) of the characteristic terminal relaxation time of the solution, which coincides with the incipient enthalpic transition.

In order to find an analogue of the phase separation of MC in synthetic polymers, we can examine the temperature dependence of the other types of HM polymers.

Hydrophobically end-capped molecules such as PEO end-functionalized with aliphatic groups (Renou *et al.*, 2009), PEO end-functionalized with alkyl groups (Winnik and Yekta, 1997) or PEO end-capped with fluorinated groups (Berret *et al.*, 2003) tend to phase separate with increasing temperature. According to Laflèche and co-workers (Laflèche *et al.*, 2003), end-capped PEO molecules with alkyl groups behave like adhesive spheres. When solutions with concentrations between 1% and 10% w/w are heated, phase separation takes place, characterized by the appearance of a clear top phase and a turbid bottom phase whose turbidity decreases with time. The bottom phase is very viscous. At large enough concentrations (20% w/w), 'soft' or 'hard' gels are observed at room temperature (in the authors' classification, hard gels are those which do not flow when they are tilted) and solutions remain clear upon heating, but they lose their elasticity and become fluid (40°C). Phase separation with these molecules has quite the opposite effect of the other thermogelling systems.

Thermo-associated copolymers were obtained by grafting randomly distributed poly (NIPAm) (Chapter 4) side chains on to PAA backbones (Durand *et al.*, 2000). It was shown that the viscosity of semi-dilute solutions measured at a fixed shear rate ($100\ s^{-1}$) increased by one to two orders of magnitude around 40°C, owing to a collapse transition of the NIPAm side chains as detected by ^{1}H NMR. To interpret the important loss of mobility of the side chains, the authors suggested that the association temperature, where viscosity increases, is correlated with the formation of NIPAm microdomains, with characteristics of a glassy state. The same backbone grafted with PEO chains does not show such an effect.

6.7 Conclusions

In this chapter, we report on a mechanism of formation of non-permanent networks through hydrophobic interactions. Most of the examples presented are obtained by modification of existing polymers or synthesis of amphiphilic polymers with various architectures. According to the degree of hydrophobicity (alkyl or fluorinated chains are the most hydrophobic ones) and the blockiness (random or block hydrophobic domains), the HM polymers self-associate at room temperature, producing varying degrees of modification of the Newtonian viscosity.

When the viscosifying effect is not sufficiently large, addition of ionic surfactants, which act to strengthen the hydrophobic associations, is an efficient way to enhance them. The concentration of surfactant has an optimum value, above which the whole solution phase separates. In most cases, when the temperature is increased, solutions containing HM polymers phase separate macroscopically. Consequently this thermogelation effect requires a subtle balance between the association of the hydrophobic groups and the solubility of the whole polymer. The Pluronic polymers exhibit this property and are recognized as good candidates for thermally viscosifying solutions, particularly at clinically relevant temperatures, and for subsequent pharmaceutical applications. Such Pluronic solutions are homogeneous and show no phase separation in the temperature range where micelles are created. Methyl cellulose is the other well-known example. Here the heterogeneity of commercially available molecules is a limitation in exploring detailed mechanisms, but the system appears to show competition between global phase separation and local association. For concentrated solutions of methyl cellulose, a different mechanism, related to the formation of the liquid crystalline state, has been suggested (Yin *et al.*, 2006). To be fully understood, this type of material needs more study, particularly based on well-synthesized samples. Thermogelling systems based on NIPAm and polyelectrolytes may suggest other, more complex microstructures and mechanisms.

References

Alexandridis, P., Hatton, T. A., 1995. *Colloids Surf., A* **96**, 1–46.
Alexandridis, P., Holzwarth, J. F., Hatton, T. A., 1994. *Macromolecules* **27**, 2414–2425.
Annable, T., Buscall, R., Ettelaie, R., Whittlestone, D., 1993. *J. Rheol.* **37**, 695–726.
Artzner, F., Geiger, S., Olivier, A. *et al.*, 2007. *Langmuir* **23**, 5085–5092.
Beheshti, N., Zhu, K., Kjoniksen, A. L., Nyström, B., 2008. *Colloids Surf., A* **328**, 79–89.
Berret, J.-F., Calvet, D., Collet, A., Viguier, M., 2003. *Curr. Opin. Colloid Interface Sci.* **8**, 296–306.
Candau, F., Regalado, E. J., Selb, J., 1998. *Macromolecules* **31**, 5550–5552.
Cohen Stuart, M. A., Fokkink, R. G., van der Horst, P. M., Lichtenbelt, J. W. T., 1998. *Colloid Polym. Sci.* **276**, 335–341.
Cox, W. P., Merz, E. H., 1958. *J. Polym. Sci.* **28**, 619–622.
Desbrieres, J., Hirrien, M., Rinaudo, M., 1998. *Carbohydr. Polym.* **37**, 145–152.
Desbrieres, J., Hirrien, M., Ross-Murphy, S. B., 2000. *Polymer* **41**, 2451–2461.

Doi, M., Edwards, S. F., 1986. *The Theory of Polymer Dynamics*. Clarendon Press, Oxford.
Durand, A., Hourdet, D., Lafuma, F., 2000. *J. Phys. Chem. B* **104**, 9371–9377.
Ferry, J. D., 1980. *Viscoelastic Properties of Polymers*. Wiley Interscience, New York.
Glass, J. E. (ed.), 1986. *Advances in Chemistry* 213. American Chemical Society, Washington, DC.
Glass, J. E. (ed.), 1989. *Advances in Chemistry* 223. American Chemical Society, Washington, DC.
Graessley, W. W., 1974. *Adv. Polym. Sci.* **16**, 1–179.
Haque, A., Morris, E. R., 1993. *Carbohydr. Polym.* **22**, 161–173.
Hvidt, S., Jørgensen, E. B., Schillén, K., Brown, W., 1994. *J. Phys. Chem.* **98**, 12320–12328.
Iliopoulos, I., Wang, T. K., Audebert, R., 1991. *Langmuir* **7**, 617.
Israelachvili, J. N., 1992. *Intermolecular and Surface Forces: With Applications to Colloidal and Biological Systems*, 2nd edn. Academic Press, London.
Jenkins, R. D., Silebi, C. A., El-Asser, M. S., 1991. Steady-shear and linear-viscoelastic material properties of model associative polymer solutions. In Schulz, D. N. (ed), *Polymers as Rheology Modifiers*. ACS Symposium Series 462. American Chemical Society, Washington, DC, p. 222.
Krieger, I. M., Dougherty, T. J., 1959. *Trans. Soc. Rheol.* **3**, 137–152.
Laflèche, F., Durand, D., Nicolai, T., 2003. *Macromolecules* **36**, 1331–1340.
Lau, B. K., Wang, Q., Sun, W., Li, L., 2004. *J. Polym. Sci., Part B: Polym. Phys.* **42**, 2014–2025.
Leibler, L., Rubinstein, M., Colby, R. H., 1991. *Macromolecules* **24**, 4701–4707.
Li, L., 2002. *Macromolecules* **35**, 5990–5998.
Lundberg, D. J., Glass, J. E., Eley, R. R., 1991. *J. Rheol.* **35**, 1255–1274.
Magny, B., Iliopoulos, I., Zana, R., Audebert, R., 1994. *Langmuir* **10**, 3180–3187.
McBain, J. W., Salmon, C. S., 1920. *J. Am. Chem. Soc.* **42**, 426–460.
Nyström, B., Thuresson, K., Lindman, B., 1995. *Langmuir* **11**, 1994–2002.
Nyström, B., Kjoniksen, A.-L., Beheshti, N., Zhu, K., Knudsen, K. D., 2009. *Soft Matter* **5**, 1328–1339.
Pham Trong, L. C., Djabourov, M., Ponton, A., 2008. *J Colloid Interface Sci.* **328**, 278–287.
Philippova, O. E., Hourdet, D., Audebert, R., Khokhlov, A. R., 1997. *Macromolecules* **30**, 8278–8285.
Regalado, E. J., Selb, J., Candau, F., 1999. *Macromolecules* **32**, 8580–8588.
Renou, F., Nicolai, T., Benyahia, L., Nicol, E., 2009. *J. Phys. Chem. B* **113**, 3000–3007.
Shedge, A. S., Lele, A. K., Wadgaonkar, P. P. *et al.*, 2005. *Macromol. Chem. Phys.* **206**, 464–472.
Tanford, C., 1991. *The Hydrophobic Effect: Formation of Micelles and Biological Membranes*, 2nd edn. Krieger Publishing Company, Malabar, FL.
Thuresson, K., Nyström, B., Wang, G., Lindman, B., 1995. *Langmuir* **11**, 3730–3736.
Volpert, E., Selb, J., Candau, F., 1996. *Macromolecules* **29**, 1452–1463.
Winnik, M. A., Yekta, A., 1997. *Curr. Opin. Colloid Interface Sci.* **2**, 424–436.
Witten Jr, T. A., Cohen, M. H., 1985. *Macromolecules* **18**, 1915–1918.
Yin, Y., Nishinari, K., Zhang, H., Funami, T., 2006. *Macromol Rapid Commun.* **27**, 971–975.
Zhou, H., Song, G.-Q., Zhang, Y.-X. *et al.*, 2001. *Macromol. Chem. Phys.* **202**, 3057–3064.

7 Helical structures from neutral biopolymers

7.1 Introduction

This chapter is devoted to gelatin and agarose, two systems which form gels via the transition from a disordered and relatively flexible coil conformation in solution at high temperatures (60–80°C) to an ordered and partly helical structure on cooling to ~30°C, provided the concentration is large enough. At these lower temperatures the network strands consist of both helical and coil-like portions. On heating, the gels 'melt' and the individual molecular chains revert to the coil state. At one level, the gel network consists of long and thin fibres which are interconnected. However, they differ fundamentally from the fibrillar networks to be discussed in Chapter 9, which are formed from globular protein precursors such as actin or the so-called acid and irreversibly heat-set 'amyloid' gels, since in these cases the strands are assembled not from molecular chains but from globular macromolecules derived from proteins.

In the present chapter, we analyse the mechanisms of network formation in gelatin, coil to triple helix, and agarose, coil to double helix to multiple strand fibres. Among the other thermoreversible multiple helix-stranded networks, carrageenan and gellan gels were discussed in Chapter 5. There, gelation occurs in the presence of various types or amounts of salts, and the role of specific ions is a major factor, which requires separate analysis. By contrast, gelatin and agarose aqueous solutions undergo a sol–gel transition mainly driven by cooling from the high-temperature form, and the gelation is not strongly influenced by ionic interactions.

Both systems are derived from natural polymers; gelatin is proteinaceous, extracted from collagen by acid or alkaline treatment, whereas agarose is a polysaccharide extracted from marine algae. Gelatin is widely used in food preparations (desserts, confectionery), technical formulations (paper coating, inks and – until digital cameras became ubiquitous – photographic film) and well-known pharmaceutical applications (hard and soft capsules). Agarose, or the cruder product known as agar, is used in similar food applications (and sometimes as a vegetarian replacement for gelatin), as a common laboratory culture medium for microbes, in the laboratory technique of gel electrophoresis and as the base of some dental impression materials.

7.2 Gelatin

Collagen is the major structural component of mammalian tendon, bone, skin and cornea, and gelatin is a name applied to degraded and denatured collagen (Ward and Courts,

1977). The main source of raw material for commercial samples of gelatin is bovine or porcine, especially bones or hides from beef, or pig skins. More recently, somewhat different gelatins extracted from fish skins have become available. We first examine the features common to all gelatins, and then clarify the differences.

7.2.1 Molecular composition and native conformation

Collagen is proteinaceous (Dickerson and Geis, 1969) and is an assembly of tropocollagen macromolecules containing a number of amino acids (Cantor and Schimmel, 1980). The proportion of each amino acid and its exact position in the chain are predetermined. These amino acids, also called residues, correspond to the monomer units in synthetic polymers; the collagen single chain, for example, is made of about 1000 residues and contains about 20 amino acids. Glycine (Gly) is the most abundant by weight, but the specific and unique feature of the sequence of amino acids in tropocollagen is that every third residue is Gly, and in mammalian sourced material the imino acids proline (Pro) and hydroxyproline (Hyp) constitute another quarter of the composition by weight. The rings of these imino acids give enhanced rigidity to the tropocollagen chain and play an important role in both the mechanical and the thermal stability of the collagen rods. The most frequent sequences found on tropocollagen are therefore –(GlyProX)– or –(GlyProHyp)–, where X are other amino acids.

7.2.2 Structure of native tropocollagen fibrils

The secondary and tertiary structures of native collagen were established in 1955, independently by Rich and Crick (1955) and by Ramachandran and Kartha (1955). The three-stranded, triple-helical model for tropocollagen was first suggested by Ramachandran and Kartha, who showed that their X-ray diffraction results could be explained by a helix comprising left-handed helical chains (secondary structure) related by a right-handed rope twist (tertiary structure) so that, overall, collagen has a coiled-coil, rope-like conformation. The Gly residues of the –Gly–X–Y– repeat sequence are placed in the core of the structure because the triple helix structure, in which the three chains are arranged about a common central axis, leaves no space in the inner core for larger lateral groups.

The imino acid hydroxyproline (Hyp) was recognized as being associated uniquely with the collagen triple-helical structure and having a role in stabilizing the structure. Moreover, the three chains are linked through inter-chain hydrogen bonds perpendicular to the common axis, although the nature and the number of these bonds is still a matter of debate. For each –Gly–X–Y– triplet, one hydrogen bond forms between the amide hydrogen atom of glycine in one chain and the carbonyl oxygen atom of residue X in an adjacent chain. Hydrogen bonds involving the hydroxyl group of hydroxyproline may also stabilize the collagen triple helix. Ramachandran's 1968 model (Ramachandran and Chandrasekharan, 1968) proposes a pattern of hydrogen bonding also involving water molecules in interstitial positions.

The left-handed helices have 10 residues per 3 turns and the pitch is approximately 0.9 nm. The right-handed superhelix has a pitch 10 times larger, around 9 nm. The

Table 7.1 Imino acid composition and denaturation temperatures of collagens from various sources.

Imino acid (wt%)	Cod	Ling	Portuguese dogfish	Megrim	Tuna	Calf skin
Pro	10.3	10.9	11.2	12.8	12.4	13
Hyp	7.1	6.9	7.5	9.1	9.2	12.2
Pro + Hyp	17.4	17.8	18.7	21.9	21.6	25.2
T_m (°C)	15	19.5	20.9	27.9	28.8	36

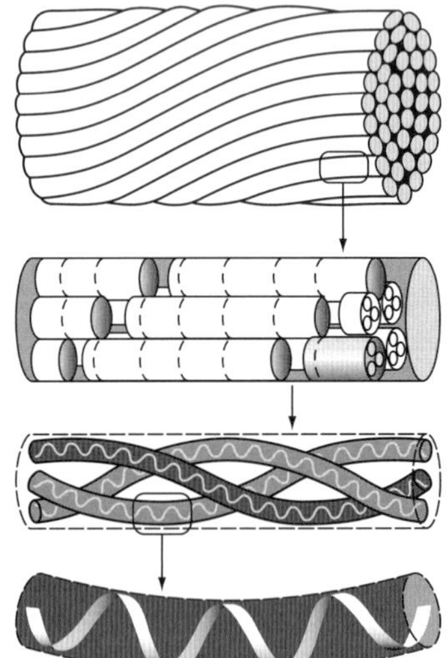

Figure 7.1 Structure of collagen fibres: (a) fibres are bundles of twisted fibrils; (b) fibrils are made of rods arranged in parallel and staggered arrays (periodic spacing 64 nm); (c) each rod is a right-handed triple helix; (d) the single collagen strand is a left-handed helix.

supramolecular arrangement of the collagen rods in fibrils is called the quaternary structure. The fibres are strengthened by covalent bonds and thus are insoluble. The hierarchical arrangement of collagen in the native state is shown in Figure 7.1. At larger scales, in tissues, the assemblies of fibres form a wide variety of patterns.

7.2.3 Amino acid composition in collagens from various sources: mammalian and fish

Mammalian collagens are the most common, but other sources have been exploited, including those from piscine sources, particularly skins. The main difference between mammalian and fish collagens lies in their imino acid composition (Table 7.1). In contrast to mammalian collagens, the content of these particular imino acids in fish collagens varies significantly in the range 17–22 wt%, while in mammalian collagens it is close to 25 wt%.

Fish are non-homoeothermic, and it was recognized that the denaturation temperatures of the various collagens parallel the natural environment temperatures of the species from

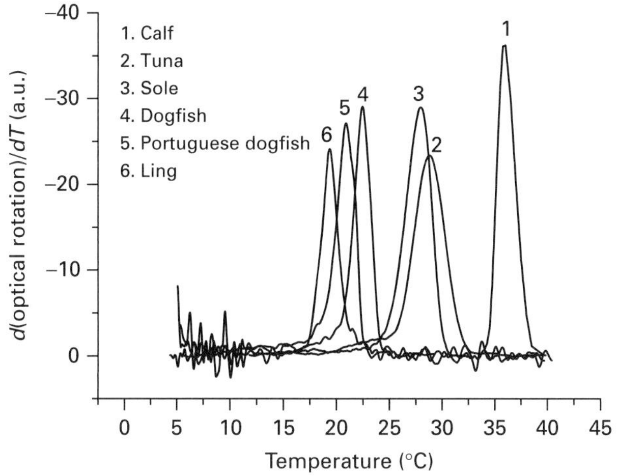

Figure 7.2 Denaturation temperatures of soluble collagen rods measured by optical rotation in dilute acidic solution. Adapted from Joly-Duhamel *et al.* (2002b) © 2002 American Chemical Society.

which they are derived. The relation between denaturation temperature and imino acid content has been established for vertebrate and invertebrate species, and is connected with the mechanism of stabilization of triple helices via the proportion of pyrrolidine residues (Harrington and Rao, 1967). This is shown in Figure 7.2, where the melting peaks or denaturation temperatures were measured by optical rotation (OR) techniques for soluble tropocollagen rods extracted under conditions which preserve their native structure. Deep-water (cold-water) fish such as cod have thermal denaturation temperatures around 15°C while those for surface-water (warm-water) fish – from rivers and lakes, and also including tuna – are much more stable (28–29°C). Warm-blooded land animals (mammals) have a constant and large amount of Pro + Hyp and a denaturation temperature around 36°C.

It is generally believed that the imino acid content affects the entropy of the 'melting' or denaturation transition: the presence of pyrrolidine rings decreases the configurational entropy of the random coil. If it is assumed that these rings undergo no conformational change at the coil–helix transition, the melting temperature (denaturation) is then related to the pyrrolidine content by

$$T_m = \frac{\Delta H}{(1-p_i)\Delta S}, \tag{7.1}$$

where p_i is the fraction of imino acids. In this simplified model, the enthalpic term is constant, while in a more refined analysis the enthalpic term per residue should decrease with the pyrrolidine content (Harrington and Rao, 1967). Good qualitative agreement is found with the first assumption, but a more quantitative estimate of T_m is obtained with the more refined analysis.

7.2.4 Collagen renaturation

Before discussing the renaturation of degraded collagen – gelatin – it is necessary to consider the essential features of the renaturation of soluble collagen. If the essential subunit, tropocollagen, is sufficiently denatured by heating, the three strands separate partially (since there may be some covalent cross-links) or completely to give the flexible polypeptide form. Tropocollagen itself is a rod-like particle of $M_w \sim 3.45 \times 10^5$ g mol^{-1}, so that each strand has $M_w \sim 110\,000$ g mol^{-1}. The length is around 300 nm and the diameter is ~1.4 nm. The temperature of denaturation depends upon the source (and thus the exact polypeptide composition), as mentioned above. However, the helix–coil transition is sharp, and the initial nucleation of the re-formed helix occurs rapidly. Such a solution is essentially a very ideal form of gelatin, and renaturation studies show a number of important features.

The kinetics of renaturation can be followed by, for example, optical rotation measurements, and this approach has been carried out by a number of workers, so here we only summarize the results. At low concentrations *c*, the kinetics is first-order with respect to *c*, increasing to second-order at higher concentrations. This is consistent with a change from intramolecular helix formation at lower concentrations to intermolecular helices at higher concentrations. Both nucleation and helix growth are slow; the nucleation presumably involves just two chains (since three-body collisions are very unlikely in dilute solutions). The nuclei have a critical size, below which helix propagation does not occur, and this has been examined by employing special samples with a very low number of residues. Estimates of the critical size lie in the range 20–40, and indeed a 36-residue peptide can renature, so the lower limit is probably realistic. Also, for such a 36-residue system, intramolecular folding is very unlikely, so the helix is formed of 3 × 36-residue strands. Helix growth is slow, but the subsequent propagation is even slower as the coil-to-triple helix 'zipping' rate is limited by the presence of *cis*-proline residues in the backbone. The subsequent reversion of these to the *trans* form allows the helix to extend only gradually so that the overall helix propagation rate is typically four to six orders of magnitude slower than for double-helical systems such as carrageenan (Chapter 5).

7.2.5 From collagen to gelatin

7.2.5.1 Molecular mass distribution

To extract gelatin from collagen, the entire structure of the fibres has to be disrupted. Both chemical and thermal treatments are required during the extraction process to break down both inter-chain covalent and hydrogen bonds. There are two processes of extraction: one uses alkali and the other acid. Generally, gelatin from bovine bones and hides is extracted with alkali and gelatin from skins (porcine, fish) with acid. The process modifies the isoelectric point of the chains, which is around pH = 5 for the basic extract and around pH = 8–9 for the acidic extract (unchanged from native collagen). The process produces gelatins with various molecular masses. Although initially a tropocollagen single strand has the known molecular mass, during extraction both

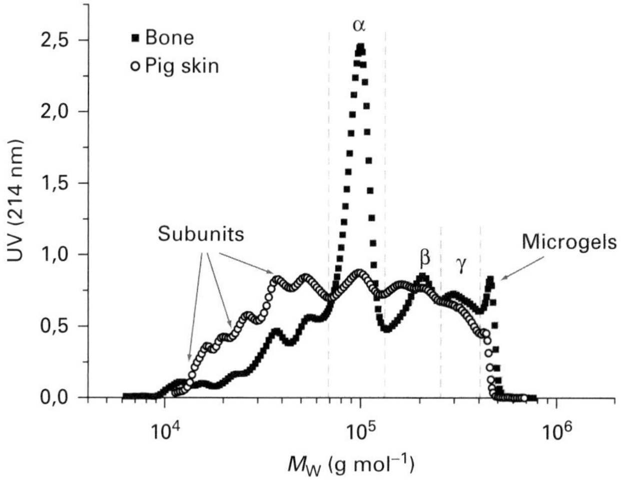

Figure 7.3 Molecular mass distribution obtained by chromatography of first extracts of gelatin from beef bone (basic extract) and pig skin (acid extract).

hydrolysis of the protein (subunits of the initial strand) and incomplete breakage of the cross-links (branched high molecular mass components) occur. The molecular mass and its distribution resulting from extraction are important parameters in determining the ability of gelatins to form a gel.

In Figure 7.3 we see that the molecular mass distribution of gelatin differs significantly depending on the source. The large central peak of the distribution, the so-called α chains, is a single tropocollagen strand; β and γ chains have, respectively, two and three such strands covalently bonded. Larger molecular masses (known as microgels) are also observed. Besides, degraded α chains appear as subunits corresponding to lower and lower molecular masses. The pig skin (first extract) has a much more even distribution, with no distinct major population. In the samples in Figure 7.3, average molecular masses are $M_w = 157\ 000$ g mol^{-1} for the bone extract and $M_w = 179\ 000$ g mol^{-1} for the pig skin extract. The indices of polydispersity are respectively $M_w/M_n = 1.89$ and 2.17, so similar average values of the molecular masses and polydispersity indices may correspond to very distinct gelatin distributions.

The molecular mass (and to a lesser extent its distribution) affects the viscosity of the solutions and is an important parameter in industrial applications. Another important factor for applications is the gel 'strength' as determined by the industry Bloom test, measured under standard conditions. Bloom strength is quoted in grams and the higher this value, the greater the modulus. Usual values range from 300 g to 50 g, with low molecular mass gelatins having a low Bloom value. Very low Bloom values are observed for fish skin gelatins, for reasons explained below, related to their gelation temperatures. In industry, gelatins with low Bloom values are used in food preparations, or technical applications such as paper manufacturing, while intermediate values are used for pharmaceutical capsules. The highest values from acidic extracts are required in confectionery and, from basic extracts, for hard capsules.

Table 7.2 Characteristic parameters of gelatin in aqueous solution at $T = 50°C$.

Parameter	Symbol	Numerical value
Radius of gyration	$\langle R_g^2 \rangle^{1/2}$	35 ± 4 nm
Average molecular mass	M_w	$(1.9 \pm 0.1) \times 10^5$ g mol^{-1}
Polydispersity index	M_w/M_n	2.3
Second virial coefficient	A_2	$(3 \pm 1) \times 10^{-4}$ mol ml g^{-2}
Flory parameter	χ_F	0.48
Statistical segment length	b	4 ± 0.5 nm
Cross-section of the chain	R_c	0.32 ± 0.1 nm
Mass per unit length	M_l	280 ± 80 g mol^{-1} nm^{-1}
Self-diffusion coefficient	D_0	2×10^{-7} cm^2 s^{-1}
Hydrodynamic radius	R_h	22 ± 2 nm
Overlap concentration	c^*	0.5 wt%
Correlation length (semi-dilute solutions)	ξ	$c = 1$ wt%, $\xi = 7 \pm 1$ nm $c = 2$ wt%, $\xi = 5 \pm 0.5$ nm $c = 5$ wt%, $\xi = 3.5 \pm 0.3$ nm

7.2.5.2 Solution properties

Gelatin is delivered as dry granules and ideally should be left for several hours in cold water (4°C) so the grains swell without dissolving. The preliminary swelling is an important step which allows the granules to dissolve fully with mild stirring at 40–50°C. Because M_w is quite low, and because the polypeptide chain itself is quite flexible and coil-like, gelatin solutions are not very viscous and are Newtonian in the range of shear rates from 0.1 to 100 s^{-1}. Gelatin in solution is a coil, so with increasing concentration the coils become entangled, and viscosity increases (Herning *et al.*, 1991; Pezron *et al.*, 1991). The persistence length of the gelatin coil is around 2 nm; since each tripeptide sequence has a length of approximately 1 nm, this persistence length corresponds approximately to six residues. The Flory parameter $\chi_F = 0.48$ is close to 0.5, as is common for water-soluble polymers. The overlap concentration c^* is close to 0.5 wt% and, with increasing concentrations in the semi-dilute range, the characteristic distance between coils (correlation length) decreases according to statistical theories of entangled solutions, as was shown by dynamic light scattering and neutron scattering experiments (Herning *et al.*, 1991; Pezron *et al.*, 1991).

Various geometrical parameters of typical gelatin coils are listed in Table 7.2.

7.2.6 Gelation mechanism

When a gelatin solution of concentration $\gtrsim 2\%$ w/w is cooled below room temperature, the protein coils undergo a coil–helix transition, and a polymer network begins to be formed. The helices are reminiscent of the triple-helical native structure of tropocollagen. The polypeptide chains only partly recover their native conformation, even when samples are annealed for hours or days, and no real equilibrium is ever reached. When the temperature is raised, the inverse helix–coil transition takes place and the gel 'melts' to give a liquid (Figure 7.4). It is therefore very important to follow the conformational

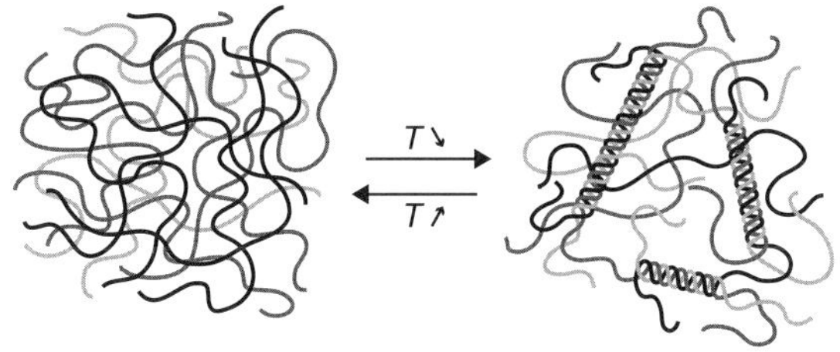

Figure 7.4. Schematic representation of the coil–triple helix transition for a semi-dilute gelatin solution.

change of the coils in solution. Although many parameters can affect the thermal 'renaturation' of triple helices, we examine below only one type of structure – the triple helix conformation – chains can adopt.

The fact that triple helices do not associate in bundles was, at one time, controversial. Godard *et al.* (1978) attributed the melting enthalpy of the gels during maturation to the formation of fringed micelle crystals, an interpretation based on generally accepted ideas for crystallization of polymer melts. Later measurements such as optical rotation demonstrated that the mechanism of renaturation is limited to the triple helix step. Moreover, X-ray fibre diffraction experiments (Pezron *et al.*, 1990) displayed scattering patterns close to native collagen, but without features related to the macro-period of the fibrils along the longitudinal axis. In the rehydrated state, the orientation of the triple helices progressively decreases with water content and the spacing between the triple helices increases, while the positions of the meridional wide-angle reflections (periodicity along the triple-helical structure) are unchanged. These experiments support the assumption that gels which usually contain large amounts of water (>50 wt%) have a network structure in which segments of triple helices are randomly oriented. Moreover, since measurements of the OR angle show a slow increase with time but apparently without limit (even when plotted on a log time axis), the proportion of residues in the ordered helical conformation must also be increasing. This suggests a considerable degree of conformational flexibility, even post-gel, and is rather unlikely to occur if the junction zones are formed of rigid crystallites.

7.2.6.1 Optical rotation (OR)

Conformational changes of certain biopolymer solutions can be detected by measuring the rotation angle of the polarization of an incident linearly polarized light beam. Far from the absorption band, the optical rotation angle changes as the beam passes through the solution. The collagen triple helices have the same optical rotation properties as the left-handed helices of the individual strands. The triple helix association gives a distortion to the chains which has very little effect on the optical rotation dispersion. Just as for carrageenans (Chapter 5), for gelatin solutions and gels the change of optical rotation angle can be related to the proportion of residues renatured or reassembled into helices (hereafter referred to as the 'helix fraction'):

$$\chi = \frac{[\alpha]_\lambda^{\text{ex}} - [\alpha]_\lambda^{\text{coil}}}{[\alpha]_\lambda^{\text{collagen}} - [\alpha]_\lambda^{\text{coil}}},$$

where χ at any time, concentration or temperature is given by

$$\chi = \frac{\text{number of residues in helical conformation}}{\text{total number of residues}}, \tag{7.2}$$

α is the rotation angle, c is the solution concentration (g cm^{-3}) and l is the optical path. The intrinsic optical rotation angle at a given wavelength λ is

$$[\alpha]_\lambda = \frac{\alpha}{c\,l}. \tag{7.3}$$

Since both $[\alpha]_\lambda^{\text{collagen}}$ can be measured for native soluble tropocollagen rods (entirely helical) and $[\alpha]_\lambda^{\text{coil}}$ for gelatin solutions entirely in the coil conformation (T=40–50°C), χ can be calculated under varying thermal protocols. Experimentally observed OR values $\alpha_\lambda^{\text{ex}}$ are of the order of degrees and can be easily detected and recorded with a computer-controlled polarimeter, so the fraction of residues in helical conformation can be accurately measured in gelatin solutions for concentrations as high as 20 wt%.

7.2.6.2 μDSC

An alternative measurement is provided by μDSC (Chapter 2). Denaturation of collagen and renaturation observed in gelatin solutions is accompanied by significant heat flows, mainly during heating and cooling ramps. An exothermic peak is observed during helix formation when cooling at a fixed rate. By applying the same thermal history in polarimetry and μDSC, a very good correlation is seen between the two techniques. A value of $\Delta H_{\text{coil-helix}}$ per gram of helix during cooling or heating ramps was found to be independent of gelatin concentration and of the molecular mass of the extract, and close to the native collagen denaturation enthalpy. An average absolute value is $\Delta H_{\text{coil-helix}} = 55 \pm 2$ J per gram of helix.

Since the helix–coil transition of gelatin is a reversible process, the local structure of a gel at any time can evolve because of continuous molecular rearrangements. Accordingly, the range of melting temperatures of the gel changes with concentration or maturation time, probably the result of annealing (elimination of loops and short sequences) and growth of helices, although the full collagen structure is never attained. It was thought possible that the enthalpy could also depend on the degree of perfection of the sequences, but this was not the case.

In mammalian gelatin gels, the melting temperature of the helices is below 36°C, while for fish skin gelatins it is below the corresponding temperatures of denaturation of the species' collagen. For gelatin films (or very concentrated gels, $c > 50$ wt%), classical DSC is an appropriate technique (Coppola *et al.*, 2008), and even for these systems the formation or melting enthalpy per gram of helix is constant and can be used in the characterization of the state of aggregation in the gels. The structure of these very concentrated gels is taken up later in this chapter.

7.2.6.3 Parameters affecting helix formation

Gelatin gels are never in true equilibrium, so that when temperature is lowered below the coil–helix transition temperature, helices start to form but never approach the 100% of

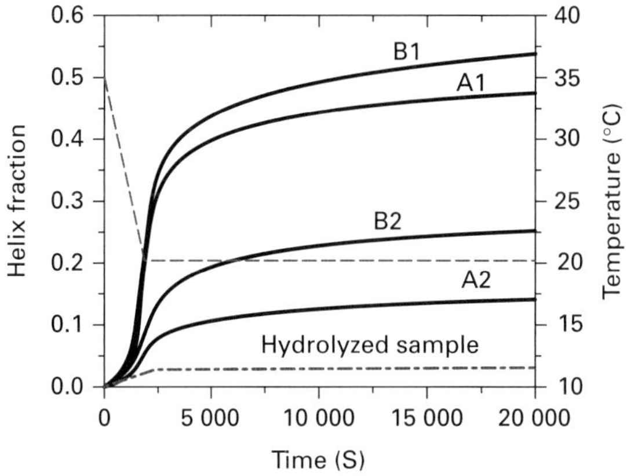

Figure 7.5 Helix formation versus time for two molecular masses of beef bone gelatin (A1: 145 700 g mol^{-1} and A2: 102 200 g mol^{-1}) and pig skin gelatin (B1: 168 500 g mol^{-1} and B2: 74 600 g mol^{-1}) at concentration $c = 4.5\%$ w/v. Also shown is a highly hydrolyzed sample ($c = 20\%$ w/v, $M_w \sim 11\,000$ g mol^{-1}) which has a very poor ability to regenerate helices. The thermal protocol is also displayed. Adapted from Joly-Duhamel *et al.* (2002a) © 2002 American Chemical Society.

native collagen. Various protocols for decreasing temperature can be used: either continuous cooling or heating at fixed rates or rapid temperature decrease similar to a quenching, followed by long-time annealing at constant temperature, or any combination that would be required to form a gel. Various parameters which affect helix formation are well established:

- Gelatin concentration: for identical protocols, the higher the concentration, the larger the helix fraction. This suggests that nucleation is favoured in concentrated solutions, but further growth may be slowed in a densely entangled system. The ultimate ratio is 60–70%, at low temperatures and long annealing times.
- Molecular mass: this is a very important parameter. The first extracts are those which exhibit the most rapid rates of growth. Figure 7.5 shows that extracts with different molecular masses from mammalian gelatins have marked differences in the renaturation kinetics at identical concentrations (4.5% w/v) and thermal treatments. Very low molecular mass samples, so-called hydrolyzed samples ($M_w < 15\,000$ g mol^{-1}), cannot recover the helical conformation even at high concentration.
- Source of gelatin sample: as already stated, gelatins from sources such as fish skin extracts have different collagen composition and their collagen denatures at lower temperatures than mammalian gelatins. Therefore their renaturation temperature is shifted to significantly lower values. A comparison of the kinetics of helix renaturation is shown in Figure 7.6.
- pH in the range 4 to 10, or salt (NaCl) at concentrations below 0.1 M, do not play an important role in helix formation for more concentrated gels (greater than a few wt%). However, in the very dilute range, effects of pH become important. Non-electrolytes

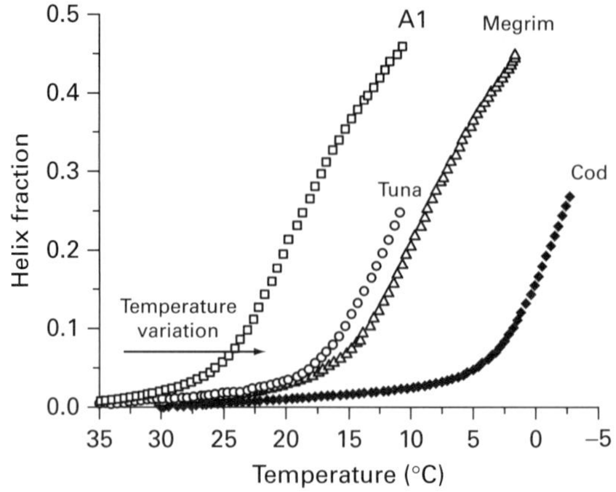

Figure 7.6 Helix formation in gelatin aqueous solution ($c = 4.5\%$ w/v) versus temperature, for gelatins with various sources. Cooling rate $-0.5°C\ min^{-1}$. Adapted from Joly-Duhamel *et al.* (2002b) © 2002 American Chemical Society.

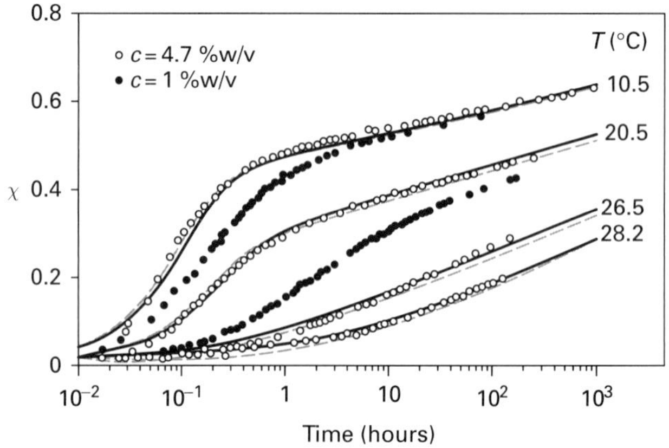

Figure 7.7 Helix fraction versus time (logarithmic scale) at various temperatures (T = 10.5, 20.5, 26.5, 28.2°C) and concentrations (1% and 4.7% w/v). Adapted from Djabourov *et al.* (1988a) with permission from EDP Sciences.

such as sugars and glycerol enhance helix formation as they decrease the solubility of gelatin. Other chemicals such as KSCN, LiBr, $CaCl_2$, urea and phenols have the power to completely prevent gelation when they are present in sufficient concentration; see, for instance, Ward and Courts (1977).

- Kinetic effects: helix formation proceeds after the nucleation step, but slows down progressively while the network is formed. An 'infinitely long' evolution characterizes helix formation in gelatin, as shown, for instance, in Figure 7.7 for various concentrations and temperatures. This slow evolution or physical ageing in gelatin gels was first reported and analysed by te Nijenhuis (1981a, 1981b).

7.2.6.4 Helix thermal stability

The coil–helix transition is reversible, but there is an element of hysteresis between the formation (renaturation) and melting (denaturation) temperatures, the extent of which depends upon the relative heating and cooling rates. Moreover, while collagen denaturation has a very narrow temperature distribution, the renaturated helices of gelatin gels present a large distribution of melting temperatures, with a clear correlation between the width of the melting peak and the helix formation temperature: after maturation, the lowest temperatures provide the largest helix fractions, but the broadest distributions.

The maximum of the peak is also related to the final cooling temperature, with the lowest gelation temperatures having the lowest melting temperatures. Figure 7.8 shows

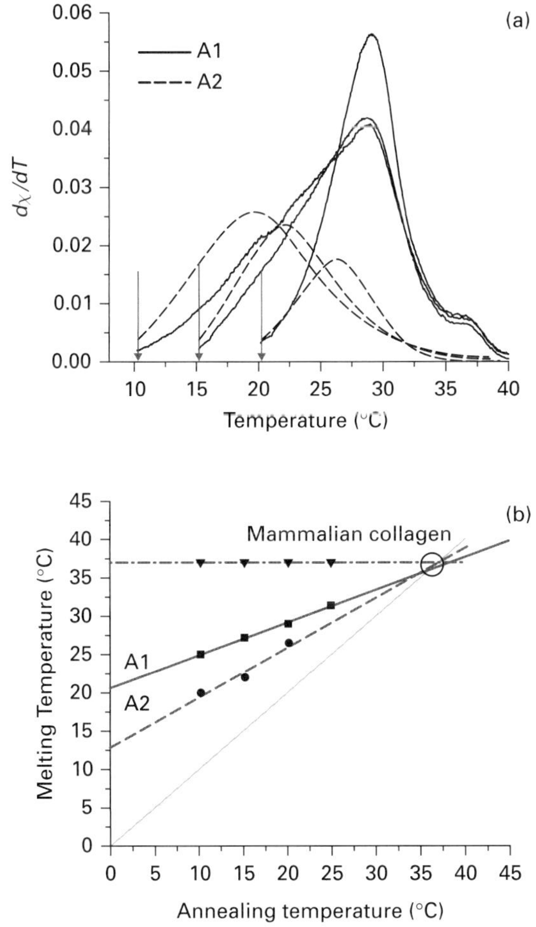

Figure 7.8 (a) Melting curves $d\chi/dT$ for samples A1 and A2 after cooling and annealing for 15 h at various temperatures, indicated by the arrows. The melting temperatures vary according to the initial temperature and depend on the molecular mass. (b) Melting temperature versus annealing temperature for samples A1 and A2. Extrapolation toward the melting temperature of mammalian collagens is shown. From Joly-Duhamel *et al.* (2002b) © 2002 American Chemical Society.

helix melting. This data was obtained from optical rotation measurements, by taking the derivative of the angle of rotation versus temperature, recorded during a slow heating ramp. Figure 7.8a shows the width of the melting peaks related to the gelation temperature for the two samples A1 and A2 of Figure 7.5. The sample with a lower molecular mass has a poorer thermal stability, a lower helix fraction and a broader melting temperature distribution.

Figure 7.8b summarizes melting temperatures versus gelation temperatures; the melting temperatures extrapolate towards the collagen melt temperature (36°C) for mammalian gelatins, if gelation (renaturation) could be achieved at the extreme limit of helix stability. The melting temperature evolution can be related to the size of the helical sequences. In crystallization theories, it is known that the size of the critical nucleus decreases with the supercooling $\Delta T = T - T_m$. After an initial nucleation step, during annealing periods crystal perfection and slow crystallization occur. In polymer melts, perfection involves an increase in melting temperature and a sharpening of the melting peaks, similar to gelatin helix perfection. The usual Zimm–Bragg model (Zimm and Bragg, 1959) predicts the cooperativity of the transition between single helix and random coil conformations based on the Ising model, so, perhaps quite reasonably, this does not explain either the broad melting peaks or the non-equilibrium behaviour of *triple* helices in gelatin gels. Thus, a more refined theory is needed in order to explain the complex features of helix formation and melting in semi-dilute gelatin solutions. Because the helices are believed to grow in one dimension (without lateral aggregation), an important simplification of the theory is expected, compared to crystallization of long polymer chains from melt or from solution.

7.2.6.5 Mechanisms of helical renaturation

The kinetics of renaturation has been studied in some detail, and over many years. As mentioned earlier, the kinetic order, as measured for example by OR, is 1 at low concentrations and increases to 2 at higher concentrations. In no case has the anticipated third-order kinetics (for a three-chain model) been observed. It has even been proposed that helix nucleation is a *bimolecular* process, involving an intramolecular β-turn and another gelatin macromolecule (Busnel *et al.*, 1989). When a third segment meets a 'kink' with the correct orientation, a triple helix is nucleated.

Network formation within an entangled solution is a complex process. Careful phenomenological analysis of triple helix renaturation during a rapid quenching of the solution from the sol state (40°C) to a fixed low temperature (below the collagen melting temperature) shows a two-step mechanism:

- a relatively fast nucleation with an exponential increase in helix fraction with a characteristic fast time of the order of 1 or 2 minutes
- a subsequent slow process in which helix growth proceeds with a logarithmic rate and a characteristic slow time, which depends on temperature.

These two steps may be independent or related to each other. The second mechanism promotes annealing and growth for the helices created during the first step, but analysis of

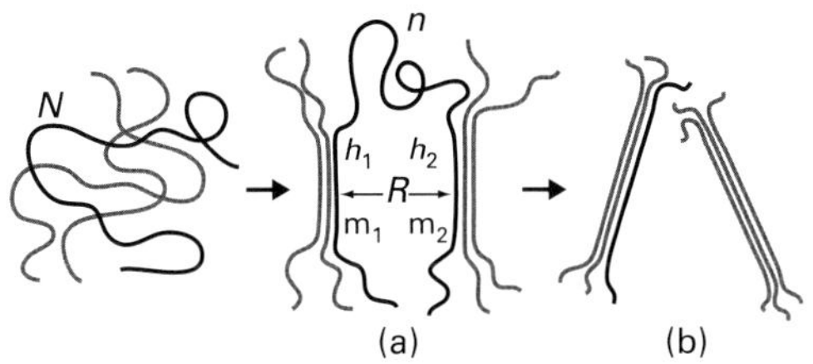

Figure 7.9 The two steps involved in the gelation process: (a) formation of a loose network in which a chain containing N statistical units bridges between two helical sequences (h_1, h_2), with n units in coil conformation and m_1, m_2 units as free ends; (b) growth of helical sequences, with the collagen rods entirely renatured at infinite time. Adapted from Djabourov *et al.* (1988a) with permission from EDP Sciences.

the kinetics does not suggest a way to discriminate between these two possibilities (Djabourov *et al.*, 1988a).

A microscopic scheme of some of the steps involved in the formation and maturation of the network of triple helices is shown in Figure 7.9.

- Helices are nucleated along individual chains, supposing, for instance, that the Pro and Hyp sequences are favourable nucleation sites. The model neglects the fact that there is a minimum stable helix length, the rate determining *cis*-proline flip and the presence of some covalent cross-links. A triple helix sequence can start to grow when helices are nucleated on segments belonging to three different chains which are closely entangled. Concentration and entanglement between chains are factors which enhance the probability of nucleation of the triple helix, as confirmed by experimental results and highlighted by this model.
- A chain is likely to be involved in several (at least two) of these sequences, creating a loose network. The first step of the kinetics can be associated with the rapid process. If a chain with N statistical units is involved in two junctions separated by a distance R, as shown schematically in Figure 7.9, R can be considered to a first approximation as a constant, depending on the local concentration. By simple statistical mechanics, we can calculate the helix fraction corresponding to this 'equilibrium' configuration.

The results for these predictions are shown in Figure 7.10

- The helix fraction χ_0 at the first step is calculated as a function of temperature. The statistical unit is taken here as a tripeptide. In this calculation, $T_0 = 767$ K corresponds to the binding energy ε per tripeptide in the helical state defined by $T_0 = |\varepsilon|/k_B$. The calculated points fit with a binding energy $\varepsilon(\text{mole}) = 6.27$ kJ mol^{-1}.
- Note that when the first step is achieved, the system is not truly in equilibrium. The second process, which has a logarithmic time dependence, could represent a superposition of different microscopic times allowing the local constraints to be released (starting with the *cis*-proline flip outlined above). In this model, the final equilibrium should correspond to the fully renatured collagen rod, where the solution becomes a suspension of independent rods. This final state is not reached within the longest observation time, and the solution remains a gel with a connected network. In the final

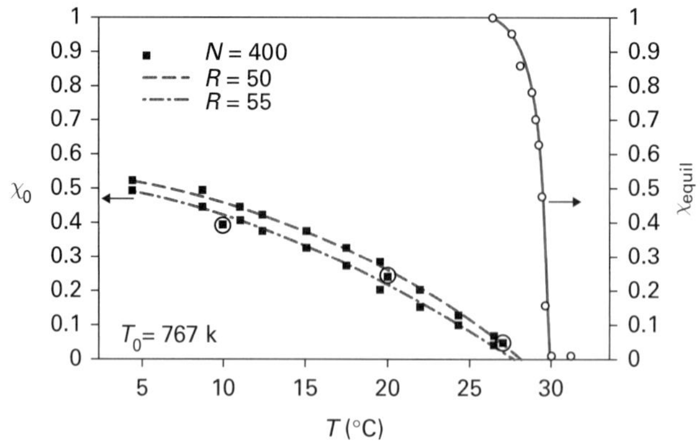

Figure 7.10 Computed values for helix fraction versus temperature. χ_0 is the helix fraction at the end of the first step, χ_{equil} is the hypothetical value reached after infinite time, if the long maturation process were to be completed. Various values for R are represented on the graph, highlighting solution concentration effects. Adapted from Djabourov *et al.* (1988a) with permission from EDP Sciences.

step, $\chi_{equil}(T)$ can be computed from the statistical model and behaves as a step-like function within a narrow temperature range, similar to the denaturation curve of native collagen. This means that the upper limit will always be lower than 100%.

The broad outline of the gelation of gelatin solutions is now accepted as being uniquely related to the renaturation of the collagen type triple helix, unlike ionic-induced gelation in polysaccharides, where the mechanisms are still debated as they touch upon on both conformation and aggregation effects (Chapter 5). A similar problem is encountered in agarose gels (see below). The conformational change in gelatin is determined by the nature and composition of the collagen. In practice, the total fraction χ of renatured helices may reach 0.75, but will still increase in time, at a fixed temperature. Indeed, since it has never been observed to reach a limit, it can be argued that the gel does not have an equilibrium helix fraction at less than 100%, and that this is a value that will never be reached for the reasons above. In practice, the helix fraction in a typical gel is around 20–40%.

7.2.7 Rheology of gelatin gels

7.2.7.1 The sol–gel transition

The rheology of gelatin gels has been widely investigated in the past; a first reference is that of Leick (1904), who measured birefringence induced under elongation and who suggested that the elastic modulus of the gels varies with the square of gelatin concentration. In 1948, Ferry published a review on protein gels (Ferry, 1948), in which he reported that the rigidity (shear modulus) of gelatin gels varies as the square of the protein concentration, and that it decreases with increasing temperature and with decreasing average molecular mass. In a subsequent paper, Ferry and Eldridge (1949) reported optical rotation measurements on solutions versus temperature, where they concluded that specific OR ‘is primarily related to intramolecular cross-links ... while a comparatively small number of intermolecular links is responsible for the rigidity of gels'. Following this, they published a paper (Eldridge and Ferry, 1954) which became a reference in the field of gel rheology, proposing their model for the elasticity of gelatin gels, relating the melting temperature of the gels to protein concentration and molecular mass. They assumed that the cross-links of the gel were formed by binary association between chains, although their model was devised while the exact structure of the gelatin junctions was still unknown, since the triple-helical structure of native collagen was not revealed until the following year (1955); see Section 7.2.2.

Since that time, the sol–gel transition has been reanalysed using a number of percolation type models (Chapter 3). To relate the small-deformation rheological properties to structural parameters, it is obviously necessary to have some model of the network structure. Unfortunately the length of the overall helical 'junction zones' is unknown, although their width (at around 1 nm) corresponds to that of the tropocollagen triple helix. The most reliable techniques for determining the helix fraction – optical rotation and DSC – are both essentially short distance scale methods, so they cannot identify the length of the helix, only the number of residues in the helical conformation.

However, because gelatin gelation is a kinetic process, it is important to follow identical paths for the structural and rheological investigations. As the physical junctions are fragile, gelation can only be investigated by limiting the strain, rather than the stress, to very small values (less than a few percent). Extensive investigations on the time and temperature dependence of dynamic moduli of gelatin gels have been performed by te Nijenhuis (1981a, 1981b), who established that an 'induction time' was necessary before the storage modulus starts to increase. He found that this time is short (1 min) at low temperatures (1–2°C), whereas it increases to about 8 h at 25°C, and found a maximum gelation temperature of ~27°C.

Such important changes of the viscoelastic moduli within a narrow range of temperatures cannot be understood without parallel investigations of the conformational changes. Figure 7.11 shows that both measurements (rheology and optical rotation) can be performed with great accuracy during ageing, after a rapid quench from hot solution to a lower temperature, between 24°C and 28°C, using two independent temperature-controlled baths. Figure 7.11a shows that the helix fraction increases at different rates, as was already mentioned in the previous section; in Figure 7.11b the increase of shear modulus with time at a fixed frequency is displayed. Obviously, there is no direct proportionality between these two measurements, so it is interesting to eliminate the time parameter and plot shear modulus versus helix fraction, for the various temperatures. This is shown in Figure 7.12.

The parameter defined by te Nijenhuis as an 'induction time' corresponds to a threshold in the helix fraction, after which the elastic modulus starts increasing. The threshold is close to $\chi_c = 0.07$ in the example shown in Figure 7.12, and is reached in 7 min at $T = 24$°C and in 214 min at $T = 28$°C. The existence of a critical threshold is predicted by all percolation type theories, but the surprising result is that all data collapses to a single curve, even though these correspond to very different rates of nucleation and growth of helices. The threshold is very low for a low-concentration system, meaning that the

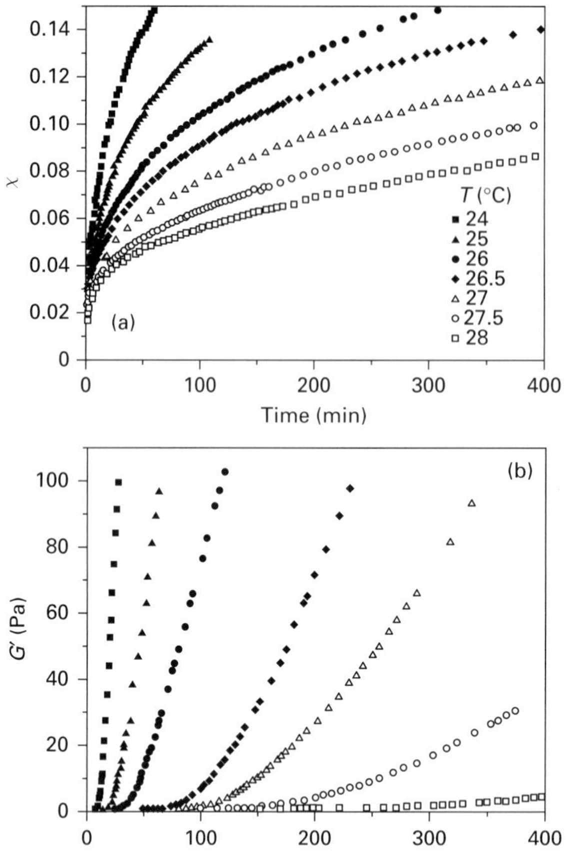

Figure 7.11 (a) Helix fraction versus time ($c = 4.7\%$ w/v) during rapid quenching at various temperatures; (b) shear modulus versus time at 0.15 Hz, at same temperatures. Adapted from Djabourov *et al.* (1988b) with permission from EDP Sciences.

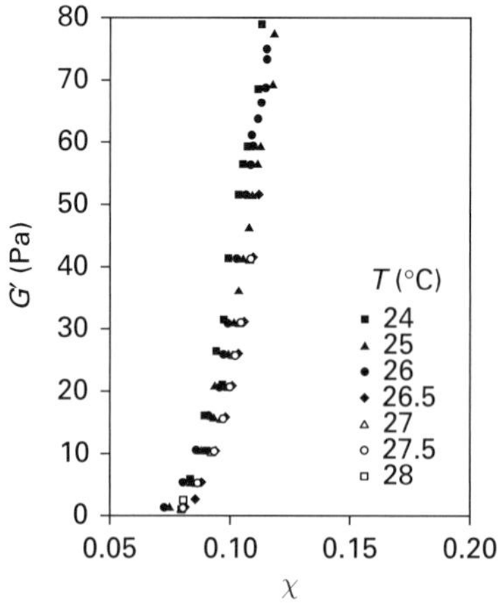

Figure 7.12 Relation between shear modulus and helix fraction derived from Figure 7.11. Adapted from Djabourov *et al.* (1988b) with permission from EDP Sciences.

elasticity is extremely sensitive to the helix fraction. As soon as the helix fraction passes beyond the threshold, the shear modulus becomes almost independent of frequency and increases progressively with helix growth. The loss modulus also increases with time before the gelation threshold, as expected. G'' is frequency-dependent but the low-frequency (Newtonian) limit G''/ω is difficult to access because G'' corresponds to a small value in this limit, and the kinetics of helix formation continues.

In all the theories discussed in Chapter 3, the elastic modulus increases with the fraction of bonds. If we assume that the helix fraction can be compared to the fraction of cross-links (bonds), a power-law dependence is expected for the elastic modulus versus distance to the gel point:

$$G' \approx \left(\frac{\chi}{\chi_c} - 1\right)^t, \quad \chi > \chi_c. \tag{7.4}$$

This is a significant assumption in itself, since what really matters is not the proportion of residues in the helical conformation but the number and distribution of helices, but it is reasonable to assume the two are proportional when $\chi \sim \chi_c$. The exponent t for elasticity defined in Chapter 3 is found here to be $t = 1.8 \pm 0.1$, in the range $\varepsilon < 0.1$ (Djabourov *et al.*, 1988b) which corresponds to that of a random network of resistors in the percolation theory, while in the classical Flory–Stockmayer approach, it should be close to $t = 3$. This still suggests that, despite the complexity of the local helical structure (Section 7.2.6), percolation type theories provide a simple framework for interpreting elasticity if the proportion of triple-helical residues is identified with randomly distributed, permanent cross-links. The relation between elasticity and structural parameters is insensitive to the kinetics of helix growth, which varies strongly with temperature.

7.2.7.2 All gelatin networks

In the case illustrated, the sol–gel transition showed a very simple correlation. On the other hand it is known that many different gelatin samples exist, with various origins and molecular masses. Various thermal treatments can also be applied and they are likely to modify the distribution of the helical sequences, as shown before. Is the storage modulus always simply related to the helix fraction? What is the effect of molecular mass, concentration or other parameters? These aspects are examined below.

Concentration dependence of the shear modulus

For many years it was known that the shear modulus increases with concentration, but it also depends on the temperature and time. Consequently the variation is not just the commonly reported c^2 dependence, and temperature also plays a role in the Eldridge–Ferry (EF) model (Chapter 3). Although gelatin gels are not in an equilibrium state, the EF approach is still useful, and is widely employed. However, it is important to point out that the commonly reported c^2 proportionality will hold only away from the critical concentration (say, $c > 5c_0$) where, in practice, the majority of experiments have been performed; as c approaches c_0 from above, the apparent exponent will increase very markedly.

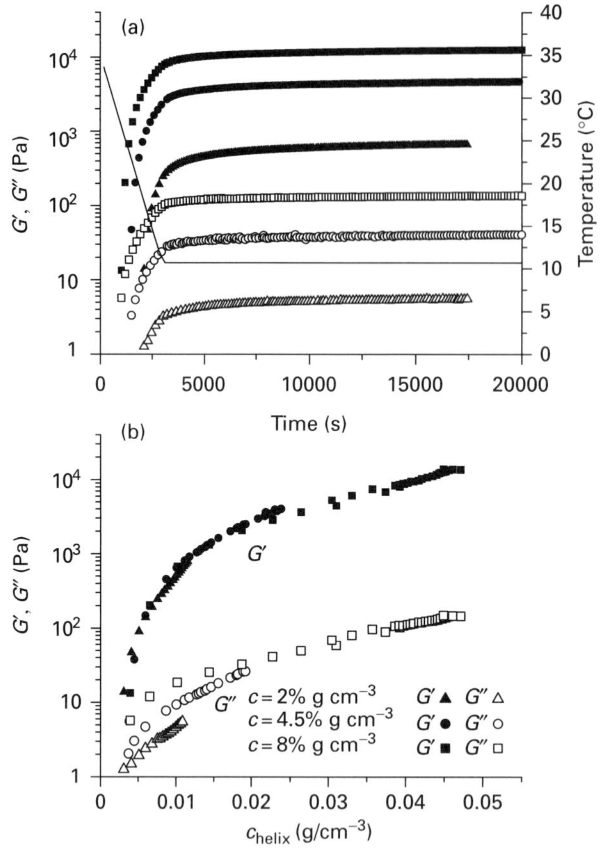

Figure 7.13 Storage and loss moduli versus time for three gelatin concentrations: 2% w/v, 4.5% w/v and 8% w/v: (a) time dependence of the shear moduli (1 Hz, sample A1) during cooling and annealing; (b) relation between the shear moduli and helix concentration during experiments shown in (a). Adapted from Joly-Duhamel *et al.* (2002a) © 2002 American Chemical Society.

The kinetics of increase of the dynamic moduli during a cooling and annealing protocol is shown in Figure 7.13a when temperature is lowered to the same final temperature at the same cooling rate, but for different gelatin concentrations. The shear and loss moduli both show a large increase during the cooling step. Over this range the modulus is indeed not far from a c^2 dependence.

In these experiments, as the concentration is a variable, helix concentrations c_{helix} should be calculated. The helix concentration c_{helix} can be simply calculated from the helix fraction χ through the relation

$$c_{\text{helix}} = \chi c_{\text{gelatin}}. \tag{7.5}$$

In Figure 7.13b, G' and G'' are plotted versus c_{helix} for three gelatin concentrations. It appears that the shear moduli are simply related to c_{helix}: for the same values of c_{helix}, very similar values of G' are observed, whereas G'' depends on the gelatin concentration. In

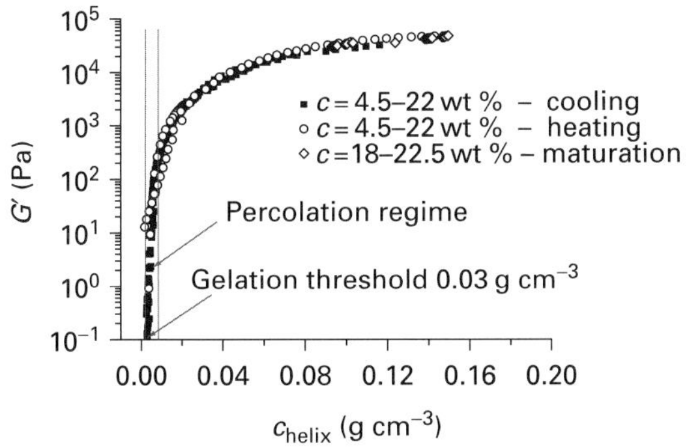

Figure 7.14 Correlation between storage modulus and helix concentration during various thermal treatments (bovine bone gelatin, $M_w = 157\ 000$ g mol^{-1}). Adapted from El Harfaoui *et al.* (2007) with permission from John Wiley & Sons.

more concentrated solutions, at equal values of c_{helix}, G'' is higher, presumably because of a larger proportion of coils and free chains, which contribute to the viscous dissipation. Also, when gelatin concentration is higher, c_{helix} can reach larger values and accordingly the storage moduli reach larger values, as shown on the plot. The threshold or critical concentration of helices is identical for the three gelatin concentrations, at $c_{\text{helix}}^{\text{critical}} \cong 0.003$ g cm^{-3}.

Thermal treatment, molecular mass, solvent effects

Helix formation is reversible, but with some hysteresis between cooling and heating. Does the correlation between storage modulus and helix concentration apply to different stages of gel formation and melting?

Figure 7.14 shows the correlation between storage modulus and helix concentration during cooling, annealing and heating steps at gelatin concentrations between 4.5% and 22% w/v. The superposition of the data is very good for all measurements of the mammalian gelatin. The plot is based on frequency-independent storage moduli, which occur later in the course of gelation for higher concentrations, after which all the moduli follow the same general trend, whether helix concentration is derived from microcalorimetric measurements or from optical rotation. Various extracts of this type of gelatin were also reported by El Harfaoui (2007) with various average molecular masses and distributions. It was shown that the correlation is also valid in this case. The correlation is also valid when the solvent is changed from water to mixtures of water and glycerol, as is often the case in pharmaceutical applications (Joly-Duhamel *et al.*, 2002a).

Source of gelatin

Collagen composition reflects the biodiversity of gelatin samples from various species. The variability in imino acid content has a direct effect on gelation temperatures, although the fundamental assembly remains the triple helix. This is an additional parameter which can be tested against data for G'. The result is shown in Figure 7.15, which, from now on, may be considered a master curve for the storage moduli of all gelatin gels. The data in Figure 7.15 includes fish skin gelatin from cold and warm seas

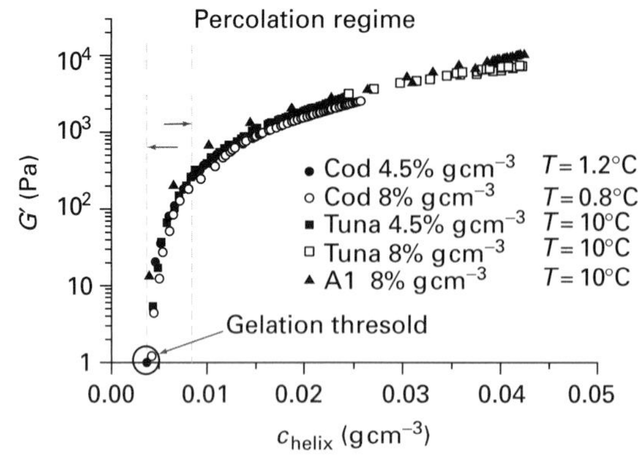

Figure 7.15 Master curve for the storage modulus of all types of gelatin networks. Adapted with permission from Joly-Duhamel *et al.* (2002a) © 2002 American Chemical Society.

(cod and tuna), which gel at different concentrations or temperatures, for some samples at temperatures as low as 1°C.

7.2.8 Elasticity in gelatin gels

If we consider the model of alternating flexible and triple-helical segments along the chain, then a first approximation would be to assume that the very flexible polypeptide segments act like an ideal rubber, and in practice a good fit can be obtained on this basis. Alternatively, if we adopt the model of a proportion of rods in a sea of coils, then a rod-like model may be more useful, as introduced below. Here we summarize both models.

7.2.8.1 Coil-like model

Using the information above and the branching theory described in Chapter 3, it is possible to obtain a very good zero-parameter fit to the modulus versus concentration behaviour of gelatin, using creep data (equilibrium compliance, $J_e = 1/G$) to provide good quality estimates (Higgs and Ross-Murphy, 1990). As noted above, at the gel point the helix fraction $\chi_c \sim 0.07$, for a gelatin concentration 4.5% w/v, so from the simple Flory gel point (Equation (3.3)) this would give an apparent *maximum* number of branches per chain (the functionality, f, as defined in Chapter 3) of $\chi_c = 1/(f-1)$, i.e. $f \sim 15$.

A more specific model may be that of Peniche-Covas *et al.* (1974), which suggests that for an n-chain model, $f = (n-1)(x-1)+1$. Here, x is the number of junction zones per chain and n (= 3 for gelatin) is the number of chains involved in this junction zone. We can also estimate the functionality if we use the fact that the minimum stable helix length is around 30 residues (Busnel *et al.*, 1989). Since the molecular mass of a peptide residue in gelatin is ~10^5 g mol^{-1}, this means that for this sample ($M_n \sim 70\ 000$ g mol^{-1}) the maximum functionality per chain should be around 20, so that, for a three-chain triple helix model, $f \sim 40$.

The latter estimate is almost certainly too high because of the presence of covalent cross-links, the need for the network to retain flexible regions etc., so a figure of $f \sim 10$–20 would be more reasonable. Still needed is an estimate for the network front factor. For a ideal rubber-like network this should lie in the range 0.5 (for a phantom network) to 1 (affine). Using these 'constraints' we find that very good fits can be obtained with f in the range 10 to 20 (Ross-Murphy, 1992). Corresponding critical concentrations are ~0.7%, whichever value of f or front factor is applied, a value which corresponds well with light scattering results.

More sophisticated analyses have been applied, particularly by te Nijenhuis, and are discussed in his review (te Nijenhuis, 1997). For example, in the appendix to that review he derives an equation which can be rewritten as

$$G_{\mathrm{eq}} \approx \frac{cRT}{M}\frac{4}{g}\left[\frac{c}{c_0}\right]^{\frac{g}{2}-1}. \tag{7.6}$$

Here G_{eq} is the equilibrium shear modulus, g is his so-called cross-link molecularity – for gelatin he assumes this will be 6 [= 3 (for a triple helix) × 2 (since he assumes the association of two such chains)] – and c, c_0 are concentration and critical concentration, respectively. If we use this value, $g = 6$, (7.6) gives the commonly observed $G \sim c^2$ dependence.

7.2.8.2 Rod-like model

The rheological data shows that the fundamental relationship between elasticity of gelatin gels and molecular properties is via the proportion of helical residues. The helices have in common an identical thickness (~1 nm) and variable lengths. Consequently the degree of perfection determined by the dispersion of melting temperatures has no influence on the rigidity.

Then two main regions can be defined: the critical region, where the modulus increases very sharply, and at higher helical concentrations, $c >> 2c_{\mathrm{helix}}^{\mathrm{critical}}$.

(a) **The critical region, where the modulus increases very sharply**

Here, according to the analysis above, the shear modulus increases as a power law with the threshold distance, $(c_{\mathrm{helix}} - c_{\mathrm{helix}}^{\mathrm{critical}})/c_{\mathrm{helix}}^{\mathrm{critical}} < 0.3$. The threshold $c_{\mathrm{helix}}^{\mathrm{critical}} \approx 0.003$ g cm^{-3} is very low:

$$G' \sim \left(c_{\mathrm{helix}} - c_{\mathrm{helix}}^{\mathrm{critical}}\right)^t, \quad t \approx 1.9. \tag{7.7}$$

This can be analysed in terms of a rigid-rod model, since, at the threshold, the volume fraction occupied by the helices (assuming a protein density of 1.44 g cm^{-3}) is only 2×10^{-3}. The percolation threshold in networks made of highly anisotropic particles, such as fibres or rods, depends on their aspect ratio (the ratio L/a between the length L and diameter a of the rods). Balberg and co-workers (Balberg *et al.*, 1984; Balberg and Binenbaum, 1987) simulated homogeneous percolation in a system of randomly oriented rods with uncorrelated contacts and established the relation

$$c_{\mathrm{critical}} L/a \approx 0.7, \quad L/a \gg 1, \tag{7.8}$$

which indicates that the contact number (average number of neighbours in contact with a certain particle) at the threshold $c_{critical}$ is of the order of 1 (actually 1.4). Networks of rod-like macromolecules show thresholds for the gel–sol transition at a low polymer concentration, $c_{critical} \approx 0.05$ wt%, in agreement with the excluded volume of the molecules.

Gels of (albeit structurally quite different) colloidal rods were also investigated experimentally by Philipse and Wierenga (1998). They also considered the case of heterogeneous networks, which leads to very low critical volume fractions at the threshold, when particles form fractal clusters.

The gelation threshold in gelatins is compatible with the formation of a loose network of rods with a large aspect ratio. As the diameter of the triple helix is around 1 nm, the length of the equivalent rods at the threshold could be around 200–300 nm. However, the gels are not made of rods with a fixed length; the helices grow step by step, increasing the helix fraction with time, but the distribution of the helical sequences is probably not identical in all circumstances. However, the elasticity behaves as if the volume fraction of helices or the total length of helices per unit volume is the only parameter which has an influence on the elasticity.

(b) **At higher helical concentrations, $c >> 2c_{helix}^{critical}$**

Here the percolating system changes towards a more homogeneous network. However, even then the helix concentration determines the elasticity of the network. The helix concentration (g cm^{-3}) can be converted to the total length of triple helices per unit volume, L_v, as if all the helical sequences were assembled end to end in one single ribbon. The ribbon creates a filamentous structure with tightly entangled strands. Transmission electron microscopy (TEM) imaging of network replicas indeed shows a filamentous structure.

In statistical analysis, if filaments adopt all orientation angles in a given volume, the nearest-neighbour distance d is related to the total length L_v of filaments by a very simple relation:

$$d = (2L_v)^{-1/2}. \tag{7.9}$$

Considering, for instance, a helix fraction of $\chi = 0.40$ for a gel with concentration $c \approx 10$ g cm^{-3}, and knowing from the collagen structure that the triple helix has an average mass per mole per unit length of 1000 g mol^{-1} nm^{-1}, we find $L_v \approx 2.4\times10^4$ μm/μm^3, which is a large value. The distance d derived from (7.9) is ≈ 4 nm, in agreement with TEM and neutron scattering investigations. Therefore, filaments are tightly entangled, so that in the range 0.01 g cm^{-3} $< c_{helix} <$ 0.05 g cm^{-3}, the calculated distance between the strands (mesh size of the network of helices) is 7 nm $> d >$ 3 nm, which is much smaller than the persistence length of collagen, $l_p \approx 170$ nm. The network of helices forms an intricate structure at scales such that they have to be considered as rigid objects ($d << l_p$), a situation comparable to that of actin gels (Chapter 9), for example.

We can then consider the gel network to consist of semi-flexible strands of persistence length l_p, of length L and linear density $1/d^2$, but further assumptions on the deformation mechanisms of the gels still need to be made. Assuming that the strands are much more rigid than the links (coil segments) that hold them together, the elasticity of such a network is then purely entropic and due to the constrained thermal agitation of the rods. Such a case has been considered by Jones and Marques (1990) and called the freely hinged network, where the elasticity is due to constrained thermal agitation of the rods.

Elasticity then scales as

$$G' \approx B k_B T L^{-1} d^{-2}, \tag{7.10}$$

where B is a prefactor and T is the temperature. To test the concentration dependence of elasticity, further assumptions must be made as to the geometry (topology, connectivity) of the network, and in particular as to the scaling of the strand length L with helix concentration c_{helix}. In the absence of such information we can make the simplest assumption, i.e. that the length of the rods L and their typical distance d scale alike. Since $d \sim c_{\text{helix}}^{-1/2}$, this leads to

$$G' \sim c_{\text{helix}}^{1.5}. \tag{7.11}$$

The moduli are extremely sensitive to the ratio L/d. We obtain a much better agreement with experimental values when the helices are considered longer than the mesh size d. In fact good agreement is found using a prefactor of 1 when we assume $L/d \sim 5$, meaning that the helices are much longer than the mesh size d. The intermediate domain between percolation and a homogeneous network is treated by a simple interpolation between the two contributions. The data for the highest helix concentrations, far from the percolation regime, is also in agreement, giving an expected exponent close to 1.5, as shown in Figure 7.16.

It is possible to illustrate the structure of such a gelatin network schematically as a collection of rigid, entangled rods (see Figure 7.17) whose total length per unit volume can be measured at any moment and is the only parameter controlling the elasticity of the

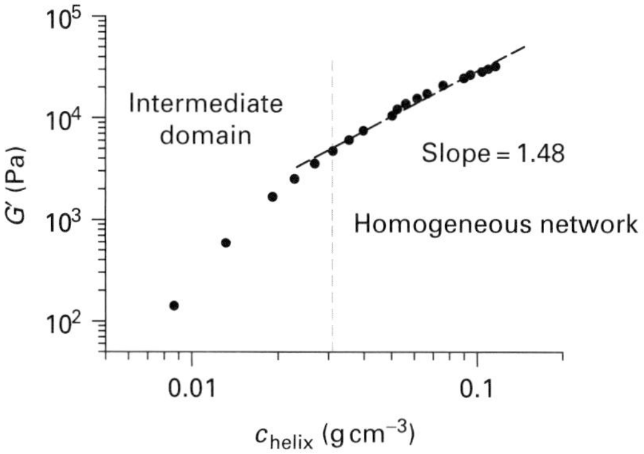

Figure 7.16 The power-law behaviour of the storage modulus versus helix concentration far from the percolation regime, after the intermediate domain. In the homogeneous network of rigid rods the exponent converges towards 1.5. Adapted from El Harfaoui *et al.* (2007) with permission from John Wiley & Sons.

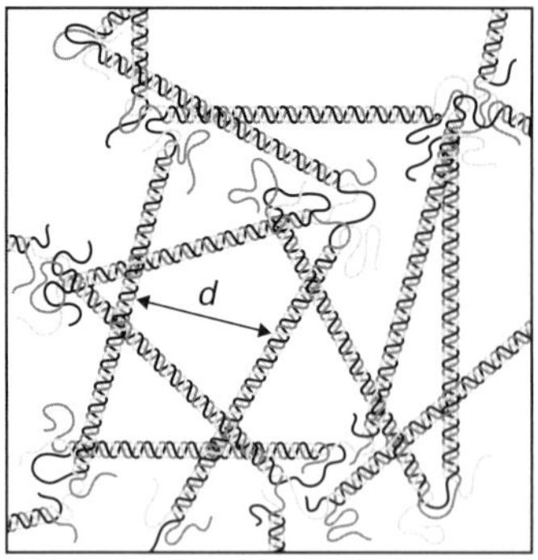

Figure 7.17. Fully developed gelatin network made of rigid rods of a typical length L separated by a distance d. The rods are connected by flexible junctions in the coil conformation. This structure is similar to TEM images, far from the gelation threshold. Adapted with permission from Joly-Duhamel *et al.* (2002a) © 2002 American Chemical Society.

network. The model supposes that the connections between rods are very flexible compared to the rigidity of the triple helix, and therefore gels can support large deformations which are reversible, even if the frame of rods is rigid by itself. The model also assumes that the length and the distance between strands are related to each other, so that there is only one independent length related to c_{helix}.

7.2.9 Mechanical properties including high strains

The vast majority of data in the literature, including that of te Nijenhuis, was obtained using the oscillatory shear method, although creep results are also interesting. One of the commonly observed features of a gelatin 'mechanical spectrum' is that there is a pronounced minimum in the G'' versus frequency spectrum for gelatin gels, but not for most other gelling systems, with a few exceptions, e.g. the E'' minimum found in agar gels (Nishinari, 1976). This is well illustrated in te Nijenhuis (1997), and this minimum suggests that, at lower frequencies, there is a maximum which reveals a long time flow process occurring at frequencies ~10^{-5} s^{-1}. This corresponds quite accurately to the retardation times measured in a series of creep experiments. Equilibrium compliances from these experiments can be estimated reasonably accurately; very high apparent creep phase viscosities ($c.10^8$ Pa s) are derived.

There are surprisingly few reliable measurements of the large-deformation or failure properties of gelatin gels. This is because even a typical 20 wt% gelatin gel is not easily self-supporting, and the commonly used but less rigorous cylindrical compression method is far more popular than the more easily analysed tensile mode. Also, relatively few papers have considered the actual stress–strain behaviour instead of more empirical (Bloom-like) units.

The work by McEvoy *et al.* (1985a) and its later extension by Bot *et al.* (1997) have concentrated on the tensile technique, however, and analysed the resultant data in terms of a rubber-like rupture method. Both groups employed the phenomenological equation due to Blatz, Sharda and Tschoegl – the so-called BST equation – which allows for a small degree of strain hardening through an exponent m, and reduces to the stress–strain behaviour from 'classic' rubber elasticity when $m = 2$:

$$\sigma = \frac{3E}{2m}\left(\Lambda^{m-1} - \Lambda^{-(m+2)/2}\right). \tag{7.12}$$

Here σ is the tensile stress, E is Young's modulus and Λ is the extension ratio. The strain to break can be treated by the Smith failure envelope method and the stress–strain behaviour by an extension of the classical stress–strain behaviour of elastomers, which allows for some high-deformation effects. Overall, results show that up to failure, which is typically at quite high strains (100% or greater, for gelatin gels of $c \gtrsim 10\%$ w/w), the stress–strain curve is surprisingly linear and ideally rubber-like, and does not show the substantial ('strain hardening') upswing seen in either experiments or theories for rod-like networks, so that experimental values of this exponent for gelatin gels are typically 2–3. Tensile strength and break strain depend markedly on strain rate, both falling as the test rate is reduced and rising again at the slowest rates. The minimum values of these quantities increase with polymer concentration.

7.2.10 Phase diagram

Although most of this chapter is concerned with aqueous gelatin sols and gels, it has been known for many years that various additives can be used to modify their behaviour, and this is particularly important in the production of pharmaceutical capsules. However, most of these are not gels *per se*, and so they are discussed only briefly.

For example, gelatin solutions with higher concentrations (around 30 wt%) are used for pharmaceutical applications for both 'hard' and 'soft' capsules. For soft capsules, various amounts of plasticizer are currently used. Films are prepared from these concentrated solutions and left to dry at room temperature under a controlled atmosphere: in the absence of a plasticizer, when the water evaporates they become brittle and make 'hard' capsules, whereas when glycerol or special grades of non-crystallizing aqueous sorbitol solutions are used as plasticizers, the capsules remain 'soft'. The water content of the films decreases during dehydration, but plasticizers are non-volatile components and so their mass does not change during film ageing. The use of plasticizers modifies T_g, the glass transition temperature of the film.

The phase diagram of very concentrated gelatin films is thus of interest, even far from the classical gel state: these films exhibit both glass and melting transition (T_m) temperatures. The ratio of plasticizer to gelatin (P/G) is an important adjustable parameter. In Figure 7.18, T_m is plotted versus gelatin concentration for P/G between 0 and 1, for films containing glycerol as plasticizer. The glass transition temperatures of the films are lower than the melting temperatures, but approach one another very closely as the films become more concentrated. An important feature appears in Figure 7.18: for gelatin concentrations below 40–50 wt%, the melting temperatures T_m of the triple helices are nearly constant (on thermograms measured at a fixed scanning rate, after a long time of maturation), in

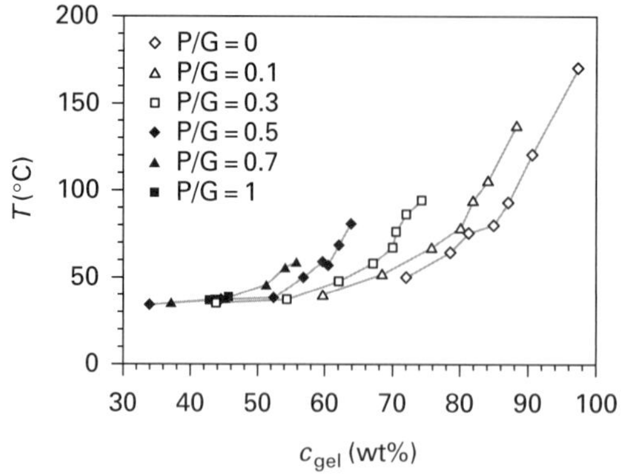

Figure 7.18 Onset of melting temperature determined by differential scanning calorimetry (DSC) of gelatin films containing various amounts of glycerol. The ratio P/G is the mass of plasticizer (glycerol) to the mass of gelatin. Reprinted from Coppola *et al.* (2008) with permission of John Wiley & Sons.

agreement with the previous description of the gels. However, when the gelatin concentration exceeds 50 wt%, T_m increases sharply, e.g. reaching 180°C at $c_{gelatin} = 98$ wt% (only the bound water is left) for non-plasticized films (P/G = 0). These very high melting temperatures are related to the formation of bundles of triple helices, which aggregate when the films start to dry and the overall volume decreases (Pezron *et al.*, 1991).

The melting temperatures for films containing glycerol increase with the ratio P/G for a given gelatin composition, as do the glass transition temperatures. Interpretation of the phase diagram of plasticized films highlights the role of hydrogen bonds between water or glycerol and the protein residues. The films are characterized by three states of aggregation according to their concentration, temperature and plasticizer content: bundles of triple helices, randomly oriented triple helices and random coils, which can be in either a glassy or a rubbery state (Coppola *et al.*, 2008).

7.3 Agarose

Agarose is a linear polysaccharide which can be extracted by treating agarophyte red algae with alkali. It consists mainly of repeated sequences of the two residues β (1→3)-linked D-galactose and α (1→4)-linked 3,6-anhydro-αL-galactose residues (Figure 7.19). Agarose is the purest material from the usual extracted commercial product agar-agar (or simply 'agar'). This product, like the carrageenan series described in Chapter 5, contains a significant amount of the sulphated form, sometimes referred to as agarose sulphate, and sometimes agaropectin. Almost all the world's production of agar is concentrated around Japan, and the article by Matsuhashi (1990) is an interesting survey of both history and applications.

Agarose itself has therefore to be separated from a mixture of crude agar-agar which also includes agarose sulphate, but this can be achieved following alkali treatment of the original algae extract, eventually leaving pure agarose. This is assumed to be chemically and electrically neutral, although it still contains some 'rogue residues' of the sulphated

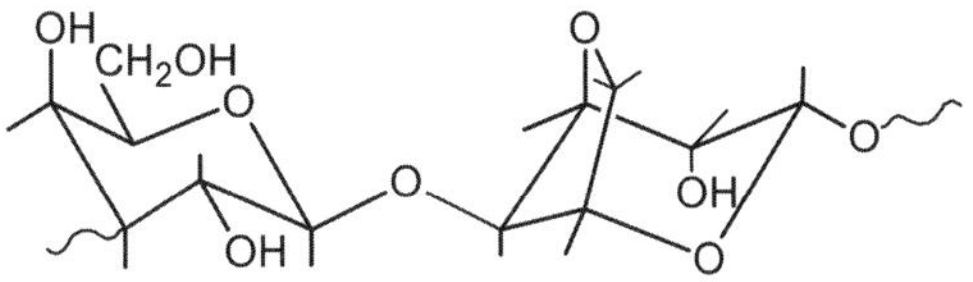

Figure 7.19 The agarose molecule consists of β (1→3)-linked D-galactose and α (1→4)-linked 3,6-anhydro-αL-galactose residues.

form (Norton *et al.*, 1986), just like the carrageenans. That said, some 'agarose' data in the literature has actually been taken using agar rather than agarose, and so care has to be taken when comparing results obtained by various workers.

As mentioned above, the partly purified agar is widely employed as a medium for microbial growth; in this application a mixture of agar and appropriate electrolyte and/or low molecular mass nutrients is dissolved by autoclaving, poured into a Petri dish and allowed to cool and gel; then the microbes of interest can be 'plated out' by gently scratching the agar plate with a needle dipped in the appropriate solution. The agar in the dish therefore serves both to encourage and sustain growth and as a solid support. From this application, some of the mechanical properties of the agarose can already be deduced, i.e. that it gels on cooling and that the resultant gel has sufficient 'strength', i.e. modulus, to act as a rigid support, but that it is brittle (easily scratched) (Matsuhashi, 1990).

7.3.1 Gelation mechanism

Early work on establishing the gelation mechanism and macromolecular structure of agarose gels (Dea *et al.*, 1972) used the technique of optical rotation to investigate what was happening. There is a large and relatively sharp change in the OR, occurring at around 35–40°C, on cooling a heated solution (100°C or higher, in an autoclave) of 0.2 wt% agarose. On reheating the solution (and 0.2 wt% is quite close to the critical gel concentration) there is a very large hysteresis and the OR does not change very significantly until the temperature exceeds 80°C. This clearly suggests that some ordered structure is formed and that, once formed, it is quite stable. However, without the help of other techniques, OR cannot establish the form of the ordered structure unequivocally, although parallels with gelatin and other gelling polysaccharides would suggest some helical structure. That said, a more quantitative re-examination of the OR and its behaviour at lower temperatures (Schafer and Stevens, 1995) strongly supports the double helix picture found from X-ray measurements and described in more detail below.

Wide-angle X-ray measurements on oriented fibres by Rees, Arnott and co-workers (Arnott *et al.*, 1974) indicated that the ordered structure was due to a double helix, so that this was the origin of the gelation process. The particular double helix proposed has 0.95 nm axial periodicity. Each chain in the double helix forms a left-handed, three-fold helix of pitch 1.90 nm and is translated axially relative to its partner by half this distance. This model accounts for the sign and magnitude of the optical rotation shift that accompanies the sol–gel transitions. However, the agarose network is described as arising by double helix formation and the subsequent aggregation of these helices into bundles. This alone

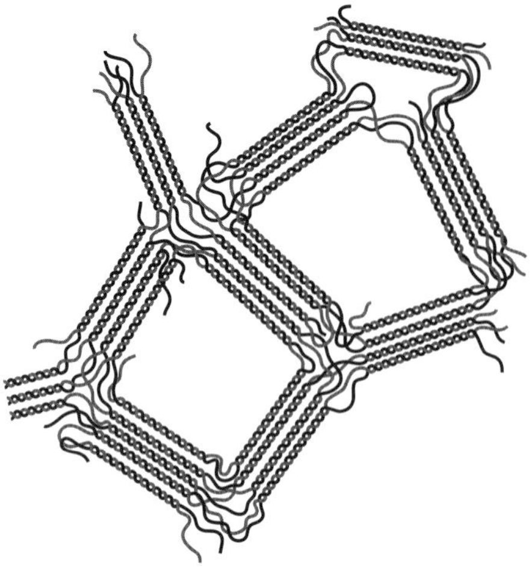

Figure 7.20. Schematic representation of agarose network structure composed of bundles of double helices. Adapted with permission from Arnott *et al.* (1974) © 1974 Elsevier.

would explain the appearance of the gel, usually slightly hazy or turbid because of the light scattered from aggregated helical regions. There is a view that the gelation of agarose is actually driven by the 'rogue residues' of the sulphated form. Without these acting to terminate the double helices, a 100% pure agarose would most likely precipitate out on cooling. However, a more detailed explanation in terms of a spinodal decomposition mechanism is described below.

Consequently the large hysteresis loop between the gelation and melting temperatures of the gels is suggested as indicating the substantial amount of helix aggregation accompanying the network building. The bundles of double helices create a rigid framework, shown in Figure 7.20. SAXR measurements support this view (Djabourov *et al.*, 1989). It is observed by this technique that the scattered intensity scales exactly with concentration between 0.1 and 5 wt% and is not sensitive to time or temperature between 34°C and 4°C, provided the sample is left to 'set' for 1–2 h. The angular distribution of the scattered intensity was found to be consistent with a model based on a fibrous structure containing rod-like fibres of variable thickness. Computed intensity curves assuming two populations of 3 and 9 nm diameters with roughly equal weights (on average 7 times as many thin rods as thick ones) are supported by TEM imaging, which mainly shows the population of thin rods. Using structural parameters (molecular dimensions and mass per unit length) proposed by Arnott *et al.* (1974), it was found that the thin fibres are consistent with a hexagonal packing of double helices, while the larger aggregates consist of assemblies of the thin fibres. The distribution of the aggregation number is likely to induce the thermal hysteresis which is well established in these gels. Because these measurements were not time-resolved, the two steps, consisting of growth of double helices followed by substantial aggregation, could not be separated in this experiment.

DSC work on agarose gels seems to have concentrated on the low-temperature behaviour and on the sol–gel transition. A low-temperature peak was detected below 0°C and associated with 'non-freezable water' (Watase *et al.*, 1988), while other studies have extended this work to examine agarose gels in sucrose and other solutions. Studies of the sol–gel transition include the work by Watase and co-workers (Watase *et al.*, 1989; Nishinari *et al.*, 1992a, 1992b). They noticed an initial endothermic peak at 75°C, shifting to higher temperatures with increasing concentration of agarose gels. An exothermic peak was seen at lower temperatures – around 30°C – and was attributed to the increase in ordered regions. More recent data collected for agarose in sucrose solutions is discussed in Section 7.3.4 below.

7.3.2 Phase behaviour

At a molecular level the helix–coil transition observed by OR and WAXS is well established, but the subsequent aggregation mechanism is still not fully clear, although it has been examined by several authors and a coherent picture is now slowly emerging. Pines and Prins (1973) were the first to report small-angle light scattering data for low-concentration gel systems (0.5–1 wt%). This showed a distinct maximum in the scattered intensity versus the scattering vector, generally ascribed to phase separation occurring by spinodal decomposition (Chapters 8 and 10), and raised the question of the mechanism of gel formation with phase separation occurring at low concentration. Since then, this point has been examined by many papers, in particular by San Biagio and co-workers (San Biagio *et al.*, 1996). They noted two domains, at low and high concentrations. When aqueous agarose solutions at concentration 0.5% w/v were quenched to 47.5°C, their experiments showed de-mixing of the sol and the sol–gel transition as two distinct phenomena, occurring on very different time scales. (Here it is important to appreciate that, although the concentration used in this work was above the expected minimum gel concentration, say, 0.2 wt% at 25°C, the temperature was also higher than normally used in preparing a macroscopic gel, so the effective minimum concentration was increased.)

Within the first 100–200 min, the sol underwent spinodal demixing as detected by several features (Chapter 10), including the occurrence of a forward-scattered light pattern (a small-angle scattering ring), the exponential growth of the scattered light intensity and the linear Cahn plot growth rates. Within this time, the sample remained liquid, with the appearance of a 'transient' viscosity peak. At later times, between 100 and 1000 min, the number of free polymer coils in the sol decreased substantially and a population of 'objects' having a diffusion coefficient two orders of magnitude smaller than that of the free polymer coils was observed. After several days, in sols of various polymer concentrations, the diffusion coefficients of these species showed an equilibrium size, with no sign of macroscopic gelation such as viscoelasticity or turbidity. Macroscopically, the sample remained a liquid. Samples containing objects of sizes 0.2–0.4 μm could be redissolved by heating, however, with the same thermal hysteresis as macroscopic gels. Such objects appeared therefore to be 'mesoscopic' (obtained by spinodal demixing), where local gelation had occurred. At higher concentrations ($c \geq 2\%$ w/v) direct gelation occurred around 70°C and no sign of spinodal demixing was observed.

The phase diagram proposed by San Biagio *et al.* (1996) for the gelation of agarose solutions is similar to the one proposed by Tan *et al.* for atactic poly(styrene) (see

Chapter 8). The authors, however, warn that in their scenario the occurrence of the coil–helix transition underlies gelation and is responsible for the thermal hysteresis, although the evidence suggests that the hysteresis reflects the subsequent aggregation. The coil–helix transition 'alters the solutes' and their interaction with water. Since double helices can be approximated as rod-like particles, the phase diagram of Flory–Onsager type might also be relevant from a qualitative point of view, but the topological constraints among helices prevent the occurrence of an actual nematic ordering (see also the discussion in Chapter 3).

Aymard *et al.* (2001) re-examined the phase diagram at a fixed concentration (2 wt%), considered as a 'high concentration' by the previous authors, with particular emphasis on the thermal history and time dependence, particularly in the range 36–43°C. From their work, it was confirmed that the cooling rate plays a major role on the state of aggregation, as first established in the early work by Dea *et al.* (1972) and in subsequent rheological studies.

Again, the growth rate of the gel modulus, as followed by small-strain dynamic measurements, was dramatically reduced above 35°C, so longer isothermal curing was required for this modulus to tend towards a pseudo-plateau value or to enter a regime of slower growth (up to 6 days at 43°C). Gelation always took place after the final temperature was reached; however, kinetic effects were not totally suppressed, within the finite long-time scale investigated. Rather than a concentration range, the temperature range 35–43°C appears as a boundary between different types of gelation. Turbidity measurements performed at various temperatures by quenching and annealing the sample, as for the rheology, also showed marked differences. Samples quenched above 35°C showed a dramatic change in turbidity, whose wavelength dependence suggested the formation of large-scale fluctuations of concentration in the system, i.e. polymer-rich and polymer-poor regions (Figure 7.21).

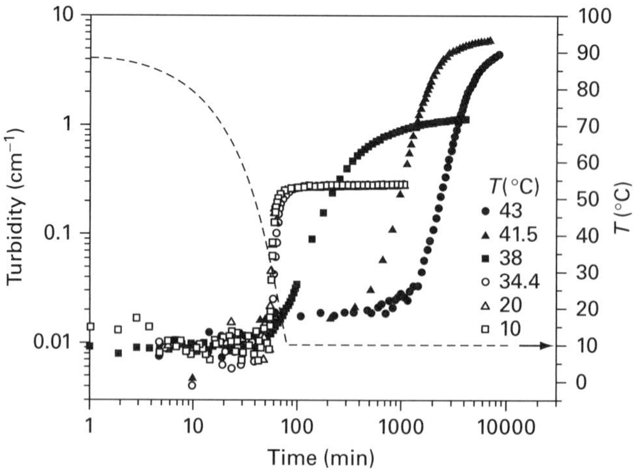

Figure 7.21 Time course of the turbidity measured at 800 nm for an agarose solution at 2 wt%, cured isothermally. The cooling temperature profile is indicated on the right-hand scale. Adapted from Aymard *et al.* (2001) with permission of John Wiley & Sons.

Solutions were transparent at high temperature. Upon cooling to $T<35°C$, a sharp increase in turbidity was observed, due to double helix formation and aggregation of helices into bundles. Major differences were observed when curing occurred at $35°C<T<43°C$. First, a marked delay in the onset of turbidity was apparent, increasing with increasing temperature. Second, the rate of increase in turbidity was much lower than for elasticity. Both effects were qualitatively similar to the apparent gelation time observed in rheology. Finally, much higher turbidity values were observed in this temperature range. The wavelength dependence of turbidity with time, at 41°C and 43°C, was interpreted in terms of heterogeneities. Assuming that the contribution of the heterogeneities dominates the scattering at the end of curing at the high temperature, the use of the Debye–Bueche expression in light scattering experiments yielded an average size of about 350 nm at 38°C and larger than 5 μm at 41.5°C and 43°C. For further confirmation, the fine structures of agarose gels with various thermal histories were also evaluated by TEM. Gels formed at 20°C appeared homogeneous at distances larger than the correlation length, similar to previous reports in the literature. At 43°C, micrographs showed a truly phase separated microstructure with polymer-rich and polymer-poor domains.

These experiments, together with previous investigations, show that gelation and phase separation are competing processes. It is likely that complete demixing would represent the thermodynamic equilibrium for agarose helices, but this is prevented by gelation, which traps the system away from equilibrium. On this basis, the existence of two temperature domains could be explained in terms of a competition between osmotic forces (in favour of a phase separation of the helical conformation) with elastic forces (which tend to prevent this). When solutions were quenched below 35°C, the rapid development of the network of aggregated double helices stopped the demixing process and a pseudo-equilibrium state was reached.

7.3.3 Mechanical properties

Gel mechanical properties have been investigated under both small and large strains or stresses. Agarose gels are easily fractured, and the linear region can only be accessed at lower strains. The vast majority of studies have been carried out on gels with concentrations below, say, 3 wt% (for high M_w starting materials), since producing homogeneous materials at higher concentrations is very difficult. That said, a recent paper on agar (note, not agarose) gels (Ayyad *et al.*, 2010) has used a novel 'hydrothermal' method to produce homogeneous materials of up to 30 wt%. However, this seems to simply involve heating the sealed sample in an oil bath up to 145°C. The authors claim that this does not degrade M_w, and the high values of moduli obtained tend to support this. Extension to higher-quality agarose samples seems to be worthwhile.

7.3.3.1 Effect of concentration

Just as for other physical gelling systems, increasing the concentration of agarose above the minimum required for gelation produces a higher-modulus product. The published data is discussed in more detail below, with both M_w and c varied. To summarize here, the minimum concentration is around 0.2–1 wt%, depending on M_w, but accessing very high concentrations is normally limited by solubility. Attempts to apply the Eldridge–Ferry melting law to these samples are also limited by both solubility and the large gap between formation and remelting. Evidence from Matsuhashi (1990) suggests that, at higher concentrations, the melting point becomes more concentration-independent. That said, some published data exists, including the extensive work by Watase and Nishinari and co-workers. This is discussed in more detail below.

7.3.3.2 Effect of molecular mass

One piece of work (Watase and Nishinari, 1983) was one of the first to investigate the M_w dependence of mechanical properties. Initially they prepared a series of agarose samples, at a fixed concentration but with a range of M_w from 3.4 to 48.5 × 10^4 g mol^{-1}, and measured the time-dependent Young's modulus $E(t)$ by stress relaxation. As expected, at low M_w there was a tendency for some stress to relax, but this was not seen for the highest M_w sample. They also measured the storage Young's modulus E' for the same samples by the longitudinal vibration method, so avoiding slippage. Values were found for all except the highest M_w samples, where solubility again appeared to be a problem. They then used Poisson's ratio (0.5) to evaluate the corresponding shear moduli, and obtained values of G' as a function of both concentration and M_w. The data, when plotted as log G' versus log c, shows the expected behaviour, that is, not a single power law but a curve with an increasing slope as c approaches the minimum concentration c_0. At the highest concentrations, the slope appears to be slightly less than 2. In a separate study, Tokita and Hikichi (1987) analysed the concentration range closer to c_0 with a power-law model and found the expected higher exponent, in their case around 4.

An attempt was made by Watase and co-workers (Watase *et al.*, 1989) to analyse the overall data using a two-dimensional branching theory model (i.e. both c and M_w fitted simultaneously). The critical concentrations found in this 'global' fit were a decreasing function of M_w, as might be predicted, but were found to lie approximately in the range 0.02–2 wt%. For the higher M_w samples, this was substantially below that found by analysing each M_w set separately (0.2–2%). In the latter case the log M_w dependence of log c_0 was close to the ideal value of −1. Perhaps unsurprisingly in view of the modulus and temperature behaviours reported above, analyses of the relaxation spectra of agarose gels of the different molecular masses reported by Watase and Nishinari (1983) showed a clear dependence on M_w, the highest M_w samples extending into the longer time region.

Work by Normand and co-workers (Normand *et al.*, 2000) has re-examined both the small- and large-deformation behaviour of agarose gels for three different M_w samples and a range of concentrations. Their small-deformation rate data has been used to superimpose the gelation curves – a method upheld by these workers for other systems – but they report an observed modulus concentration exponent of 2.1, although the usual curvature is seen. In fact a more useful comparison comes from their initial large-deformation measurements, and they report a slope of log E (measured under both tension and compression, although the exponents for the same samples differ slightly) versus log c of 2.6 at low concentrations and 1.5 at the highest concentrations. More detailed discussion of these results is given in Section 7.3.3.5.

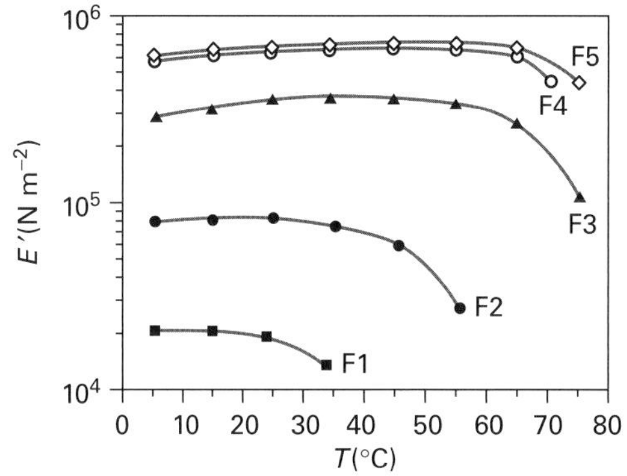

Figure 7.22 Temperature dependence of storage modulus E' for five agarose fractions with M_w from 3.4×10^4 g mol^{-1} (F1) to 48.5×10^4 g mol^{-1} (F5). Reprinted with permission from Nishinari and Watase (1993) © 1983 Springer.

7.3.3.3 Effect of temperature

Nishinari and co-workers (Nishinari *et al.*, 1992a, 1992b) investigated the temperature dependence of the modulus – mainly the storage Young's modulus E' – over the range 5°C to 80°C. For lower M_w samples, this range tended to exceed the melting temperature, but generally the modulus showed a slight increase before the expected decrease in the melting range. Perhaps surprisingly in view of the structure, these moduli showed rubber-like behaviour, i.e. proportionality to absolute temperature in the lower temperature range (Figure 7.22). These results can be analysed in terms of both entropic and enthalpic contributions to the elasticity, with the former dominating at lower temperatures and the latter at higher temperatures. The melting behaviour, particularly as investigated by DSC, is amenable to treatment by the 'zipper model' (Chapter 3).

7.3.3.4 Other characterization work

In solution, the overlapping concentration c^* for agarose was estimated by Aymard *et al.* (2001) to be 0.6 wt% and by San Biagio *et al.* (1996) to be 1.5 wt%; both values are also in the range of gelatin c^* (0.5 wt%). The gelatin chain is more flexible than agarose; therefore c^* should be lower in agarose at equivalent molecular masses. The gel mesh size determined by Aymard *et al.* (2001) in the homogeneous state of gels at $c=2$ wt% is about 100 nm, while at these concentration it is only a few nm in gelatin. This indicates that there are greater void spaces in agarose gels than in gelatin gels.

7.3.3.5 Large-deformation behaviour of agarose gels

The data here is also somewhat limited, and few workers have used a rigorous geometry set-up so that the actual stress and strain are obtainable. Here we summarize results obtained by a number of workers (Watase and Nishinari, 1983; McEvoy *et al.*, 1985b; Normand *et al.*, 2000). The Watase results were obtained with the same stress relaxation apparatus used in the small-deformation work, and they studied the various M_w samples, but at a fixed concentration (2.4 wt%). In all cases failure occurred at comparatively small strains, but at the highest M_w the strain to fail extended to *c*.20%. This is, of course, much lower than found for gelatin, confirming the more brittle nature of this material, but up to this point the stress–strain curve is close to linear, i.e. there is no indication of yield behaviour.

The McEvoy work used the 'racetrack' sample method described above for gelatin, for 1 wt% and 2 wt% samples. Results are in good agreement with the Watase results, as far as the stress–strain behaviour and ultimate properties are concerned. From the deformation rate–dependent failure envelope, strain to break fell in the range 8–30%. Data for gelatin at approximately the same c/c_0, the appropriate comparison in this case, shows failure occurring at much higher strains (40–100%).

The small-deformation work by Normand *et al.* (2000) was mentioned above. For large-deformation measurements in compression they used pre-moulded cylinders and lubricated the test bed; in tension they used dumbbell-shaped cutters applied to gelled sheets. Data for the two deformation modes gives fair agreement in terms of initial modulus, but failure of cylindrical samples in compression is always ambiguous. In tension their results compare reasonably well with those described above, with a linear stress–strain trace up to the point where failure occurs. They comment that strain to fail is largely invariant with concentration at *c*.40 wt%, but regrettably they did not report measurements of failure at various rates, or construct the relevant failure envelope.

7.3.4 Solvent effects

The state of aggregation can be effectively modified by changes of the solvent, as was first proposed by Watase *et al.* (1990): upon addition of sucrose, agarose gels lose their turbidity and eventually become transparent, presumably partly a refractive index matching effect and partly structural, as discussed below. These phenomena were first investigated in detail by Nishinari and co-workers (Watase *et al.*, 1990; Nishinari *et al.*, 1992a, 1992b). In more recent experiments reported by Normand *et al.* (2003), large amounts of sucrose were added in water, and calorimetric and rheological measurements were performed for gels with various molecular masses and concentrations. A distinct exothermic peak, generally attributed to double helix formation, was observed during setting (cooling) of 2 wt% solutions. As sucrose was increased from 0 to 60 wt% the enthalpy remained almost constant at $\Delta H = 10$–11 kJ mol^{-1} per residue, but decreased by 20% at 70 wt% of sucrose. This effect can be ascribed either to a sudden decrease of the enthalpy of the coil–helix transition or, more likely, to a decrease in the helix fraction. Increasing sucrose concentration in the solvent up to 60 wt% leads to an increase in the apparent gelation temperature, the rate of gel formation and the elastic modulus obtained at long times. Within this range of sucrose concentrations, sucrose seems to accelerate the gelation process. However, when sucrose concentration exceeds 60% the reverse trend is observed and the gelation process seems to slow dramatically. A monotonic increase in the loss modulus at long times is observed with increasing sucrose content, in the binary solvent. This is consistent with previous observations, which report that sucrose reduces the quantity

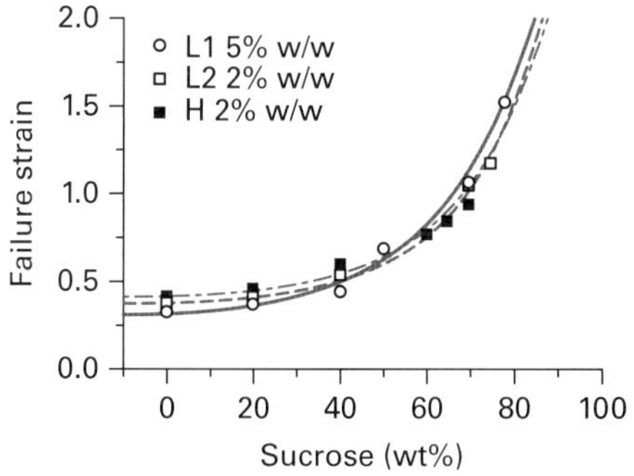

Figure 7.23 Dependence of the failure strain on sucrose concentration for various molecular masses (11.6×10^4 g mol^{-1} for sample L1, to 16.2×10^4 g mol^{-1} for sample L2 and 16.2×10^4 g mol^{-1} for sample H) and two concentrations. Adapted with permission from Normand *et al.* (2003) © 2003 Elsevier.

of helical aggregates, increases the number of cross-links and decreases the length of the helices (Watase *et al.*, 1990; Nishinari *et al.*, 1992a, 1992b).

Agarose helices are known to be stabilized by hydrogen bonds with water molecules, some occupying the inner part of the helix (Foord and Atkins, 1989) as in collagen (although we note that their model for agarose was single helical). The reduction of available water molecules decreases their contribution to the stabilization of the helices, which explains the decrease in the coil–helix conversion degree. While the aggregation number of the double helices decreases, the gels can be deformed to a much higher extent, an effect ascribed to increased internal flexibility of the ordered chains, possibly as a consequence of the decrease in cross-sectional thickness.

The effect of sucrose on the strain at failure is very important: increasing the sucrose concentration leads to a transition from brittle gels (about 40% strain at rupture) to highly deformable gels (about 200%), somewhat unusually for polysaccharide gels; see Figure 7.23. As the sucrose concentration increases, the strain at failure increases as well, although mainly above 60 wt% sucrose when a larger proportion of coils are in the non-helical conformation.

7.4 Comparison between helical type networks

To conclude, it is interesting to compare the small-strain limit shear moduli of gelatin and agarose, both of which, in aqueous solution, can be considered as fibrillar type networks constructed of helices. Such a comparison was made by Clark *et al.* (1983), and is reported in Figure 7.24 over a wide concentration range, up to 25 wt% in the case of gelatin. The plot highlights the fact that agarose gels have a higher modulus than gelatin at low concentrations. For instance, at $c \approx 2$ wt% the modulus is about 40 Pa for gelatin and 35 kPa for agarose.

As has been pointed out in Chapter 3, the shapes of modulus versus concentration curves are very similar but, as can be judged from Figure 7.24, they differ in both

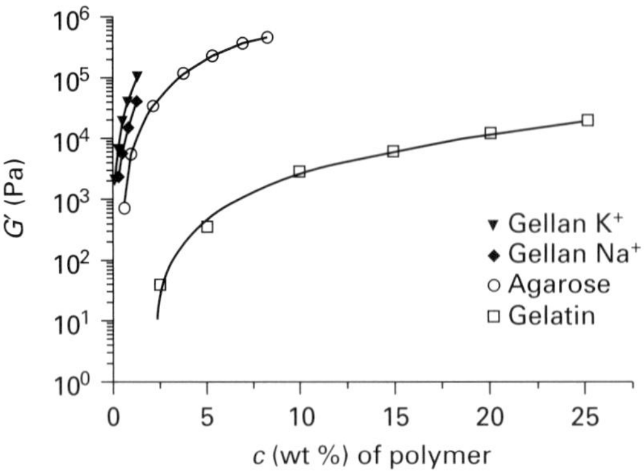

Figure 7.24 Comparison between storage moduli for agarose, gelatin and gellan at various polymer concentrations. Adapted with permission from Clark *et al.* (1983) © 1983 American Chemical Society, and from Milas and Rinaudo (1996) ©1996 Elsevier.

horizontal and vertical scaling factors. The horizontal factor largely reflects the value of the critical concentration c_0, which in this case is around 0.17% w/w for agarose and 1.2% w/w for gelatin. As discussed in Chapter 3 and elsewhere, the value of c_0 is by no means universal – for example, it is lower for a sample of mammalian gelatin formed at 5°C than for the same sample formed at 20°C, and will also depend upon the cooling regime.

However, even with appropriate horizontal scaling, the data for gelatin and agarose will never superimpose because it appears that the modulus of an agarose gel is substantially greater at the same value of c/c_0. In most approaches this can be attributed to differences in the fibrillar structure of agarose and gelatin, as we discuss below. For more rigid network structures such as agarose, elasticity is very unlikely to be entropic. In other words, the contribution per cross-link comes from bending and twisting contributions, which raise the enthalpy of the system. Consequently we adopt the term 'enthalpic' and, in one approach, assign its contribution to a generalized front factor, sometimes denoted a, which for agarose is estimated to be $\gtrsim 10$ (Clark and Ross-Murphy, 1987).

Because of the careful work on gelatin described above, the generalized behaviour of shear modulus on helix concentration is now firmly established. However, without further hypotheses this does not tell us what we need to know to establish the concentration of *junction zones*, and without this no a priori estimate of the shear modulus can be made simply from the available structural information. The contribution of the local rigidity of junctions in the elasticity of rubber-like networks is not elucidated.

The mesh sizes for agarose and gelatin gels at equal concentration ($c \sim 2$ wt%) are, respectively, 100 nm and only a few nanometres; the latter contains an entangled network with both triple helices and coils. The fibrils in agarose gels are rigid and non-entangled; therefore, without modification the models proposed for gelatin should not apply to agarose. In the latter case, the rigidity of the network comes from both:

- long and rigid rods with variable thicknesses, constituted from the aggregation of double helices

- junctions between the rods which are probably much less flexible than gelatin loops or coils. The non-helical content of agarose gels is probably very low, although this is not discussed in the literature, and may be related to defects in the structure of the polysaccharide ('rogue residues'). Both of these 'stiffening' (i.e. enthalpic) effects will tend to contribute to the greater modulus of agarose gels even if the data is replotted in terms of c/c_0.

For comparison, data for another helical type network, the gellan gel presented in Chapter 5, is shown in Figure 7.24. The data shows the storage moduli derived from Young's modulus E with the simple assumption $G = E/3$ at room temperature, and after 24 h maturation (Milas and Rinaudo, 1996). E values were calculated from the linear region of the stress–strain diagram, for gels cut into small cylinders. The counterions from commercial samples were carefully replaced and the results shown in Figure 7.24 contain either K^+ or Na^+ counterions with $M_w = 2.5 \times 10^5$ g mol^{-1}. The authors also explain the role of cooling history on gel properties. It is clear that under these conditions very high shear moduli were observed in the gellan networks compared to the other helical networks. Here the gellan concentration did not exceed 1.3 wt% but the moduli are extremely high, K^+ counterions giving the highest values.

Most helical type networks show very high moduli, compared for instance to chemically cross-linked gels, and ionic effects give additional rigidity to the helical aggregates. In the case of gelatin, if we choose the master curve (Figure 7.15), then helix concentration is the only parameter that controls the storage modulus of the gels. In this case the gelatin plot is shifted to lower concentrations because the minimum concentration of helices is about 0.3 wt%. For agarose and gellan gels, OR data suggests that the helix fraction is very high, so the polymer concentration is very close to the helical concentration. Again, however, whether the high modulus from the high helix fraction is due solely to the same factors as for agarose or whether the effect of ions is simply to increase the concentration of junction zones is currently too difficult to establish.

7.5 Conclusions

A gelatin network is made of portions of triple-helical rods of constant cross-section and variable length, and along the chain contour there are more flexible coiled sequences. The conformational change from ordered to disordered (triple helix to coil) occurs cooperatively, at a specific temperature. Taking into account the complexity of gelatin samples, "a specific temperature" cannot be a single temperature. Such a gel is never at equilibrium, and experimentally the growth in helix fraction continues indefinitely, as can be observed by several techniques. There is only a small degree of thermal hysteresis between gel formation and gel melting temperatures, even though this depends on many factors. Overall, the parameter controlling the gel elasticity is the helix concentration.

By contrast, an agarose network is made of aggregated fibrils of double helices, with variable aggregation numbers and/or variable cross-section of fibrils. There is no clear indication of the cross-linking mechanism, but it is most likely to be bridging by individual chains participating in various helical sequences. Again, the conformational change is observed to occur quite suddenly over a narrow range of temperatures, but currently there is no clear way to separate the two aggregation stages, helical formation and lateral aggregation during gelation. A very large hysteresis appears between gel formation and melting temperatures of the gels. The state of aggregation, however, can be modified by changes in the solvent composition (sucrose). Elasticity is controlled by concentration and molecular mass in aqueous solution. Yet another example of a helical network is provided by gellan, which apparently shows an additional enhancement of the shear modulus from ionic effects (Chapter 5). The mechanism of increase of the shear modulus in this case is not well known and may be due either to a larger size of aggregates or to a loss of their flexibility.

References

Arnott, S., Fulmer, A., Scott, W. E. *et al.*, 1974. *J. Mol. Biol.* **90**, 269–272.

Aymard, P., Martin, D. R., Plucknett, K. *et al.*, 2001. *Biopolymers* **59**, 131–144.

Ayyad, O., Muñoz-Rojas, D., Agulló, N., Borrós, S., Gómez-Romero, P., 2010. *Soft Matter* **6**, 2389–2391.

Balberg, I., Binenbaum, N., 1987. *Phys. Rev. A* **35**, 5174–5177.

Balberg, I., Anderson, C. H., Alexander, S., Wagner, N., 1984. *Phys. Rev. B* **30**, 3933–3943.

Bot, A., Wientjes, R. H. W., deHaas, K. H., 1997. *Imaging Sci. J.* **45**, 191–196.

Busnel, J. P., Morris, E. R., Ross-Murphy, S. B., 1989. *Int. J. Biol. Macromol.* **11**, 119–125.

Cantor, C. R., Schimmel, P. R., 1980. *Biophysical Chemistry, Part I: The Conformation of Biological Macromolecules*. W. H. Freeman, New York.

Clark, A. H., Ross-Murphy, S. B., 1987. *Adv. Polym. Sci.* **83**, 57–192.

Clark, A. H., Richardson, R. K., Ross-Murphy, S. B., Stubbs, J. M., 1983. *Macromolecules* **16**, 1367–1374.

Coppola, M., Djabourov, M., Ferrand, M., 2008. *Macromol. Symp.* **273**, 56–65.

Dea, I. C. M., McKinnon, A. A., Rees, D. A., 1972. *J. Mol. Biol.* **68**, 153–172.

Dickerson, R. E., Geis, I., 1969. *The Structure and Action of Proteins*. Harper & Row, New York.

Djabourov, M., Leblond, J., Papon, P., 1988a. *J. Phys. France* **49**, 319–332.

Djabourov, M., Leblond, J., Papon, P., 1988b. *J. Phys. France* **49**, 333–343.

Djabourov, M., Clark, A. H., Rowlands, D. W., Ross-Murphy, S. B., 1989. *Macromolecules* **22**, 180–188.

Eldridge, J. E., Ferry, J. D., 1954. *J. Phys. Chem.* **58**, 992–995.

El Harfaoui, N., Djabourov, M., Babel, W., 2007. *Macromol. Symp.* **256**, 149–157.

Ferry, J. D., 1948. Protein gels. In *Advances in Protein Chemistry*. Academic Press, New York, pp. 1–78.

Ferry, J. D., Eldridge. J. E., 1949. *J. Phys. Chem.* **53**, 184–196.

Foord, S. A., Atkins, E. D. T., 1989. *Biopolymers* **28**, 1345–1365.

Godard, P., Biebuyck, J. J., Daumerie, M., Naveau, H., Mercier, J. P., 1978. *J. Polym. Sci., Part B: Polym. Phys.* **16**, 1817–1828.

Harrington, W. F., Rao, N. V., 1967. Pyrrolidine residues and stability of collagen. In Ramachandran, G. N. (ed), *Conformation of Biopolymers*. Academic Press, New York, pp. 513–531.

Herning, T., Djabourov, M., Leblond, J., Takerkart, G., 1991. *Polymer* **32**, 3211–3217.

Higgs, P. G., Ross-Murphy, S. B., 1990. *Int. J. Biol. Macromol.* **12**, 233–240.

Joly-Duhamel, C., Hellio, D., Ajdari, A., Djabourov, M., 2002a. *Langmuir* **18**, 7158–7166.

Joly-Duhamel, C., Hellio, D., Djabourov, M., 2002b. *Langmuir* **18**, 7208–7217.

Jones, J. L., Marques, C. M., 1990. *J. Phys. France* **51**, 1113–1127.

Leick, A., 1904. *Ann. Phys.* **319**, 139–152.

Matsuhashi, T., 1990. Agar. In Harris, P. (ed.), *Food Gels*. Elsevier Applied Science, London, pp. 1–51.

McEvoy, H., Ross-Murphy, S. B., Clark, A. H., 1985a. *Polymer* **26**, 1483–1492.

McEvoy, H., Ross-Murphy, S. B., Clark, A. H., 1985b. *Polymer* **26**, 1483–1492.

Milas, M., Rinaudo, M., 1996. *Carbohydr. Polym.* **30**, 177–184.

Nishinari, K., 1976, *Jap. J. Appl. Phys.* **15**, 1263–1270.

Nishinari, K., Watase, M., 1993. *Rep. Prog. Polym. Phys Jpn.* **36**, 661–664.

Nishinari, K., Watase, M., Kohyama, K. *et al.*, 1992a. *Polym. J.* **24**, 871–877.

Nishinari, K., Watase, M., Kohyama, K. *et al.*, 1992b. *Rep. Prog. Polym. Phys Jpn.* **35**, 169–172.

Normand, V., Lootens, D. L., Amici, E., Plucknett, K. P., Aymard, P., 2000. *Biomacromolecules* **1**, 730–738.

Normand, V., Aymard, P., Lootens, D. L. *et al.*, 2003. *Carbohydr. Polym.* **54**, 83–95.

Norton, I. T., Goodall, D. M., Austen, K. R. J., Morris, E. R., Rees, D. A., 1986. *Biopolymers* **25**, 1009–1029.

Peniche-Covas, C. A. L., Dev, S. B., Gordon, M., Judd, M., Kajiwara, K., 1974. *Faraday Discuss. Chem. Soc.* **57**, 165–180.

Pezron, I., Djabourov, M., Bosio, L., Leblond, J., 1990. *J. Polym. Sci., Part B: Polym. Phys.* **28**, 1823–1839.

Pezron, I., Djabourov, M., Leblond, J., 1991. *Polymer* **32**, 3201–3210.

Philipse, A. P., Wierenga, A. M., 1998. *Langmuir* **14**, 49–54.

Pines, E., Prins, W., 1973. *Macromolecules* **6**, 888–895.

Ramachandran, G. N., Kartha, G., 1955. *Nature* **176**, 593–595.

Ramachandran, G. X., Chandrasekharan, R., 1968. *Biopolymers* **6**, 1649–1658.

Rich, A., Crick, F. H. C., 1955. *Nature* **176**, 915–916.

Ross-Murphy, S. B., 1992. *Polymer* **33**, 2622–2627.

San Biagio, P. L., Bulone, D., Emanuele, A., Palma-Vittorelli, M. B., Palma, M. U., 1996. *Food Hydrocolloids* **10**, 91–97.

Schafer, S. E., Stevens, E. S., 1995. *Biopolymers* **36**, 103–108.

te Nijenhuis, K., 1981a. *Colloid Polym. Sci.* **259**, 522–535.

te Nijenhuis, K., 1981b. *Colloid Polym. Sci.* **259**, 1017–1026.

te Nijenhuis, K., 1997. *Adv. Polym. Sci.* **130**, 1–252.

Tokita, M., Hikichi, K., 1987. *Phys. Rev. A* **35**, 4329–4333.

Ward, A. G., Courts, A. (eds), 1977. *The Science and Technology of Gelatin*. Academic Press, London (UK).

Watase, M., Nishinari, K., 1983. *Rheol. Acta* **22**, 580–587.

Watase, M., Nishinari, K., Hatakeyama, T., 1988. *Food Hydrocolloids* **2**, 427–438.

Watase, M., Nishinari, K., Clark, A. H., Ross-Murphy, S. B., 1989. *Macromolecules* **22**, 1196–1201.

Watase, M., Nishinari, K., Williams, P. A., Phillips, G. O., 1990. *J. Agric. Food Chem.* **38**, 1181–1187.

Zimm, B. H., Bragg, J. K., 1959. *J. Chem. Phys.* **31**, 526–535.

8 Gelation through phase transformation in synthetic and natural polymers

8.1 Introduction

Physical cross-links can be created when solutions undergo a phase transformation (Keller, 1995), but the gels arising specifically from such a phase transformation do not usually correspond to a stable state. However, phase transformations tend to lead to network formation under specific circumstances:

- when phase transformation is incomplete: for instance, a crystalline polymer fraction coexists with an amorphous part
- when a newly formed macroscopic phase does not develop, owing to kinetic constraints
- when connectivity is achieved either between individual molecules (chains leading from one junction site to another) or by the structured phase morphology, not specifically related to the phase transformation itself
- when phase diagrams are needed to understand the gel formation but are insufficient to predict this gelation
- when, overall, physical gels arising through phase transformation are in a non-equilibrium state.

At its final stage, liquid–liquid phase separation would generate two fluid layers; in a liquid-to-crystal phase transition the final state would be a crystal with macroscopic dimensions. In order to obtain a gel, the final stages of the phase separations are not reached and, instead, the phase separations are arrested at a certain point where the network is formed. Every case reported in this chapter illustrates a different situation, although the majority of the examples examined are of synthetic polymers.

8.1.1 Crystallization

Since crystallization was regarded initially as a principal route for physical gelation, specifically in the case of synthetic polymers in organic solvents, we first recall the conditions for crystallization to occur from a solution. In the case where polymer and solvent are miscible in all proportions at high temperatures, under supercooling conditions crystals can form below the liquid-to-crystalline solid phase limit. These crystals from supercooled solutions normally appear as suspended particles, giving a turbid appearance to the solution. The state of maximum stability is that of a macroscopic crystal, and the usual polymer crystallization leads to a crystal–amorphous structure in a single-component system (polymer melt) in a non-equilibrium state. In the gels, the non-equilibrium 'structure' expands throughout the solvent. Such systems, not being in equilibrium, eventually collapse by expelling the solvent (the so-called syneresis effect), but the non-equilibrium state can be long-lived, thus retaining the appearance of a gel state for weeks or months. The explanation for the growth restriction on new crystals, and the limitation of size for the existing crystals, is a major question which has to be solved to understand the persistence of the gel state.

In block copolymers, the sequences that can crystallize under certain circumstances form the network junctions and the rest of the chain the connecting paths. Gelation should then proceed until all the crystallizable portions of the copolymer have formed crystal junctions, and so the gel state would approach true equilibrium for the micro-structure. This is indeed observed for micelle formation in aqueous solutions of some amphiphilic copolymers under certain conditions (Chapter 6). In a homopolymer the factors that can influence the amount of crystallizable material are the tacticity of the chain, type of solvent, temperature and time (since this is a non-equilibrium state).

Tacticity, which can be measured directly using proton or ^{13}C NMR spectroscopy, refers to polymers with one non-hydrogen substituent (R) attached to the carbon chain per monomer unit. In isotactic macromolecules all these substituents are located on the same side of the macromolecular backbone; in syndiotactic macromolecules they have alternate positions along the chain; in atactic macromolecules all Rs are placed randomly along the chain. In principle atactic polymers cannot crystallize, but this is not always the case. Chain tacticity will be thoroughly examined for a number of gel systems in this chapter.

8.2 'Crystallization'-induced gelation: poly(vinylchloride) (PVC) gels

Poly(vinylchloride) is the third most widely produced commercial plastic, after poly(ethylene) and poly(propylene). PVC is widely used in construction because it is cheap, durable and easy to assemble. It can be made softer and more flexible by the addition of plasticizers, the most widely used being phthalates. In this form, it is used in clothing and upholstery and to make flexible hoses and tubing, flooring and electrical cable insulation.

PVC is usually produced by radical polymerization at 50°C, and under these conditions the material is *atactic*. Using various heating protocols between −60°C and +90°C during synthesis, Ceccorulli *et al.* (1977) investigated their influence on PVC tacticity and degree of crystallinity. They observed that changes of tacticity produced changes of glass transition temperature (T_g) in the bulk material, and showed that *partially syndiotactic* samples have a higher T_g than nearly atactic samples. Such samples are synthesized at low temperature (low-temperature PVC) and have a higher degree of crystallinity. After extensive thermal cycling of the bulk material around 230°C, the difference in T_g persisted between low- and high-temperature PVCs, while the crystalline content was melted out.

The formation of PVC gels is strongly related to tacticity. PVC physical gels are generally placed in the class of crystallization-induced gelation processes. In order to compare gel properties obtained from various investigations, it is important to specify the polymer tacticity, molecular mass and solvent. Increasing the proportion of syndiotactic sequences, the crystallizable chain portions of the polymer, themselves analogous to the copolymer sequences mentioned before, are expected to enhance the elasticity of the gels. Indeed, even in very early investigations it was noticed that PVC can form gels with relatively high modulus in a large variety of solvents. Reports on the gelation of PVC date back to the late 1940s (Stein and Tobolsky, 1948; Alfrey *et al.*, 1949; Walter, 1954).

Guerrero and co-workers (Guerrero *et al.*, 1980; Guerrero and Keller, 1981) reported that the crystallinity of PVC gels was strongly influenced by polymer–solvent interactions. Their results suggest that the gelation ability of various PVC compositions was not only related to stereo-regularity, but also to this polymer–solvent interaction, i.e. the hydrogen bonding ability of the solvent. The physical junctions of PVC gels could therefore arise from two different origins:

(a) crystalline junctions related to the degree of stereo-regularity
(b) polymer–solvent complexes.

The former are the most stable, melting at high temperatures, while the latter melt at low temperatures. Most studies, especially those using X-ray diffraction (Chapter 2), were carried out on dried gels and stretched films (either wet or dried); such processing may induce rearrangements of the initial structure of the gel, and should therefore be confirmed by *in situ* measurements of the gel properties. In order to characterize the gel state, Mutin and Guenet (1989) synthesized samples with variable proportions of syndiotactic sequences (different temperatures of synthesis) and used different solvents. High-temperature PVC contained 0.33 syndiotactic groups and low-temperature PVC 0.39. The authors first noticed that, in the same solvent, highly syndiotactic PVC possesses a much higher elastic modulus (five times higher) than atactic PVC, consistent with assumption (a).

These workers established a temperature–concentration phase diagram for high-temperature PVC (HTPVC) gels in diethylmalonate (DEM) gels aged for 24 h, in which the two types of gels could be identified (Figure 8.1).

Their Gel I systems have a range of thermal stability which extends from +50°C to 100–150°C, above which they melt out and the solution returns to a liquid-like (sol) state, while Gel II systems exist between −50°C and +50°C. According to the authors, Gel I contains only syndiotactic microcrystals and melts at high temperature (gel terminal melting), this temperature being higher for PVC with long syndiotactic sequences. Gel II contains two types of junctions, syndiotactic microcrystals and independently an organized form of the polymer–solvent complex.

The transition from Gel II to Gel I occurs at a constant temperature, which argues in favour of the formation of a complex with a fixed solvent/monomer ratio. Mutin and Guenet propose the formation of a bridge between two carbonyl groups C=O of the diethyl ester and two H–Cl–C groups which involve the methyne hydrogen of two

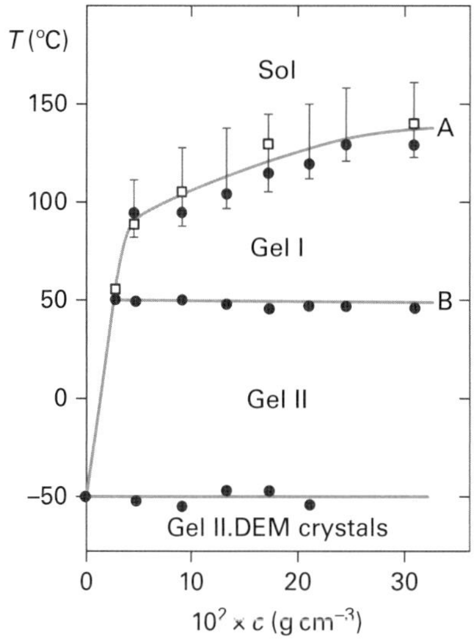

Figure 8.1 Temperature–concentration phase diagram of HTPVC in dimethylmalonate gels aged for 24 h at 20°C: differential scanning calorimetry (DSC) experiments at 20°C min^{-1} (•); ball-drop method at 2°C min^{-1} (□). Adapted with permission from Mutin and Guenet (1989) © 2002 American Chemical Society.

different PVC chains, and where the interaction has an electrostatic origin. As a consequence, gel elasticity should be enhanced by the polymer–solvent complex.

Although several authors agreed with this general picture, there are a number of points which are still unclear. First of all, DSC showed no clear endotherm (melting peak) when the HTPVC gels were heated, either in diethylmalonate or other solvents in well-controlled conditions, at the limit of 100–150°C, around the melting temperature expected for the crystalline component. Depending on sample preparation, the thermograms observed by Mutin and Guenet were quite different. (i) The thermogram of a freshly prepared gel always contained one endotherm, corresponding to gel melting as measured independently by the ball-drop method (data shown in Figure 8.1). This temperature was independent of the solvent, but extrapolation of the melting area of the peaks at zero heating rates gave a zero melting enthalpy! (ii) Reheating a gel that had been molten and cooled to 20°C in the DSC pan showed no endotherm. Further, this endotherm did not reappear, even after several days of ageing at room temperature. In addition, no formation exotherm could be detected, unlike with other gelling systems. In Figure 8.1 the thermal limits of existence of Gel I were confirmed by the ball-drop method, the method detecting the softening of the elastic gel above 100°C. However, the crystalline fraction was difficult to quantify. WAXD diffraction measurements by Hong and Chen (1998) on PVC with a syndiotactic content of 0.30 to 0.35 gelled in bromobenzene (BrBz) and dioxane (DOA) showed only amorphous peaks, indicating that the size of the microcrystals was too small to produce a crystalline diffraction pattern in the X-ray intensity curves. However, the peak of amorphous scattering in PVC–DOA gel was much sharper than that in PVC–BrBz gel, implying that PVC–DOA gel has a thicker network structure as a result of a larger degree of polymer aggregation, i.e. a heterogeneous network structure.

Scherrenberg *et al.* (1993) performed SANS experiments on PVC–tetrahydronaphthalene-d_{12} (TDN) gels, a solvent considered as a plasticizer for PVC. SANS patterns exhibit a distinct interference maximum but, surprisingly, no dependence of the interference maximum on the volume fraction plasticizer or the polymer tacticity was observed. This, the authors suggested, could indicate an inhomogeneous distribution of plasticizer in the gel. Under uniaxial deformation, plasticized PVC with a high syndiotactic content (0.50–0.58) and a high concentration (50% w/w) shows a distinct 'butterfly' scattering pattern which indicates the formation of concentration inhomogeneities as a consequence of the random distribution of the crystalline tie-points. Nevertheless, WAXS experiments on the stretched plasticized PVC samples showed no orientation of the crystalline and/or non-crystalline structure. The authors concluded that there is a 'superstructural order' in plasticized PVC, directly associated with the network-like mechanical properties.

The question of gel morphology and its implications for the solid-like behaviour remains difficult. Electron microscopy images obtained by critical-point drying of PVC gels by Cho and Park (2001) with mono- or di-ester type solvents (dibutyl phthalate (DBP) or butyl benzoate (BB)) in both cases showed an extended three-dimensional network of a fibrillar type (fibres with diameters 30–50 nm), in contrast with the assumption of point-like junctions inferred from the amorphous X-ray scattering patterns. In order to elucidate the origin of elasticity of PVC gels, it is interesting to compare DSC and rheological measurements performed on the same system. The work of Cho and Park allows this comparison.

Figure 8.2 shows G' and $\tan\delta$ versus temperature in a heating ramp, where the sample was prepared by dissolution at 160°C and was stabilized at room temperature. First, it is noteworthy that, after a long ageing period of 7 days, gels containing the diester solvent DBP show a softening transition near 50°C whereas no such transition is observed in monoester solvents or for fresh diester gels. The DSC traces corresponding to the diester and monoester gels are shown in Figure 8.3: indeed, at the larger heating rate in DSC, very likely to shift the transition temperatures compared to the low heating rates used in Figure 8.2, there is a small endothermic peak in the temperature range 60–70°C.

In the rheological experiments, the modulus decreased at around 50°C but $\tan\delta$ was very small, meaning that G'' is very low. Assuming that this peak corresponds to the melting of the polymer–solvent complex or Gel II–Gel I transition in Figure 8.1 (with dimethyl malonate as solvent), in the next range of temperatures, between 60°C and 130°C, the gel elasticity would result only from crystalline junctions and should be stable until the crystals themselves melt at high temperatures. Figure 8.2 shows that the modulus decreases progressively, while the crystals do not melt before 140°C, corresponding to the conditions for fully dissolving PVC. The value of $\tan\delta$ is low and almost constant, thus both G' and G'' decrease. The softening in this region is much enhanced; following the first wave of softening at low temperatures (60°C) (factor of 2), G' is lowered overall by a factor of 200, i.e. more than two orders of magnitude. It is therefore necessary to try and understand why the modulus changes. If there is no crystal melting spread over this very large range of temperatures (at least 70°C), it means that G' has a

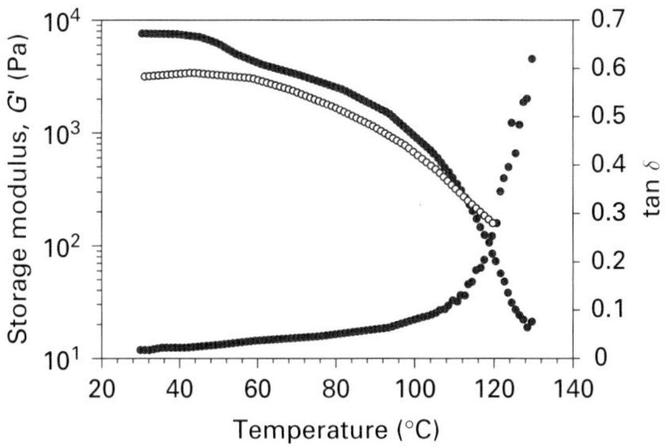

Figure 8.2 Storage modulus and tan δ versus temperature for aged (•) and fresh (○) gels of PVC–DBP, $c = 20$ wt%, with a heating rate of 1°C min^{-1}. Reproduced from Cho and Park (2001) with permission from John Wiley & Sons.

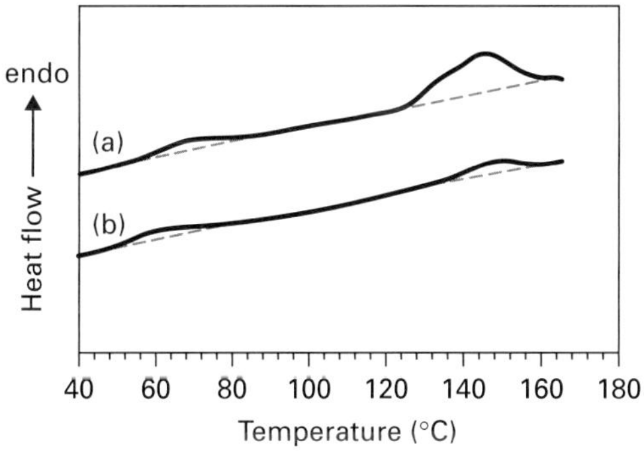

Figure 8.3 DSC thermograms of aged gels at $c = 20$ wt% for (a) PVC–DBP and (b) PVC–BB, with a heating rate of 20°C min^{-1}. Reproduced from Cho and Park (2001) with permission from John Wiley & Sons.

different contribution from crystalline junctions in this range. At the end of the ramp, tan $\delta = 0.5$, so the system still has an important elastic component. Examining EM images of the dried gels, for gels prepared at room temperature an important structuring of the solution appears in which the fibrillar structure extends throughout the whole volume, as discussed before.

Electron micrographs were prepared by critical-point drying techniques (Figure 8.4) using CO_2, which, according to the authors, allows drying of the gels without distortion or alteration of their structure. Electron micrographs show that the gels' morphology consists of a three-dimensional fibrous network in DBP or BB solvent. The mesh size is about 200–300 nm and fibres have diameters of 30–50 nm. The gel morphology deduced from the EM images is similar for both solvent types and with ageing; thus fresh and aged gels have similar structures. Here BB is considered as a good solvent for PVC.

It is not clear which mechanism of aggregation – crystal or solvent complex formation – can be associated with this particular structure. It could also indicate an incipient phase separation after rapid quenching of the solution from high temperature (160°C) to

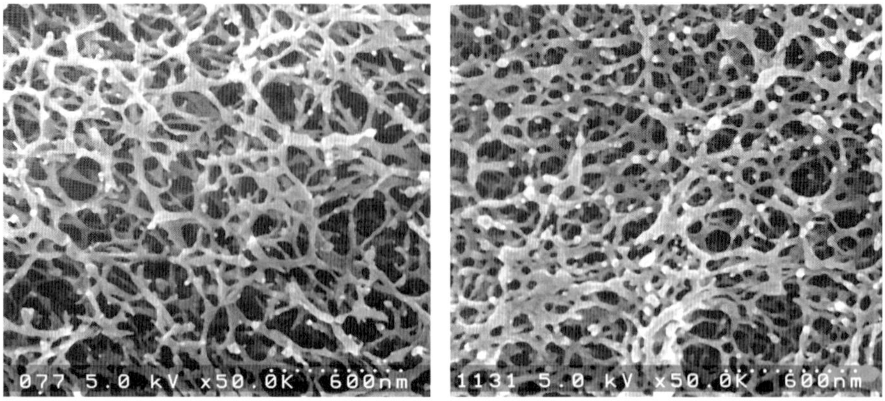

Figure 8.4 Electron micrographs of dried gels prepared by the critical-point drying technique. Left: PVC–DBP, $c=20$ wt%. Right: PVC–BB, $c=20$ wt%. Reproduced from Cho and Park (2001) with permission from John Wiley & Sons.

room temperature. During heating, although the enthalpy in DSC experiments indicates no measurable signal for junction dissociation, the rheology of the solution shows an important change, which can be ascribed to a progressive vanishing of the initial structure of the solution at low temperature. If the fibrillar structure is related to the polymer–solvent complex formation (or to a phase separation), then fibrils would be the polymer-rich phase, containing tightly bound solvent. The solvent is freed when the temperature is raised, but the local association and high polymer concentration within the fibres persist and disappear only progressively on thermal agitation at high temperature (no mechanical action being applied in the above protocol).

There is a large temperature lag between the melting of local structures of the polymer-complex type (around 50–60°C by DSC) and the full recovery of the homogeneous solution back to 160°C. The frequency dependence of the moduli should be explored in this range of temperatures to establish the progressive transition from a fibrillar network to a homogeneous solution. The time dependence of the moduli should also be investigated. The contribution to the storage modulus of the 'crystalline' domains still existing at the end of the ramp (120°C), just before melting out, appears to be much less important (10–100 Pa) than was first thought. Along the same lines as these observations, Barendswaard *et al.* (1999), using a combination of ^{13}C solution NMR and ^{13}C solid-state NMR, determined the degree of 'crystallinity' for PVC–DOP samples versus temperature. They reported that the fraction of so-called 'rigid' PVC, i.e. rigid on the time scale of the single pulse NMR experiment, as represented by the broad resonances, gradually decreased with increasing temperature. As the authors commented, here the term 'crystallinity' should be treated with some caution since, in the case of polymers such as PVC where crystallites can be very small, the value obtained for the degree of 'crystallinity' can be technique-dependent.

Indeed, in Figure 8.5 the percentage of rigid PVC as determined from the area of the broad NMR peak (^{13}C) decreases linearly with temperature in the range 90–180°C. This variation of a rigid component versus temperature is unusual for the melting

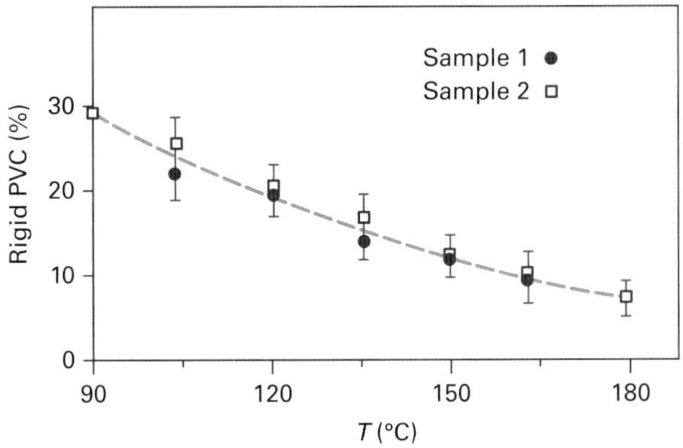

Figure 8.5 Percentage of rigid PVC determined by ^{13}C NMR for PVC–DOP gels versus temperature. Sample 1: $M_w = 60 \times 10^3$ g mol^{-1}; sample 2: $M_w = 85 \times 10^3$ g mol^{-1}. Adapted with permission from Barendswaard *et al.* (1999) © 1999 American Chemical Society.

behaviour expected for 'small crystallites', since these should not be distributed over such a large range of temperatures. The shape of the curve (linear decrease) does not bear the features normally associated with cooperative phenomena, for instance a sigmoidal or a step-like profile which would correspond better to melting of crystallites of a well-defined average size.

8.2.1 Rheological measurements

A series of papers has dealt with the rheological properties of gels using various dissolution protocols (te Nijenhuis and Winter, 1989; Aoki, 2001; Kakiuchi *et al.*, 2001; Watanabe *et al.*, 2001). te Nijenhuis and Winter (1989) dissolved PVC directly into bis(2-ethylhexyl) phthalate (DOP) and studied gelation by cooling solutions at 80°C and observing ageing effects at various frequencies (for concentrations around 10% w/w). In the other papers, the authors started the PVC dissolution at room temperature (about 30°C) in a good solvent, 70–90% tetrahydrofuran (THF), which contained varying amounts of DOP. The evaporation of THF from each solution proceeded at room temperature and lasted from 1 to 2 weeks depending on the polymer concentration. After the THF had completely evaporated, transparent liquids or solid films of just PVC–DOP were obtained. After about 2 weeks, the weight of each sample became constant. Reproducibility of rheological measurements was considered to be an indication of a stable equilibrium state. No traces of THF were left in the solutions.

Li and co-workers (Li and Aoki, 1997, 1998; Li *et al.*, 1997) studied stabilized solutions heated from room temperature to 40°C and measured mechanical spectra versus frequency for various concentrations and molecular masses of PVC. They applied the Winter–Chambon criteria (Chapter 3) to identify a gel–sol transition corresponding to concentrations at 40°C where the softening transition takes place. At this temperature they determined power laws with respect to concentration, around the critical PVC concentration. Unfortunately, these investigations cannot supply any information about the mechanism of gel formation. DOP is a plasticizer for PVC but is not a good solvent. For gels prepared in THF–DOP mixtures it is likely that liquid–liquid phase separation had taken place during THF evaporation in the presence of the poor solvent, but the authors claim that the gels were transparent. The size of the phase separated domains thus remained small compared to the wavelength of visible light.

Lozinsky *et al.* (2000) chose to heat the evaporated initial solutions up to 120°C, then cool them rapidly to room temperature and follow the increase of viscoelasticity with time. They therefore followed the build-up of the structure (sol–gel transition) with time and concluded that equilibrium values of G' and $\tan\delta$ for the aged samples were in good agreement with the corresponding values of the as-prepared samples at room temperature. This means that the two protocols used by Li *et al.* (1997) and by Lozinsky *et al.* (2000) lead to the same rheological state: the same PVC–DOP structure builds up either during THF evaporation at room temperature or after heating at 120°C and cooling at room temperature. The samples were aged for 45 to 180 days and the authors found the samples stabilized after 1 month, in agreement with other publications (Walter, 1954; te Nijenhuis and Dijkstra, 1975; Dorrestijn *et al.*, 1981). The ultimate values, although close to each other, are not identical to the as-prepared gels, and the ageing period is surprisingly long considering these are solutions with concentrations of about 10% w/w.

Aoki *et al.* (1998) and Barendswaard *et al.* (1999) used the Winter–Chambon criteria, and determined the gel–sol transition temperature to be $T = 54$°C for a PVC of concentration of 70 g L^{-1} (7% w/v) or $T = 82$°C for the same PVC at a concentration of 100 g L^{-1}.

Reviewing this series of experiments on PVC gels, it is difficult to relate the gel–sol transition to any thermodynamic solution state, since there is such a range of concentrations and temperatures where it can be observed. The question of the existence of a real equilibrium state remains totally open. In reality, the viscoelastic components reported at any time or temperature probably depend on the whole thermal history. The local heterogeneities created during evaporation of the good solvent (THF) or during a rapid quenching of a hot solution to room temperature (DOP solvent) do not allow any interpretation of the gelation process in terms of a purely molecular mechanism. The G' values, which are indeed very high in these gels ($c.10^4$ Pa for $c = 20$% w/v) at room temperature, appear to be frequency-independent after sufficient ageing times and at low temperatures (Aoki *et al.*, 1998), indicating long-lived junctions or phase morphologies, in contrast to the labile networks observed in hydrophobic associations (Chapter 6). It must be noticed also that, unlike polysaccharide or gelatin gels (Chapters 5 and 7), PVC gels have been studied at relatively high concentrations (around 80%). In PVC gels various types of solvents point towards subtle solubility properties and complex formation.

8.3 Gelation in the absence of crystallization

8.3.1 Atactic poly(styrene) in *trans*-decalin: phase separation and glass transition

Poly(styrene) is a vinyl polymer. Structurally, it is a long hydrocarbon chain with a phenyl group attached to every other carbon atom.

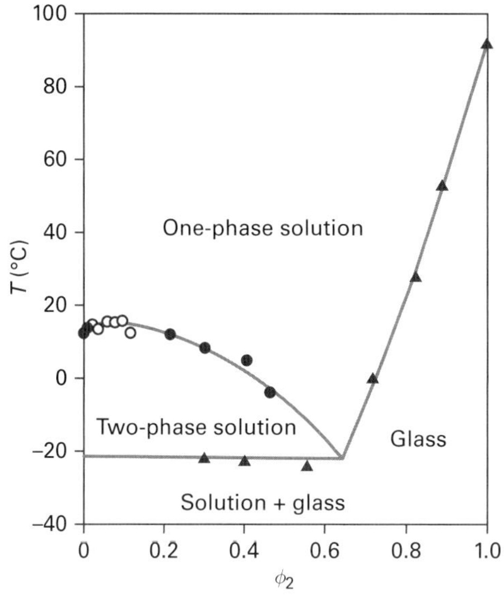

Figure 8.6 Phase diagram of aPS-TD obtained during cooling: by visual determination of the turbidity (○); demixing onset temperature measured by DSC (●); glass transition temperature determined by DSC (▲). Adapted from Arnauts *et al.* (1993) with permission from John Wiley & Sons.

Atactic poly(styrene) (aPS) has no order with respect to the side of the chain on which the phenyl groups are attached. Moderately concentrated solutions of aPS in *trans*-decalin (TD) solidify on cooling. This solidification results from the interference between liquid–liquid demixing (LL demixing) and a glass transition (T_g). In a first step, the LL demixing transforms the solution into an opaque, complex, finely dispersed two-phase system. This initial, metastable situation has to evolve towards a macroscopic two-phase system in which two solutions with different concentrations are in equilibrium with each other (Arnauts *et al.*, 1993).

The phase diagram in Figure 8.6 was established by optical and calorimetric techniques, using a narrow molecular mass distribution sample.

No demixing was observed for $c > 60\%$ w/w. Calorimetric investigations measured T_g, which increased with polymer content. Details concerning the spinodal line in the phase separated system are given in Figure 8.7, and this locus was determined by light scattering techniques. Solutions of aPS-TD exhibited an upper critical solution temperature (UCST) at 15°C and the critical concentration appeared to be situated around $c = 6\%$ w/w.

The region below the binodal in the phase diagram of a two-phase system exhibiting an UCST can be divided into two parts, namely a metastable region between the binodal and the spinodal and an unstable region below the spinodal (Chapter 10). In the metastable region, minimization of the free energy of the system requires that large changes in concentration occur about the average in neighbouring regions and in the phases. In this region, phase separation proceeds through a nucleation and growth mechanism (Wellinghoff *et al.*, 1979). Isolated particles nucleate and continue to grow with the

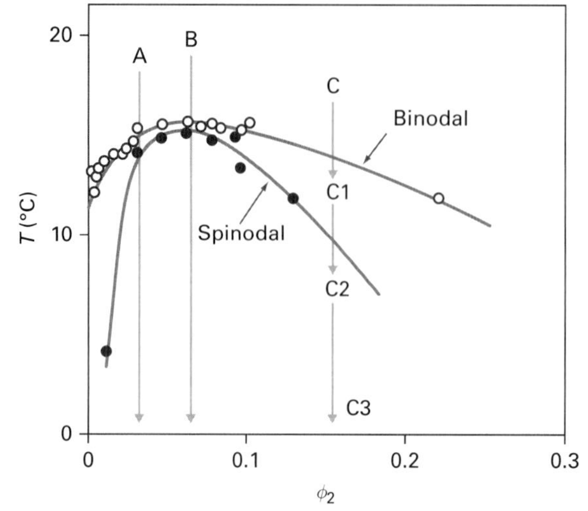

Figure 8.7 Binodal (○) and spinodal (●) lines for the system aPS-TD. Reproduced from Arnauts *et al.* (1993) with permission from John Wiley & Sons.

same internal composition until they are large enough to scatter light. Typically, polymer solutions will exhibit this type of behaviour near the θ-temperature. If the mobility of the system is sufficiently low and if the solution is quenched rapidly to a temperature below the spinodal line, nucleation and growth can be avoided, and the solution can phase separate by a spinodal mechanism. Provided the volume fraction of the minor component is high enough, the microphase morphology will consist of two continuous interconnected phases separated by a characteristic distance. In the initial stages of microphase separation, the same characteristic interphase distance will be maintained with increasing time. If one of the phases is a polymer-rich phase, the morphology will be frozen in, as soon as *the polymer concentration increases* enough to cause the *glass transition* of this phase to be below the temperature of phase separation. Interesting morphologies can be obtained when the solvent is eliminated from the demixed samples.

8.3.1.1 Morphology investigations

For the system aPS-TD (Arnauts *et al.*, 1993), three different situations have been followed through the phase diagram, indicated in Figure 8.8; samples were cooled to −40°C before morphological observations. Cooling of a solution with $c = 4.4\%$ w/w into the binodal region resulted in the formation of spherical, glassy particles in turbid, freely flowing solutions, which showed a tendency to agglomerate into larger domains and eventually deposit on the bottom of the tube (Figure 8.8a). More concentrated solutions ($c > 5$–10% w/w) phase separated into opaque gels at *c.*10°C. A solution with the critical polymer concentration (6% w/w) demixed by a spinodal mechanism, resulting in the formation of the typical spinodal morphology (Figure 8.8b). At this overall polymer concentration, material continuity of the *glassy phase* was achieved and this phase was no longer deposited on the bottom of the tube. The morphology represented in

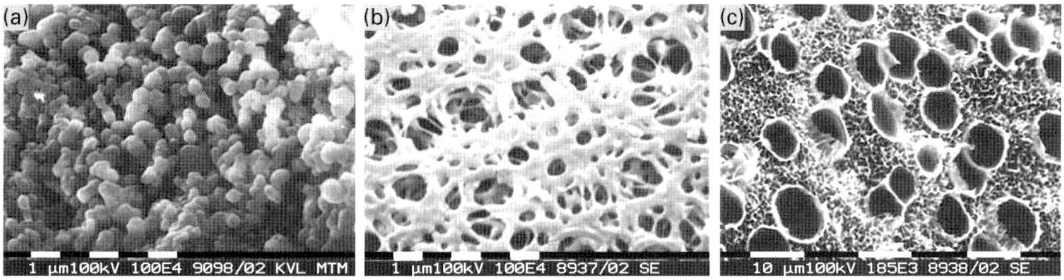

Figure 8.8 Scanning electron micrographs of dried aPS-TD samples with various concentrations: (a) $c = 4.4$ wt% (white bar is 1 µm); (b) $c = 6.7$ wt% (bar is 1 µm); (c) $c = 15.9$ wt% (bar is 10 µm). Reproduced from Arnauts *et al.* (1993) with permission from John Wiley & Sons.

Figure 8.8c was obtained in two steps ($c = 16\%$ w/w). First, the solution was cooled into the binodal region and left there overnight (Cl). Under these conditions, the concentrated phase formed a continuous matrix. In the second step, the solution was quenched to 9°C (C2), which induced spinodal demixing in the concentrated phase. Then the concentrated phase was vitrified (C3) and the solvent eliminated, and the image shows a complex, porous material.

8.3.2 Atactic poly(styrene) in carbon disulphide: gels induced by conformational changes

Wellinghoff *et al.* (1979) investigated aPS in carbon disulphide (CS_2) and determined the melting temperatures visually by gelling the solutions in tilted test tubes, increasing temperature and observing when the gel in the non-vertical test tubes began to flow under its own weight and formed a level meniscus. Gel 'melting' points (gel to 'mobile transition') of aPS in CS_2 were determined for various molecular masses: as the molecular mass increased from 4000 to 670 000 g mol^{-1}, the gel melting points increased from −25°C to +5°C. The authors then considered the gelation mechanism(s) of these solutions. Crystallization was definitely excluded in the aPS-CS_2 systems since high-sensitivity DSC measurements on these solutions revealed neither endothermic nor exothermic peaks. Additionally, the IR spectra of aPS-CS_2 solutions showed no significant change upon cooling to −78°C. The remaining possibility is that quenching the solution into the unstable region yielded a fine dispersion of polymer-rich regions whose T_g was sufficiently low to pin different polymer chain ends in the same glassy particle. The small size of these domains (<100 nm, as indicated by the gel clarity) and the rapidity of their formation (<1 min) suggested, according to (Wellinghoff *et al.*, 1979), that they developed by the spinodal mechanism.

Tan *et al.* (1983) prepared aPS gels in various solvents by dissolving the polymer at room temperature at the desired concentration in sealed tubes and cooling them slowly. Among other solvents, they also investigated CS_2; here the (Flory) θ-temperature was estimated to be around −73°C. They performed DSC measurements and used two simple mechanical methods to determine the sol–gel transition: the tilting-tube and falling-ball methods.

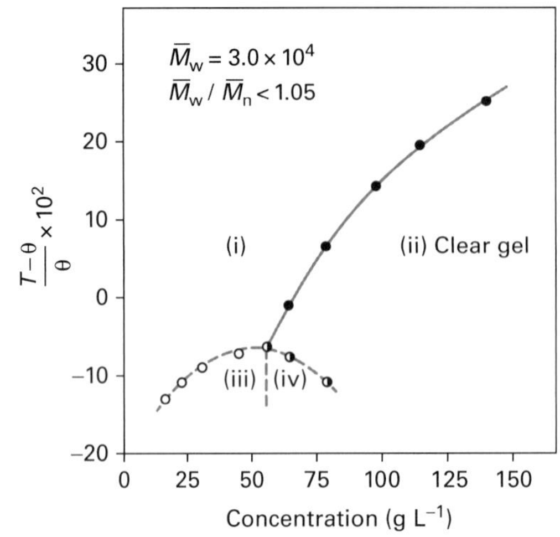

Figure 8.9 Phase diagram of narrowly dispersed aPS ($M_w = 3.0 \times 10^4$ g mol^{-1}) in CS_2: (i) one-phase solution; (ii) one-phase gel; (iii) two-phase solution; (iv) two-phase gel. Adapted with permission from Tan *et al.* (1983) © 1983 American Chemical Society.

The phase diagram of narrow M_w distribution aPS in CS_2 is given in Figure 8.9. Tan *et al.* (1983) investigated other solvents including nitropropane, and the corresponding phase diagram is shown in Figure 8.10. The temperature scale (ordinate) in each figure is centred around and normalized by the θ-temperature of the corresponding M_w of aPS and the solvent species. The binodal line in nitropropane is broader than in CS_2 and is shifted to higher polymer concentrations. These investigations depart from previous ones because the gelation transition in these two solvents is *distinct* from the phase separation boundary. It is interesting to note that the gelation curve for CS_2 in Figure 8.9 is located well above the θ-temperature and is therefore not related to the binodal line, and it intersects the binodal above the upper critical solution concentration. In nitropropane (Figure 8.10), a continuous gelation curve is observed across the binodal regime. Consequently, a turbid gel can be formed in nitropropane either from a turbid solution or from a transparent gel. Upon cooling a one-phase solution containing 190 g L^{-1} or less, the solution becomes a two-phase turbid solution which then transforms into a two-phase turbid gel. On the other hand, at 270 g L^{-1} a one-phase solution forms a one-phase gel, which then becomes a turbid gel when cooled further.

Tan *et al.* (1983) showed that stable thermoreversible gels were obtained from all M_w investigated, and in a large number of solvents. As the temperature was lowered, when gels formed from relatively dilute solutions they exhibited syneresis, followed by phase separation, indicated by turbidity. Once formed, the gels were insoluble in an excess of the same solvent. The gelation temperature T_{gel} and gel melting temperature T_m of aPS gels in the transparent gel region were found to be identical; gels aged for 2 weeks remained transparent and no change in T_m occurred. From these observations with $T_{gel} = T_m$, and the absence of any detectable time-dependent phenomena, the authors concluded that the aPS gel system had reached an *equilibrium state*.

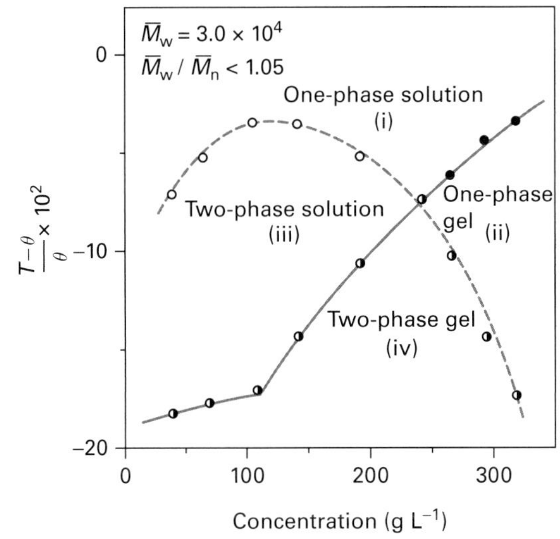

Figure 8.10 Phase diagram of a narrow M_w distribution aPS in nitropropane. Adapted with permission from Tan *et al.* (1983) © 1983 American Chemical Society.

High-sensitivity calorimetric measurements were carried out to evaluate the heat associated with the sol–gel transition. A transition was detected with DSC cooling scans, corresponding to the same temperature as T_{gel} measured by the tilted test tube method. From the exotherm associated with this transition, a 'heat of gelation' of the order of 1–2 J g^{-1} was calculated, which is a very low value, probably undetectable by earlier authors. The phase diagrams (Figures 8.9 and 8.10) for aPS indicate that gelation occurs as a distinct transition and that the phenomenon is *not caused by phase separation*. In this case, gelation behaviour is determined by polymer–solvent interactions, but does not correlate in a simple way with the binodal curve. Gel formation does not appear to correlate with the glass transition of the highly plasticized polymer either, as was the case for aPS-TD. The critical gelation concentration (in our notation, c_0) is strongly dependent on M_w. The authors also conclude that chain overlap is a necessary condition for gel formation; since the 'heat of gelation' is so small, this may be reasonable.

The gelation behaviour of the polydisperse aPS-CS_2 system was compared with two narrow M_w distribution samples, one of a similar M_w and the other of a similar M_n. The narrow fraction with similar M_w exhibits a sol–gel transition at higher temperature and has a lower c_0 than the polydisperse system. On the other hand, the polymer with similar M_n shows very comparable gelation behaviour, i.e. almost identical sol–gel transition temperature and c_0. Presumably the low-M_w tail of the polydisperse polymers remained in solution at the sol–gel transition and hence did not influence the gelation behaviour.

In order to further clarify the mechanisms of gelation of aPS in CS_2, Yanxiang and Deyan (1997) investigated solutions by IR spectroscopy. They examined the possibility of a *conformational change* in the polymer inducing gelation, and made some interesting observations. In the IR spectra of aPS, the spectral region 500–600 cm^{-1} has been found to be especially sensitive to conformational behaviour of the polymer chains, and perhaps also to intra-chain and inter-chain interactions. Based on studies of syndiotactic poly(styrene) (sPS), which could crystallize in the helical *ttgg* (*trans-trans-gauche-gauche*) conformation, it was assumed that the appearance of the band with a peak at 572 cm^{-1} indicated the formation of regular *ttgg* sequences. Therefore, the 572 cm^{-1} peak was one of the conformational sensitive bands.

The IR spectrum of aPS-CS_2 gel, dried as a film, was measured, and it was almost the same as that of normal aPS film, but a peak at 572 cm^{-1} was clearly observed which did not exist in the IR spectra of normal aPS film. This band did not disappear even after one month at −26°C. This suggested that, in such a film, some syndiotactic segments became segregated in a helical conformation, but short stereo-regular sequence length in the aPS prevented the formation of long regular sequences of the *ttgg* conformation or crystallization, because in its IR spectra only one of the *ttgg* conformational sensitive bands, at 572 cm^{-1}, could be detected. However, the formation of sequences of a particular conformation suggested that, compared to glassy polymers, some more ordered structures existed in the non-crystallizable polymer. Thus, in such films, ordered structures could be formed through stacks of some stereo-regular segments adopting the *ttgg* type conformation, and these were strongly dependent on the interaction between CS_2 and polymer segments. The dynamics of interaction was studied by measuring IR spectra at increasing temperature. The solvent escaped from the bulk film, the regular *ttgg* conformation transferred to the less regular *gauche tg* conformation and the molecular chains relaxed. Independently of these studies, Izumi *et al.* (1995) performed wide-angle neutron diffraction (WANS) experiments to identify the local structure in the aPS-CS_2 gels. They observed highly ordered structures far below the gelation temperature. However, the gel near the gelation temperature contained less distinct order.

The overall picture that emerges is that gelation of aPS in CS_2 is ascribed to intermolecular associations of chain segments including syndiotactic sequences and solvent molecules (similar to the polymer solvent complexes mentioned for PVC). François *et al.* (1986) adopted a different procedure for investigating this system: using DSC they examined aPS-CS_2 gels which had been kept in tightly sealed pans, annealed for *several days* at 40°C and then kept for at least *2 months* at room temperature to ensure thorough homogenization. Some DSC measurements were repeated a month later to test homogenization. All of the DSC experiments were consistent with the existence of structures possessing some degree of order. Their work shows that the structures responsible for the physical cross-links form a stoichiometric compound. From their temperature–concentration phase diagram, François *et al.* (1986) found a fixed polymer concentration of $c = 0.47$ g cm^{-3} (40% w/w) of intermolecular associations, and concluded that physical gelation in CS_2 arose from the formation of a polymer–solvent compound whose stoichiometry was defined between $c = 40\%$ and 50% w/w. Consequently the maximum proportion of physical junctions should be formed in this concentration range. Even if other authors do not agree with this particular stoichiometry, there is clear evidence that the nature of the solvent favours polymer complex formation and conformational change.

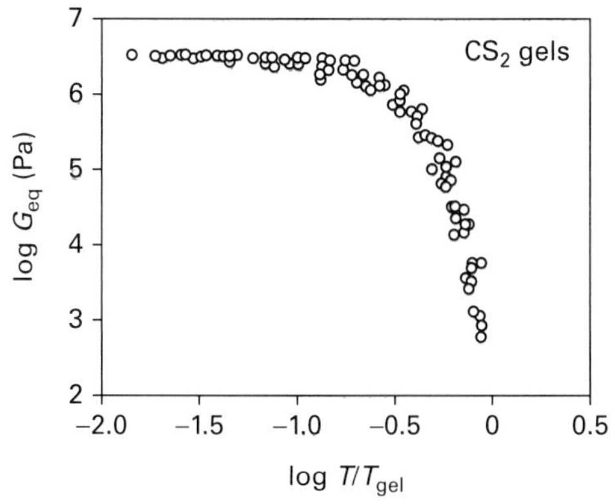

Figure 8.11 Master curve of aPS-CS_2 gel modulus versus temperature for various molecular masses and concentrations. Reprinted with permission from Koltisko *et al.* (1986) © 1986 American Chemical Society.

8.3.2.1 Rheology of transparent gels

Rheological properties were carefully measured by Koltisko *et al.* (1986). Narrow polydispersity aPS ($M_w/M_n < 1.2$) samples with M_w ranging from 2 to 900 kg mol^{-1} were investigated. Simple shear tests were performed on gels as a function of temperature (between +25 and −100°C), molecular mass and concentration. Solutions were introduced into a coaxial cylinder instrument and gels were formed *in situ* by lowering the temperature of the apparatus. Stress–strain measurements were conducted by attaching the inner cylinder to the crosshead of a standard Instron tensile testing device. Measurements of the shear modulus G_{eq} were performed by cooling the gel to the desired temperature, allowing the gel to equilibrate for 10 min and applying the stress. The modulus–temperature curves are similar in shape for all concentrations. As temperature decreases below the gelation threshold temperature, the modulus increases over a 40°C temperature range and then levels off at a value that depends on the concentration. Molecular masses of 35 kg mol^{-1} and higher exhibit the same low-temperature plateau modulus over which the gel modulus is constant. The width of the plateau depends on molecular mass, and, like the rubbery plateau, as the molecular mass increases the temperature range becomes broader. The data shown in Figure 8.11 corresponds to transparent gels, above the binodal line. It follows that the mechanism of gelation may be similar to that describing the rubbery plateau in a temperature region over which physical associations between chains are allowed to form.

The values of the modulus obtained with various reduced temperatures (T/T_{gel}) and a fixed M_w indicate that the modulus depends on the square of the concentration over the entire temperature range. In order to construct the master curve in Figure 8.11, which includes data for various concentrations ($M_w = 900$ kg mol^{-1}), the shift factors for the effective modulus G_{eff} were defined by

$$G_{eff} = G_0 \left(\frac{\rho}{c}\right)^2, \tag{8.1}$$

where G_0 is the shear modulus at this MW in bulk, c is the concentration and ρ is the polymer density. The master curve represents the modulus–temperature relationship for gels of 900 kg mol^{-1} aPS with no solvent. The plateau modulus of 3×10^6 Pa is close to the 10^5–10^6 Pa range for aPS in the rubbery plateau region and is significantly below the 10^9 Pa modulus of aPS in the glassy state.

In the Flory–Rehner model for an ideal rubber, the modulus should depend only on the molecular mass of the chain between junctions; for gels, this corresponds to the molecular mass M_a between association points. According to this, the simplest form of rubber elasticity theory,

$$G = \frac{cRT}{M_a}\left(1 - \frac{2M_a}{M}\right). \tag{8.2}$$

If (8.2) is obeyed, M_a should be independent of M.The experiments on aPS show that M_a is indeed largely invariant with M for a reduced temperature value corresponding to the plateau region. However, the average value of M_a is considerably smaller than the calculated entanglement molecular mass of various concentrations, so it is apparent that the physical associations that form the gel network are not simply entanglements, in agreement with structural investigations by IR and neutron scattering.

The calculation of M_a also suggests a lower molecular mass limit for gelation of aPS, and this was subsequently confirmed by observation. A strong temperature dependence of M_a was established. For all but the lowest molecular masses, M_a increases rapidly with T, reflecting the rapidly decreasing modulus. In terms of rubber theory, this corresponds to the gradual melting out of associations. Neutron scattering measurements by Izumi *et al.* (1995) confirmed that the formation of physical associations is favoured by lowering the temperature. The authors proposed a refined model for the molecular mass dependence of the plateau modulus, using a statistical approach taking account of the amount of polymer not forming part of the gel network. The apparent molecular masses M_a' are larger than the M_a found using the average approach of (8.2). The discrepancy between M_a and M_a' may be a result of the simplifying assumption that the gel network structure is the same as in cross-linked rubbers; the structure of gels is not necessarily homogeneous since they contain multi-chain associations.

The results of Koltisko *et al.* (1986) therefore encourage further measurements on thermoreversible aPS in the transparent gel region, along the lines adopted for conventional rubbers, including specific physical associations. The lifetime of the associations is probably sufficiently long that the end of the rubbery plateau (relaxation of the shear modulus) is not reported in currently published experiments. All of these theories also assume that, as in ideal rubber theory, elasticity is purely entropic, which is unlikely to be justified.

Additional rheological data is available from the work of Xie *et al.* (1990), who investigated gelation time as a function of temperature, polymer concentration and M_w of aPS in CS_2. A master curve was found for the gelation time–gelation temperature curves at various polymer concentrations, by shifting the curves along the temperature axis. However, the gelation point, where solutions lost their fluidity, was determined by the approximate tube tilting method. According to the phase diagrams established by Tan *et al.* (1983), the gels investigated here were also in the transparent gel region.

It is worth mentioning that François *et al.* (1986) found no molecular mass effect on the melting points determined by DSC, unlike those measured mechanically by Koltisko *et al.* (1986) for this type of gel. The differences in experimental conditions, in particular the time scale of observation of the gel state adopted in the DSC study, make it difficult to compare the two experiments. However, because of the discrepancy between the various results, the statement by Tan *et al.* (1983) that the aPS gels are in equilibrium should also be reconsidered.

8.4 Stereo-complexation and conformational changes: isotactic and syndiotactic PMMA gels

Polymethylmethacrylate (PMMA) is a member of the (vinyl) acrylate family with an extra methyl group attached to the α-carbon. PMMA is a strong, hard and transparent material with the advantage of being more transparent than glass, and so replaces glass in some applications.

Of the structural parameters affecting PMMA aggregation in solution, the most important are the configurational and conformational structure of PMMA and the mobility of the various PMMA functional groups. A diad containing two α-carbons of equal configuration is described as isotactic; a diad formed by two α-carbons of unequal configuration is described as syndiotactic. By the term 'configurational structure' of PMMA we mean the distribution of isotactic and syndiotactic diads in the polymer. Purely *i*-PMMA consists of diads with a linear backbone structure with all methyl ester groups situated on one side of the chain, whereas the purely *s*-PMMA consists of a linear backbone structure where these groups alternate on either side of the chain. The overall configurational structure of PMMA can be described most simply by determining the content of syndiotactic or isotactic diads. This result alone yields no information on the distribution of diads in longer sequences, but strongly affects the aggregation of PMMA. In PMMA, conformational isomers can be generated by rotation about three types of single bonds: (1) the bonds between the methylene group and the quaternary carbon, CH_2–C; (2) the bonds between the quaternary and carbonyl carbons, C–CO; and (3) the bonds between the carbonyl carbon and oxygen, CO–O. Rotation about bond 1 (CH_2–C) determines the conformational structure of the chain backbone, and rotation about bond 2 (C–CO) determines the orientation of the ester groups with respect to the chain. Rotation about bond 3 (CO–O) does not need to be considered in studies of the conformational structure of PMMA because, in all simple methyl esters, the ester group has been found to have a planar structure, with the ester methyl group in *cis* orientation with respect to the carbonyl; therefore, it may be assumed that this structure of the ester group is also preserved in PMMA. The structures generated by rotation about CH_2–C bonds are often described in terms of *trans* (*t*) and *gauche* (*g*) forms, even if they deviate somewhat from the strictly staggered structures. The conformational structures of the PMMA chain in the energetically most favoured forms of *i*- and *s*-PMMA differ very little; in both cases, in the most favoured form the chain structure can be approximated as *tt* (the structure with the α-CH_3 group coplanar with the ester group and oriented away from the carbonyl).

Watanabe *et al.* (1961) reported that, in some solvents, mixing solutions of *s*-PMMA and *i*-PMMA led to the formation of aggregates, which yielded pronounced changes in the rheological properties of the mixture and led to gel formation.

Indeed, long stereo-regular sequences of PMMA have a strong tendency for mutual interaction (Spevacek and Schneider, 1987), and sequences of both equal stereo-regularity (*i*–*i* and *s*–*s*) and differing stereo-regularity (*i*–*s*) are capable of this. Consequently, aggregates of stereo-regular PMMA can be divided into three types: (1) self-aggregates of *i*-PMMA; (2) self-aggregates of *s*-PMMA; and (3) the so-called stereo-complex formed by the mixing of solutions of *i*- and *s*-PMMA.

These three types of aggregates differ in their thermal stability and the extent of aggregation in the given solvent. *i*-PMMA self-aggregates exhibit the highest thermal stability and *s*-PMMA self-aggregates the lowest, with aggregates of the stereo-complex in between. In the literature, the greatest attention has been paid to aggregates of the stereo-complex and to *s*-PMMA self-aggregates. Information about the self-aggregates of *i*-PMMA is scarce because the content of these aggregates is always relatively low, irrespective of the type of solvent. Self-aggregates of *i*-PMMA decompose at temperatures close to the melting temperature of crystalline *i*-PMMA, which might indicate that the aggregates in solution are small crystallites.

Self-aggregates of *s*-PMMA and stereo-complexes of PMMA exhibit a number of common features. Bosscher *et al.* (1982) performed X-ray scattering on stretched films, cast from solutions. They found that the X-ray pattern of the stereo-complex resembles the pattern of *s*-PMMA more than that of *i*-PMMA, and assumed that the syndiotactic chain in the stereo-complex has a conformation comparable to the conformation of *s*-PMMA after solvent-induced crystallization of *s*-PMMA. The isotactic chain in the stereo-complex takes positions similar to those occupied by the solvent molecules in pure crystalline s-PMMA. The stereo-complex was found to be a double-stranded helix in which an isotactic chain with a small radius is surrounded by a syndiotactic chain with a large radius, in such a way that the requirement $i/s = 1/2$ can be satisfied.

One of the decisive factors in the self-aggregation of *s*-PMMA and in stereo-complex formation is the stereo-regularity of polymer chains. A stable, ordered structure is formed only in those cases where the *length of the interacting sequences* is longer than some minimum length, but the actual values of this minimum length depend on the solvent. In the case of the stereo-complex of PMMA, it was possible to classify the solvents as strongly complexing (minimum length of associated *s*-sequences ~3–4 monomer units) and weakly complexing (~10 monomer units). At the same time, the thermal stability of the aggregates is determined by stereo-regularity: the temperature of melting of aggregates formed by long stereo-regular sequences is higher than for aggregates formed by short sequences. The solvent has a less pronounced effect on the thermal stability of the aggregates. It was found that ester groups play an important role in the interactions of stereo-regular sequences leading to aggregation, both with self-aggregates of *s*-PMMA and with the stereo-complex of PMMA. The mobility of ester groups in the aggregates is strongly hindered. After removal of the solvent from solutions or gels where self-aggregates of *s*-PMMA or stereo-complex of PMMA exist, the structure generated during aggregation is preserved even in the solid state, similar to the case of aPS-CS_2 presented above (Spevacek and Schneider, 1987). Up until this time, structures of the double helix type had only been considered for biological macromolecules (e.g. carrageenans and, agarose; see Chapters 5 and 7). Stereo-regular PMMAs may be the first synthetic polymers in which double helices generated by physical interactions have been observed.

A more detailed analysis has been published for the *self-aggregates* of *s*-PMMA in o-xylene and toluene, and a two-step mechanism more clearly identified. On cooling solutions of *s*-PMMA from high temperature (85°C) to room temperature in o-xylene, a transparent gel is immediately formed (Berghams *et al.*, 1987). On standing, no changes in its optical characteristics were observed, while heating resulted in a transition to a transparent solution. The corresponding enthalpy change was measured by DSC and increased with increasing tacticity. The gel formed on cooling was very brittle at any concentration, so stretching was not possible. When solutions of *s*-PMMA were transformed into a gel, changes in the IR spectrum were also observed. The most interesting frequencies were those at 843 and 860 cm^{-1}, which are ascribed to the $-CH_2-$ rocking vibration and correspond, respectively, to *tt* and *tg* conformations. An increase of the 860 cm^{-1} peak at the expense of absorption at 843 cm^{-1} is characteristic for a transition from a random-coil (predominantly *tg*) into a regular all-*trans* conformation, as observed by Berghmans *et al.* (1994) in toluene gels. Theoretical calculations designate this slightly deformed *tt* conformation as the energetically most stable form on which the helix conformation of the polymer chain is based. In Figure 8.12 the intensity ratio of

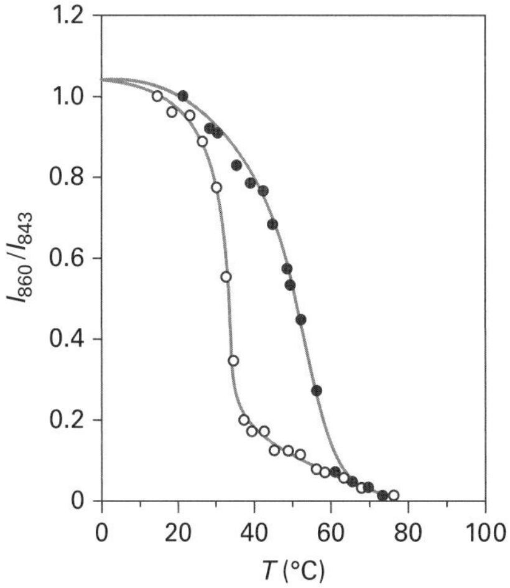

Figure 8.12 IR intensity ratio I_{860}/I_{843} as a function of temperature during cooling (○) and during heating (●) for *s*-PMMA in toluene, $c = 10$ wt%. Reprinted with permission from Berghmans *et al.* (1994) © 1994 American Chemical Society.

these two signals (*tt*/*tg* or I_{860}/I_{843}) has been used as a quantitative measure of the degree of coil–helix transformation.

This IR transition occurs during cooling in the same temperature domain as that observed in DSC experiments. The IR data indicates that the amount of conformational change increases with decreasing gelation temperature and reaches an *equilibrium value* at every gelation temperature (during cooling). Consequently, the proportion of regular sequences available for network formation increases correspondingly. Network formation represents the second step in the gelation: it is responsible for the hysteresis observed between cooling and heating in both DSC and IR measurements, and it occurs at a finite rate which depends on the concentration of ordered sequences. This is further confirmed by rheological measurements.

8.4.1 Rheology of the stereo-complex gels

Aggregation is a very complex process, depending on many factors, but the effect of time is of great importance. The formation of 'primary' aggregates, which is relatively rapid and which is preceded by an even more rapid change of *conformational structure* of the stereo-regular sequences, is followed by the very slow process of so-called 'secondary aggregation', with steadily increasing particle size. This process has been detected by various physical methods, albeit with different sensitivities.

The rate of gelation increases with decreasing temperature because of the increasing number of regular sequences present. Instantaneous gelation occurs at temperatures where at least 50% of all the transformable sequences have a regular conformation, for instance room temperature. At higher temperatures, gels are formed at a measurable rate, as deduced from rheological observations. The first step is apparently dominated by an equilibrium constant with a position that changes with temperature. The second step is not at equilibrium since, as a consequence of intermolecular associations, a temperature hysteresis is observed.

The viscoelastic behaviour of gels of the PMMA stereo-complexes in o-xylene ($c \sim 5$–12% w/w) was studied by Pyrlik and Rehage (1975) using rheological techniques. They showed that G' increases with time and depends strongly on temperature. After prolonged ageing, G' and G'' depend only weakly on frequency. The results also depend on the thermal history of the solutions: fast gelation occurs when the solutions are heated at a temperature not sufficiently high to destroy the primary nuclei of stereo-complexation. (It was found necessary to heat the gels at 145°C, the boiling point of o-xylene, to fully erase the thermal history.) G' values can reach values of the order of 10^4 or 10^5 Pa at lower temperatures, from room temperature down to −20°C (Pyrlik and Rehage, 1975).

In Figure 8.13 the macromolecular component of the gel was a mixture of 5 parts *s*-PMMA and 4 parts *i*-PMMA, giving a ratio of triads of about 1:1. Using o-xylene as solvent, the polymer concentration was 12.6% w/w (Pyrlik and Rehage, 1975). The authors also observed that syneresis accompanies the gelation process of these thermoreversible gels. Stress relaxation at constant strain is observed after long periods of time, indicating that the physical network can rearrange slowly (10^3 to 10^4 s), as shown in Figure 8.14 for a low initial modulus (~15 Pa) gel in *s*-PMMA.

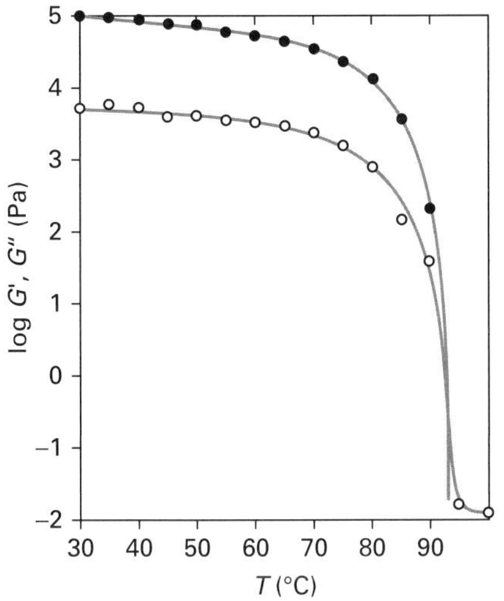

Figure 8.13 Storage (●) and loss (○) moduli of stereo-complexes of PMMA in o-xylene at $c = 12.6$ wt%. From Pyrlik and Rehage (1975).

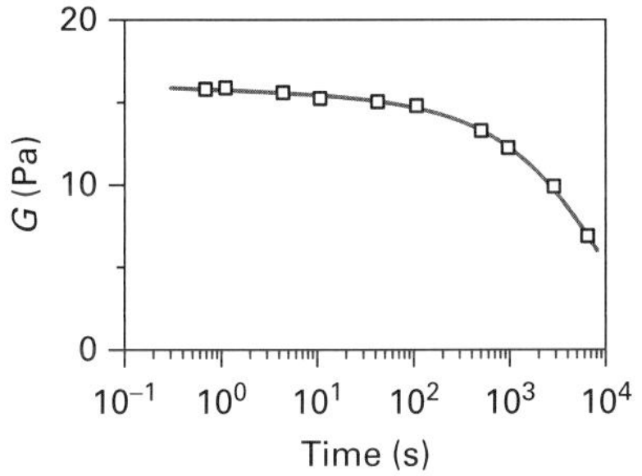

Figure 8.14 Stress relaxation modulus measured at a fixed strain (0.5%) versus time for *s*-PMMA–toluene gel with $c = 10$ wt%. From Berghmans *et al.* (1994) © 1994 American Chemical Society.

8.4.2 The two-step mechanism in *s*-PMMA solutions

IR, DSC and rheological experiments performed on progressive cooling of solutions of *s*-PMMA point towards a two-step mechanism: a very fast intermolecular conformational change, followed by a further intermolecular association. Gels formed at a measurable rate (e.g. cooled to an intermediate temperature of, say, 46°C, annealed for a time and then cooled again at 40°C) show measurable time-dependence effects related to the progressive association of the primary structure (single helix) into more aggregated structures. Annealing also has an important effect on the value of the enthalpy change on melting and reflects the contribution of intermolecular associations. Berghmans *et al.* (1994) illustrated the two-step mechanism for gelation of *s*-PMMA in toluene, as shown in Figure 8.15.

In studies of the self-complexation of *s*-PMMA, Berghmans *et al.* (1994) found no firm evidence for the formation of a double helix at the very beginning of the

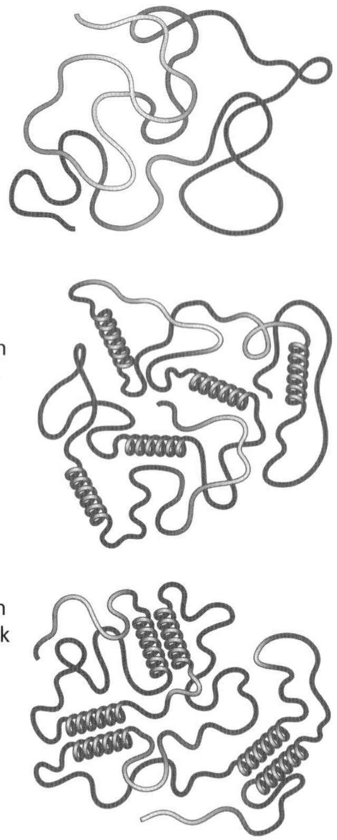

Figure 8.15 Two-step mechanism of gelation for *s*-PMMA: (a) hot solution; (b) formation of single helices; (c) initial formation of the network by aggregation of helices. Adapted with permission from Berghmans *et al.* (1994) © 1994 American Chemical Society.

process, as assumed by other authors. Instead, the occurrence of a two-step gelation mechanism suggested the initial formation of a single helix. According to these authors, the formation of double helices in the first step would lead to a much faster intermolecular network formation, as the physical cross-links could already be formed in this step.

The main conclusion by Berghman's group, drawn from these studies on stereoregular PMMA solutions, is that thermoreversible gelation occurs in two steps, by a mechanism quite different from the crystallization gelation systems. Generally this last type of gelation is nucleation-controlled, occurring at a certain degree of undercooling (crystalline cross-links, with a fringed micellar structure) (Berghmans *et al.*, 1979). Here a random coil diffuses from the surroundings (melt or solution) on to the growing crystal surface and is laid down in a regular manner. Consequently, the polymer chains acquire their regular conformation only when entering the crystal lattice. The change in molecular conformation takes place during the formation of the crystal, and the helix gets its stability from its incorporation into the crystal lattice. In their review article, Spevacek and Schneider (1987) supported a different point of view based on primary and secondary crystallization concepts, with the initial step being the formation of the double-helical structure of the self-aggregates of *s*-PMMA. It is not possible at present to further discriminate between these approaches. Finally, gelation of atactic PMMA (*a*-PMMA) in 1-butanol and cyclohexanol (Vandeweerdt *et al.*, 1991) can be explained by liquid–liquid demixing interfering with a glass transition, similar to the case of aPS in TD.

8.5 Cryogels of poly(vinyl alcohol) (PVA)

8.5.1 Mechanisms of gel formation

PVA is a water-soluble synthetic polymer with a high degree of hydrolysis (98–99%) and is considered to be atactic. Thermoreversible PVA physical cryogels, prepared by repeated cycles of freezing and thawing of an aqueous solution of the polymer, have captured the attention of both academic and industrial researchers because of their potential applications in many fields (Inoue, 1975; Peppas, 1975). Because they are biocompatible, PVA hydrogels are suitable for a variety of biomedical and pharmaceutical applications (e.g. artificial tissues, contact lenses, controlled-release devices for drug delivery). Freeze/thaw PVA hydrogels show high mechanical strength and good elastic properties, since they can endure large deformations upon stretching, and recover their original shape and dimensions on release of this tension.

PVA can be dissolved in water at high temperature (96°C) but, when cooled at room temperature, solutions ($c \sim 10\%$ w/w) do not gel and remain transparent when left in a sealed tubes for more than 1 month (Ricciardi *et al.*, 2005). Elastic properties of cryogel films are obtained only by subjecting the polymer aqueous solutions to several repeated freeze/thaw cycles, consisting of 20 h freezing steps at −22°C followed by 4 h thawing steps at room temperature. PVA cryogels undergo syneresis upon application of repeated freeze/thaw cycles, expelling some water on their surface. However, they acquire good mechanical properties, keep a high water content, are stable at room temperature and can retain their original shape. The elasticity and tenacity of the gel increases with the number N_c of cycles, as indicated in the stress–strain curves in Figure 8.16. With increasing N_c, the original cryogel (with a thickness of 1.5 mm) becomes more turbid and almost non-transparent above $N_c = 3$, suggesting the development of a microphase separated structure (Yokoyama *et al.*, 1986). In parallel with changes of elasticity, crystalline reflections become more distinct with N_c.

Traces of ice crystal growth are left as pores in the cryogel, as suggested by scanning electron micrograph of a xerogel (solvent-free gel) obtained from 5% w/w cryogel, shown in Figure 8.17.

It is seen that the pores, of order 10 μm, are linked together linearly, and it was established that they are oriented along a direction nearly normal to the freezer plate. Under freezing, the ice crystals grow along the direction of temperature gradient, the PVA-rich solution phases being segregated around them, and the gelation proceeds in the segregated solution phases, which then form the continuous porous gel skeleton.

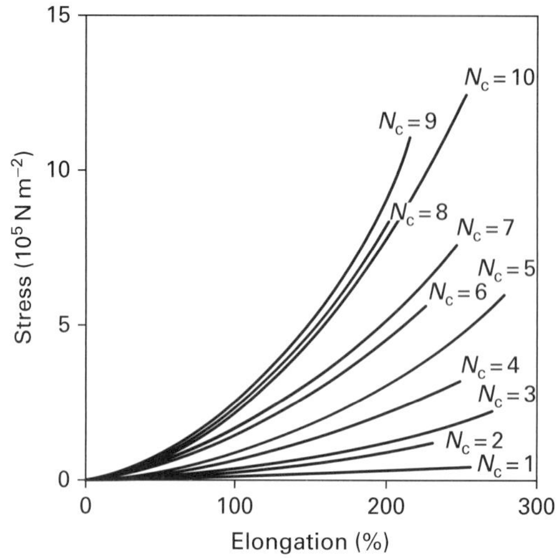

Figure 8.16 Stress–strain curves for $c = 15$ wt% PVA cryogels during N_c consecutive freeze/thaw cycles. Adapted with permission from Yokoyama *et al.* (1986) © 1986 Springer.

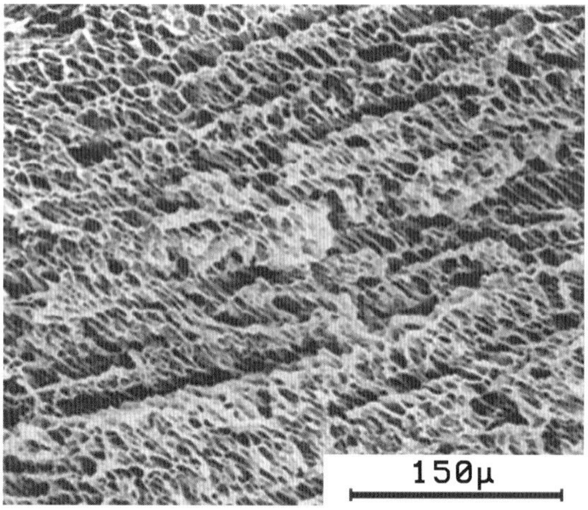

Figure 8.17 Scanning electron micrograph of a xerogel obtained from 5 wt% PVA cryogel ($N_c = 10$). Adapted with permission from Yokoyama *et al.* (1986) © 1986 Springer.

Yokoyama *et al.* (1986) considered that the repeated freezing stages are nothing but repeated recrystallizations, and regarded the cycling as a kind of refinement process for ice crystals. They therefore expected that polymer chains would be rejected from the growing ice crystals as an 'impurity' whenever the PVA–water system was frozen. Thus the concentration of PVA in the PVA-rich solution phase would increase with the freezing cycle number, and in the PVA-poor solution phase would decrease, giving rise to the porous structure. They observed that the pore size in the xerogel increased on lowering of the concentration of PVA. However, they reported shrinkage of the gels (for $N_c > 3$–4) when transformed into xerogels for EM observations. This made it difficult to determine the exact size of the porous channels in the initial cryogel state (shrinkage is about 40% in diameter for a disc shape). Nevertheless, in the original cryogel it may be considered that the pore size (or the size of ice crystals) increased when the concentration was lowered. This trend is consistent with a mechanism in which diffusion plays a major role since, on a lowering of the concentration of PVA, it will become easier to exclude chains from the ice crystal growth front, and a larger ice crystal size is then expected. When these EM images of PVA gels are compared to EM images of PVC gels obtained using the critical-point drying technique (Figure 8.4), we can see that the structure in the cryogels is much more heterogeneous and the morphology much coarser – the pores are larger and the polymer network structure is very thick. That said, the drying stage may itself have partially generated this coarsened structure.

The microstructure of these peculiar physical gels has been investigated by many authors using various techniques. Willcox *et al.* (2000) performed DSC measurements. As Figure 8.18 shows, when N_c increases, a melting peak appears at 78°C and persists, while a low-temperature shoulder at 71°C to 72°C emerges and grows. The integrated heat of fusion for the composite peak increases with N_c: as N_c goes from 1 to 24 for 'non-aged' samples, the heat of fusion rises from 0.63 to 8.2 J g^{-1} (the normalizing mass is that of the polymer). By comparing with the bulk heat of melting for PVA (150 J g^{-1}), the gel crystallinity can be inferred to increase from 0.4% to 5% of the total PVA in the system. However, the crystals in the gels are small and highly hydrated, so this crystallinity should be regarded as a crude, lower-bound estimate. In addition, the melting of bulk PVA occurs at 250°C, nearly 170°C above the endotherms under discussion. On ageing, the low-temperature shoulder grows and the higher-temperature peak shifts upward to 79–83°C, so eventually a distinct second peak in the range 67–72°C replaces the shoulder. The overall area under the two peaks saturates at 22–24 J g^{-1}, irrespective of N_c, suggesting that the crystallinity itself saturates at ~15% of the total PVA content.

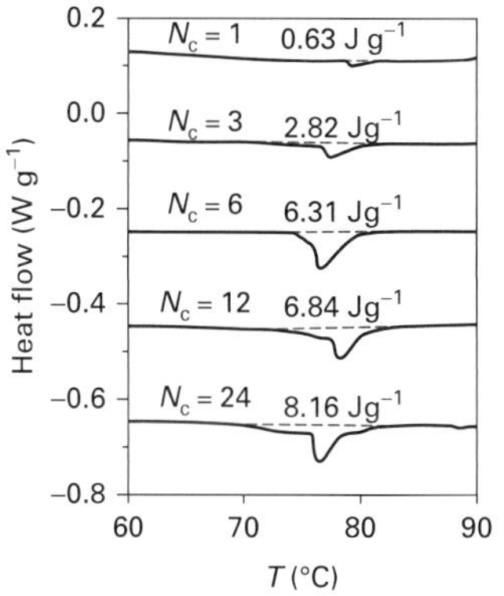

Figure 8.18 Effect of the number of cycles N_c on the endotherms of freshly prepared PVA cryogels by DSC heating at a rate of 10°C min^{-1}. Adapted from Willcox *et al.* (2000) with permission from John Wiley & Sons.

Opinions in the literature differ as to the driving force for the spontaneous gelation of aqueous PVA solutions at room temperature. However, the role of crystallinity appears to be widely accepted. For some authors (Ogasawara *et al.*, 1975, 1976; Watase and Nishinari, 1985; Wu *et al.*, 1991), liquid–liquid phase separation is the primary event, with crystallization in the polymer-rich phase reinforcing this domain as a secondary effect, while others contend that junction zones directly crystallize from the solution phase. Another opinion is that both conjectures are true, depending on gelation conditions (Komatsu *et al.*, 1986). The second hypothesis (direct crystallization) is supported by essential observations from the literature. Water is known to be a good solvent for PVA near and above room temperature, partially explaining why PVA melting temperatures are deeply depressed in the gel. PVA gels lose mechanical properties at 60–80°C, roughly corresponding to the DSC endotherm associated with the melting of PVA crystals. This superposition strongly suggests that crystallites hold the gel network together.

The gelation of PVA solutions by freeze/thaw processing is represented schematically in Figure 8.19 (Willcox *et al.*, 2000). After the first temperature cycle, network formation can be attributed to the formation of cross-links by kinetically frustrated crystallization. As more freeze/thaw cycles are applied, or as the gel is aged, additional PVA crystallization reinforces the original gel network. Reinforcement mainly involves the formation of a new, secondary class of crystallites, but the primary crystallites created in the first quench may also grow. The average crystallite size remains small, in the range of a few nanometres, and the gel mesh spacing of tens of nanometres changes little. Figure 8.19e also shows how EM imaging may change the real gel structure after removal

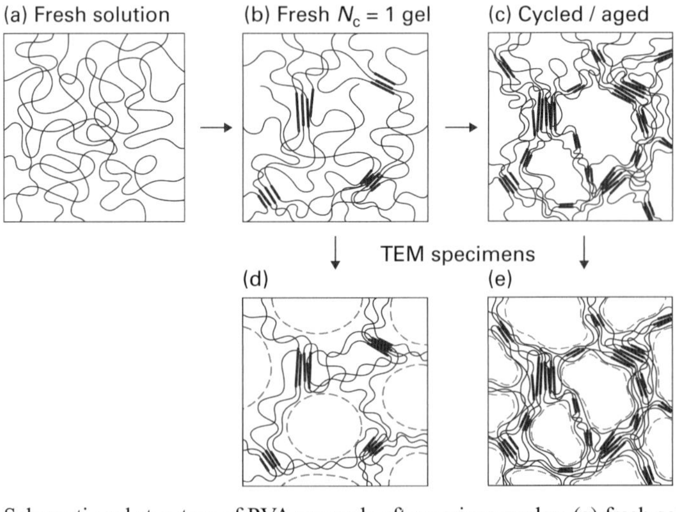

Figure 8.19 Schematic gel structure of PVA cryogels after various cycles: (a) fresh solution; (b) primary crystallites appearing during the first temperature cycle; (c) secondary crystallites develop after cycling/ageing; (d) TEM observations by removal of the frozen water in fresh gels; (e) TEM obervations of cycled/aged gels with large pores, obtained after collapse of the amorphous chains. Adapted from Willcox *et al.* (2000) with permission from John Wiley & Sons.

of water (xerogels) by collapsing the amorphous chains. The following sections deal with the swelling or drying and rehydration of xerogels, important aspects of cryogel applications.

8.5.2 Swelling

Hassan and Peppas (2000) examined the long-term morphology changes (i.e. over several months) of PVA on *swelling* in water at 37°C, relative to preparation conditions. They observed that, during swelling, a significant fraction of PVA chains were not incorporated into the overall crystalline structure, and therefore dissolved into solution. The crystalline structure of PVA gels was examined in terms of the overall degree of crystallinity as derived from DSC experiments in open pans. During these experiments the samples started to dehydrate around 100°C, whereas melting of PVA crystals started around 230°C. The degree of crystallinity was corrected for the residual water content and calculated for samples with various freezing/thawing cycles and after swelling. Parameters investigated included the number of cycles, the PVA molecular mass, the initial concentration in solutions and the time of swelling. Approximately a third of the initial crystallinity was lost in the first day of swelling, but enhanced stability was created upon increasing the number of freezing/thawing cycles. Although an increased number of cycles did not necessarily increase the overall degree of crystallinity, it appeared that repeated cycles did reinforce those crystals that already existed. For $N_c = 7$, approximately 20% more of the chains participated in the overall crystalline structure. The samples treated for $N_c = 3$ showed a much more swollen structure than those treated for $N_c = 5$ or 7.

8.5.3 Drying and rehydration of cryogels

Ricciardi *et al.* (2004) carried out an extensive X-ray diffraction study on PVA dried gels and PVA cryogels obtained by rehydrating the dried (kept in air) samples or obtained by a slow drying process, by storing for 2 months in sealed vials at room temperature. The results of their analysis confirm that highly stable PVA cryogels, with a water uptake higher than 80%, may be obtained from freeze/thaw cycles of the initial solutions. They showed that the porous structure formed during freeze/thaw cycles was not greatly altered upon drying. In fact, during the successive rehydration steps, rehydrated gels almost completely recovered the volume, shape and physical properties of the as-formed freeze/thaw PVA cryogels, as if they were a permanently cross-linked network. Thus, they showed that the outstanding physical and mechanical properties of freeze/thaw PVA cryogels may be preserved for a long time by drying the samples immediately after the preparation, and then restored when needed upon rehydration of the dried samples. The X-ray diffraction profiles of PVA gels may be considered as arising from the sum of three contributions: free water, crystalline PVA aggregates and swollen amorphous PVA. The results also support the hypothesis that pores are mainly occupied by water and that the porous walls consist of swollen amorphous PVA, while the crystalline domains act as knots of the gel network, validating the schematic structure proposed in Figure 8.19. The

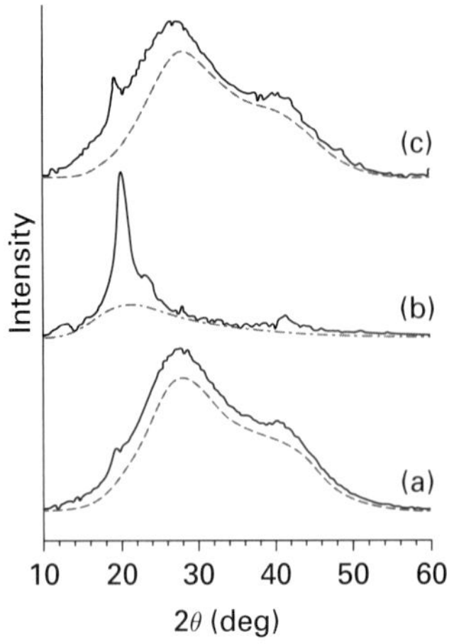

Figure 8.20 X-ray powder diffraction profiles of PVA gel after $N_c = 9$ cycles, in as-formed (a), dried (b) and rehydrated (c) states. The X-ray powder diffraction profile of solvent (liquid D_2O) (dashed curves) and of amorphous PVA (dash-dotted curves) are also shown. Adapted with permission from Ricciardi *et al.* (2004) © 2004 American Chemical Society.

degree of crystallinity and the size of the crystals increased with the number of freeze/thaw cycles and the ageing time (in agreement with previous investigations). For instance, the apparent size of the crystallites increased by 3–4 nm as the number of cycles went from $N_c = 1$ to 10. WAXS patterns of *melt-crystallized PVA films* were also reported, and exhibited sharp crystalline reflections, typical of crystalline PVA with a degree of crystallinity around 64% (a surprisingly high value for a polymer which lacks stereo-regularity, but well established in the case of atactic PVA). Xerogels also exhibit a high degree of crystallinity (up to 50% after $N_c = 9$ cycles).

Comparison of the diffraction patters for as-formed, dried and rehydrated gels is shown in Figure 8.20. Whereas in the hydrogels the main contribution to the diffraction pattern arises from the free water molecules (Figures 8.20a and c), in the dry state (Figure 8.20b) for rehydrated gels, sharp diffraction peaks are clearly observed. This figure well illustrates the difficulties in studying hydrated PVA gels, or in general any network of physical gels, compared to the completely amorphous (sol) or dry state.

In a more recent publication the same authors (Ricciardi *et al.*, 2005) propose a more detailed scheme for the organization of phases inside the polymer-rich domains. They provide detailed information in the medium-range (mesoscopic) length scales by using small-angle neutron scattering (SANS), but this does not alter previous conclusions. They suggest that during the first freeze/thaw cycle the homogeneous solution undergoes phase separation, leading to the formation of a polymer-rich and a polymer-poor phase, accompanied by crystallization of PVA in the polymer-rich phase. The polymer-rich phase was modelled as a collection of polydisperse and homogeneous spherical particles, interacting via hard-sphere potentials. The SANS intensity of the hydrogel is dominated by particle–particle interference. The volume fraction of hard spheres (particles) roughly corresponds to the volume fraction of crystallites in the gel (from X-ray diffraction). The hard-sphere interaction radius R_{HS} corresponds to the average distance between the crystallites. The values of R_{HS} are of the order of 25 nm, whereas the average size of crystallites (R ~3 nm) is close to the value determined by X-ray diffraction. The size polydispersity of the of crystallites is remarkably small, close to 1.1 at $N_c = 9$.

8.6 Cryogels from polysaccharides

Cryogels from polysaccharides have been investigated for a number of years, although most work has been performed on the galactomannans, particularly locust bean gum (LBG). Galactomannans are essentially linear macromolecules with a backbone of β (1→4)-linked D-mannopyranose (Man, M) residues to which are attached (1→6)-linked D-galactopyranose (Gal, G) residues. In LBG the mannose/galactose (M/G) ratio is *c.*4:1 (Chapter 10). However, the mechanism is not restricted to these, and cryogels have been prepared from a number of other polysaccharide systems, including xanthan, curdlan, certain β-glucans and starches. In all relevant cases it seems the water–polymer solution is, in thermodynamic terms, a poor solvent system. This implies that polysaccharide–polysaccharide interactions are relatively favourable. Under equilibrium conditions this would result in either liquid–liquid phase separation or precipitation, but, as we have seen for the synthetic systems, the process may be frozen, literally, into a non-equilibrium state. Below we survey some of these systems.

Perhaps the most studied system is LBG in water (McCleary *et al.*, 1981; Tanaka *et al.*, 1998; Doyle *et al.*, 2006), and here gels form as a result of extended cryogenic treatments (freezing, frozen storage, thawing), whereas at positive temperatures the same solutions remain as fluids during prolonged storage (only after 2 or 3 months do they transform into very weakly structured systems). The mechanism seems to be quite straightforward in that the formation of ice crystals raises the concentration of the polysaccharide and so drives polymer–polymer interactions that they become largely irreversible. When the initial polymer concentration exceeds a certain critical value, which often depends on the particular type of polysaccharide and of the freezing/thawing conditions, formation of freeze-induced gels is observed. Gels are thermo-reversible in nature; they can be melted upon heating and will re-form a cryogel during the next freeze/thaw cycle. The cryogels resemble soft sponges when the initial polysaccharide concentration is about 0.05% w/v, and they release free liquid when lightly compressed; they evolve towards cellular textures at concentrations around 0.75% w/v. The formation of non-spongy and more homogeneous cryogels only occurs when more concentrated solutions (2.5–3% w/v) are used. These materials, in contrast to the less concentrated sponge-like structures, do not release free liquid under moderate compression after cryogenic treatments.

McCleary and co-workers (McCleary *et al.*, 1981) studied a range of galactomannans produced by enzymatic (α-galactosidase) treatment of guar (M/G ratio around 2:1) to convert it to galactose-lean samples. When the M/G ratio lay in the range 80:20 to 85:15, freeze/thaw gels could be produced, whereas at lower galactose content (90:10) a precipitate was formed. Lozinsky *et al.* (2000) provided evidence for the participation

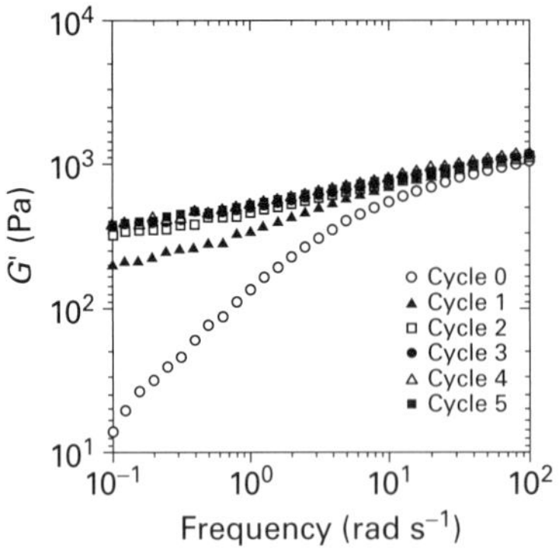

Figure 8.21 G' versus radial frequency for 3% w/w LBG gels after each freeze/thaw cycle (−25°C, followed by heating to 25°C), as indicated. From Tanaka *et al.* (1998) with permission from John Wiley & Sons.

of hydrogen bonds in the formation of LBG cryogels by preparing water solutions containing a chaotropic agent (urea). They observed a decrease in the mechanical strength with a rise in urea concentration, which could be explained by a reduction in the level of hydrogen bonding in these gels due to the chaotropic action of urea. Thus, the junction zones in LBG cryogels are likely to be created by hydrogen bonds between the hydroxyl groups of the galactomannan chains. It was suggested that the gelation process occurs as in PVA, in a two-phase medium, composed of the frozen solvent polycrystals and an unfrozen microphase. The macropores in cryogels are usually open and interconnected. Tanaka *et al.* (1998) showed that G' increases with the number of freeze–thaw cycles, as shown in Figure 8.21. The freeze/thaw behaviour of LBG systems is rather similar to the cryostructuring of starch polysaccharides, and the extreme dependence of the properties of LBG cryogels on the frozen storage temperature is similar to that of PVA. Hence, the same basic mechanisms may be responsible for the formation of non-covalent polymeric cryogels from various types of polymers, either synthetic or natural biopolymers.

Curdlan is a rather unusual microbial polysaccharide in that structurally it is a β (1→3) glucan, and can be produced by several strains of *Agrobacterium* and certain strains of *Rhizobium* (Nishinari, 1988). On heating to around 60°C and then cooling it forms a gel, which on reheating to higher temperatures (>80°C) then forms irreversibly. When either the so-called low- or high-set gels are subjected to freeze/thaw cycling, the gel modulus is increased and the amount of syneresis is decreased. Another class of β–glucans are those with mixed (1→3)(1→4) linkages derived from cereals. These are commonly extracted from oats, although they can also be found in barley and wheat. It has been known for many years that these form pasty products when heated solutions are cooled. Lazaridou and Biliaderis (2004) prepared a number of samples in water, typically 1–3% w/w, by heating them to 85°C and then subjecting them to repeated freeze/thaw cycles (N_c = 3–12). For the higher concentrations, soft cryogels could be produced even after $N_c = 3$ cycles, but the effect was magnified by increasing N_c.

Work by Giannouli and Morris (2003) has shown that xanthan can also form freeze/thaw gels. They prepared some of these by heating aqueous xanthan solutions (0.2–2% w/w) to 80°C , freezing rapidly to −20°C, holding for 24 h and then allowing the frozen mass to warm to 5°C before testing. The resultant gels were relatively stable, but could be melted out by heating in the range 20–40°C. According to the authors, the cryogels have a 'soft, spreadable texture, similar to the consistency of normal jams and marmalades'. However, recovery of structure appeared to occur almost immediately after spreading.

From this it may be concluded that a number of (perhaps all) thermodynamically poor aqueous polysaccharide solutions can be converted into cryogels, either by shock cooling or by employing freeze/thaw cycling. What is perhaps more interesting is to investigate how stable these gels are. For example, if they are left with in an excess of water at room temperature, do they then redissolve? In other words, are they merely trapped as a high-concentration solution that can then redisperse with time, as appears to be the case for LBG, or do they show equilibrium swelling behaviour? This hypothesis appears not to have been widely tested.

8.7 Conclusions

The review of the mechanisms of gelation of well-known, widely used polymers presented in this chapter illustrates the very complex behaviour that physical gels can exhibit. The nature of the solvent and the chain tacticity both play a crucial role in the ability to create a network and to give solutions with interesting mechanical properties. However, the various mechanisms of gelation are very difficult to classify within a clear thermodynamic framework since they include crystallization, phase separation, glass transition or more subtle effects of local conformation, stacking of chains and polymer solvent complexes. Results from the abundant literature suggest that none of these gels is in equilibrium. The contradictory results stress the major influence of the whole history of the sample from dissolution to gel preparation and further thermal treatments. This non-equilibrium, 'frustrated' structure in these and other physical gels is very difficult to characterize from either a theoretical or an experimental point of view. For none of the gels examined in this chapter do rheological measurements link to structural parameters. The experimental observations show the influence on the viscoelastic moduli of polymer concentration, temperature, molecular mass, solvent, thermal history and whole processing. At this stage of investigations, no-one is able to relate rheological parameters directly to topology, morphology or microstructure of the physical gels. However, morphology obviously has a major effect; for example, by creating different morphologies via processing, initially poor mechanical properties of a PVA cryogel can be greatly improved. This ability to modify material properties by various treatments applied to physical gels reminds us of the traditional metallurgy approach. We believe that this trend will develop very much in the future.

References

Alfrey, T., Weiderhorn, N., Stein, R. S., Tobolsky, A. V., 1949. *Ind. Eng. Chem.* **41**, 701–703.

Aoki, Y., 2001. *Macromolecules* **34**, 3500–3502.

Aoki, Y., Li, L., Kakiuchi, M., 1998. *Macromolecules* **31**, 8117–8123.

Arnauts, J., Berghmans, H., Koningsveld, R., 1993. *Makromol. Chem.* **194**, 77–85.

Barendswaard, W., Litvinov, V. M., Souren, F. *et al.*, 1999. *Macromolecules* **32**, 167–180.

Berghams, H., Donkers, A., Frenay, L. *et al.*, 1987. *Polymer* **28**, 97–102.

Berghmans, M., Govaerts, F., Overbergh, N., 1979. *J. Polym. Sci., Part B: Polym. Phys.* **17**, 1251.

Berghmans, M., Thijs, S., Cornette, M. *et al.*, 1994. *Macromolecules* **27**, 7669–7676.

Bosscher, F., Ten Brinke, G., Challa, G., 1982. *Macromolecules* **15**, 1442–1444.

Ceccorulli, G., Pizzoli, M., Pezzin, G., 1977. *J. Macromol. Sci., PartB: Phys.* **14**, 499–510.

Cho, K., Park, S. H., 2001. *Macromol. Symp.* **166**, 93–102.

Dorrestijn, A., Keijzers, A. E. M., te Nijenhuis, K., 1981. *Polymer* **22**, 305–312.

Doyle, J. P., Giannouli, P., Martin, E. J., Brooks, M., Morris, E. R., 2006. *Carbohydr. Polym.* **64**, 391–401.

François, J., Gan, J. Y. S., Guenet, J. M., 1986. *Macromolecules* **19**, 2755–2760.

Giannouli, P., Morris, E. R., 2003. *Food Hydrocolloids* **17**, 495–501.

Guerrero, S. J., Keller, A., 1981. *J. Macromol. Sci., PartB: Phys.* **20**, 167.

Guerrero, S. J., Keller, A., Soni, P. L., Geil, P. H., 1980. *J. Polym. Sci., Part B: Polym. Phys.* **18**, 1533–1559.

Hassan, C. M., Peppas, N. A., 2000. *Macromolecules* **33**, 2472–2479.

Hong, P. D., Chen, J. H., 1998. *Polymer* **39**, 711–717.

Inoue, T., 1975. Gelled vinyl alcohol polymers and articles therefrom. U.S. Patent 3 875 302, 7 August 1975.

Izumi, Y., Suzuki, J., Katano, S., Funahashi, S., 1995. *Phys. B (Amsterdam)* **213**, 724–726.

Kakiuchi, M., Aoki, Y., Watanabe, H., Osaki, K., 2001. *Macromolecules* **34**, 2987–2991.

Keller, A., 1995. *Faraday Discuss.* **101**, 1–49.

Koltisko, B., Keller, A., Litt, M., Baer, E., Hiltner, A., 1986. *Macromolecules* **19**, 1207–1212.

Komatsu, M., Inoue, T., Miyasaka, K., 1986. *J. Polym. Sci., Part B: Polym. Phys.* **24**, 303–311.

Lazaridou, A., Biliaderis, C. G., 2004. *Food Hydrocolloids* **18**, 933–947.

Li, L., Aoki, Y., 1997. *Macromolecules* **30**, 7835–7841.

Li, L., Aoki, Y., 1998. *Macromolecules* **31**, 740–745.

Li, L., Uchida, H., Aoki, Y., Yao, M. L., 1997. *Macromolecules* **30**, 7842–7848.

Lozinsky, V. I., Damshkaln, L. G., Brown, R., Norton, I. T., 2000. *Polym. Int.* **49**, 1434–1443.

McCleary, B. V., Amado, R., Waibel, R., Neukom, H., 1981. *Carbohydr. Res.* **92**, 269–285.

Mutin, P. H., Guenet, J. M., 1989. *Macromolecules* **22**, 843–848.

Nishinari, K., 1988. Food hydrocolloids in Japan. In Phillips, G. O., Wedlock, D. J., Williams, P. A. (eds), *Gums and Stabilisers for the Food Industry* **4**. IRL Press, Oxford, pp. 373–390.

Ogasawara, K., Nakajima, T., Yamaura, K., Matsuzawa, S., 1975. *Prog. Colloid Polym. Sci.* **58**, 145–151.

Ogasawara, K., Nakajima, T., Yamaura, K., Matsuzawa, S., 1976. *Colloid Polym. Sci.* **254**, 982–988.

Peppas, N. A., 1975. *Makromol. Chem.* **176**, 3433–3440.

Pyrlik, M., Rehage, G., 1975. *Rheol. Acta* **14**, 303–311.

Ricciardi, R., Auriemma, F., De Rosa, C., Lauprêtre, F., 2004. *Macromolecules* **37**, 1921–1927.

Ricciardi, R., Mangiapia, G., Lo Celso, F. *et al.*, 2005. *Chem. Mater.* **17**, 1183–1189.

Scherrenberg, R., Reynaers, H., Mortensen, K., Vlak, W., Gondard, C., 1993. *Macromolecules* **26**, 3205–3211.

Spevacek, J., Schneider, B., 1987. *Adv. Colloid Interface Sci.* **27**, 81–150.

Stein, R. S., Tobolsky, A. V., 1948. *Text. Res. J.* **18**, 302–314.

Tan, H., Moet, A., Hiltner, A., Baer, E., 1983. *Macromolecules* **16**, 28–34.

Tanaka, R., Hatakeyama, T., Hatakeyama, H., 1998. *Polym. Int.* **45**, 118–126.

te Nijenhuis, K., Dijkstra, H., 1975. *Rheol. Acta* **14**, 71–84.

te Nijenhuis, K., Winter, H. H., 1989. *Macromolecules* **22**, 411–414.

Vandeweerdt, P., Berghmans, H., Tervoort, Y., 1991. *Macromolecules* **24**, 3547–3552.

Walter, A. T., 1954. *J. Polym. Sci.* **13**, 207–228.

Watanabe, W. H., Ryan, C. F., Fleischer, P. C., Garrett, B. S., 1961. *J. Phys. Chem.* **65**, 896.

Watanabe, H., Osaki, K., Kakiuchi, M., Aoki, Y., 2001. *Macromolecules* **34**, 666–670.

Watase, M., Nishinari, K., 1985. *J. Polym. Sci., Part B: Polym. Phys.* **23**, 1803–1811.

Wellinghoff, S., Shaw, J., Baer, E., 1979. *Macromolecules* **12**, 932–939.

Willcox, P. J., Howie Jr, D. W., Schmidt-Rohr, K. *et al.*, 2000. *J. Polym. Sci., Part B: Polym. Phys.* **37**, 3438–3454.

Wu, W. L., Kurokawa, H., Roy, S., Stein, R. S., 1991. *Macromolecules* **24**, 4328–4333.

Xie, X. M., Tanioka, A., Miyasaka, K., 1990. *Polymer* **31**, 281–285.

Yanxiang, W., Deyan, S., 1997. *Polym. Bull.* **39**, 633–638.

Yokoyama, F., Masada, I., Shimamura, K., Ikawa, T., Monobe, K., 1986. *Colloid Polym. Sci.* **264**, 595–601.

9 Colloidal gels from proteins and peptides

9.1 Introduction

One of the most interesting and scientifically richest class of physical gels are those formed from proteins and polypeptides (Doi, 1993; Clark, 1998; Foegeding, 2006). Here we distinguish 'true' gels from loose colloidal aggregates formed, for example, by coagulation and collapse of otherwise stable colloids, by the extra requirement that the 'particles' cannot be redispersed simply by agitation; in other words, that the interaction between particles is of the order of 10kT or more. The simplest common example of this class of materials is the easily recognized gel formed by boiling an egg. Classically, all of these networks were once thought of as being formed from the polypeptide backbone itself (i.e. through complete unfolding), much in the way fibrous networks of denatured collagen are formed. However, it is now appreciated that they are usually constructed in a multi-step process from 'partially folded still globular entities' (known as the 'string-of-beads' interpretation).

In the distinction first made explicit by Hermansson and co-workers in the 1980s, these may be divided into coarse-stranded (sometimes simply referred to as 'particulate' or colloidal) gels, i.e. those having a heterogeneous and phase separated appearance; and fibrillar gels, i.e. those formed of more uniform assemblies of fine strands (Stading and Hermansson, 1991; Langton and Hermansson, 1992; Stading *et al.*, 1992). The latter are now often referred to as 'amyloid', because such fine-stranded materials have been studied very extensively over the last few years, revealing many apparent similarities with the amyloid structures found in disease conditions such as Alzheimer's.

We will discuss the latter systems later, as this is an active area of research. However, since the fibrils are formed from particles, they can still be regarded as 'particulate' at one level. Moreover, the emphasis seen in some of the literature might suggest that all relevant work has been performed in the last decade, but this is certainly not the case. Indeed highly fibrillar insulin gels were known and studied in the 1950s (Koltun *et al.*, 1954), and again by Clark and co-workers in the 1980s (Clark *et al.*, 1981a, 1981b). Later in this chapter, we include a brief description of the very specific ordered fibrous assemblies produced by actin, tubulin and fibrinogen, although the biology of these is rather specific and we cannot really consider them to be full members of this class. As with the other physical gels in this volume, the principal aims are to investigate phenomena over a range of length scales and if possible to establish descriptive models. Finally we introduce, albeit briefly, those gels formed from specially synthesized peptides – again a very topical subject.

What most of the simple particulate and fibrillar systems have in common is that they are formed from proteins in aqueous (or electrolyte) solutions that have become partly denatured, generally irreversibly, by some physical or chemical treatment. This distinguishes the particulate gels from those prepared by treating a protein with strong denaturant such as 8 M urea (van Kleef *et al.*, 1978). This treatment reduces the protein to its flexible polypeptide backbone, and such materials, not discussed here, have much more in common with chemical gels.

It is now appreciated that in, for example, simple heat-induced denaturation, the protein size and shape is only mildly perturbed. Instead some of the hydrophobic groups, which at ambient temperatures remain buried in the protein core, become exposed above some minimum unfolding or denaturation temperature. This ‘hydrophobic effect’ leads to aggregation, forming either the fine-stranded networks or the amorphous particulate structures of the physical gel (Clark and Ross-Murphy, 1987, 2009; Clark, 1998; van der Linden and Foegeding, 2009). In succeeding sections we discuss the various mechanisms and their structural and macroscopic consequences.

9.2 Colloidal gels formed from partially denatured proteins

9.2.1 Structural aspects

The production of gels from solutions of globular proteins usually requires a degree of unfolding of the protein as a first step. The details of secondary and tertiary structure of globular proteins, and the mechanisms of folding into a specific globular structure, are outside the scope of this volume, but it is now clearly established that the various globular structures are stabilized by a combination of α-helix, β-sheet and other secondary structure components, further held together by hydrogen bonding between non-adjacent peptide residues.

As mentioned above, it was once assumed that protein gels were formed only if this entire secondary and tertiary structure was disrupted, as in the urea example above. However, structural work carried out by a number of groups some 30 years ago, using X-ray diffraction, Fourier transform infrared (FTIR) and electron microscopy, established quite clearly that this was not the case. The picture instead is that, under appropriate mildly denaturing conditions, a protein will partly denature via various intermediates (Chiti and Dobson, 2006) and then partially renaturate in a slightly different manner to create non-covalent interactions between various proteins, and this is what subsequently leads to gelation.

Most of the published work involves either serum albumen, particularly from bovine sources (BSA), or the milk protein β-lactoglobulin (β-Lg) (Donald, 2008). This is not because these are interesting or particularly special in their own right, but more pragmatically because they can be obtained in reasonably pure form at comparatively low cost. For the class of experiments of interest to a gel researcher, grams – even tens of grams – of material are required. In much of this chapter, then, because of the amount of published work we will naturally tend to concentrate on gels formed from these two proteins.

Little of the generic science depends on which of these two is studied. What does matter, however, is the detail of the preparative conditions. This includes temperature (and heating rate), pH, ionic strength (and specific concentration of cations), and even the particulate batch number if the protein is obtained from a commercial source. This explains why much of the work has been to try and deliver results in a 'scaled' form so that comparison within and between resultant gels can be made, without becoming stalled by practical detail.

From all of the published work, the overall picture is that heating a sufficient concentration of protein – typically ~10% w/w – to just above the denaturation temperature – typically 60–80°C, depending upon other conditions – causes partial denaturation to take place. This leads to the formation of inter-protein β-sheet, giving rise to a polymeric network, the modulus of which is in turn accentuated by more prolonged heating, and/or by cooling back to room temperature (Clark, 1998). In this picture, the 'strands' of the network are typically a few protein diameters wide – around 5–10 nm – so these differ substantially from the so-called 'molecular networks' of gelatin and most other physically aggregated gel network strands, which are typically 1 nm in width (Figure 1.1).

Spectroscopic methods can be used to investigate the detailed structures of aggregates formed during the heating and gelling of protein solutions. As mentioned in other chapters, changes in structure at a molecular level (native-to-denatured) can be monitored by optical techniques such as optical rotatory dispersion (ORD), circular dichroism (CD), FTIR and Raman spectroscopy (Clark and Ross-Murphy, 1987). The loss of the native structure during aggregation and the development of new ordered structure (such as β-sheet) are usually significant, and CD and ORD have been used to calculate the relative proportions of various secondary structures present in both native and denatured forms. A limitation of the CD method is that suitable measurements can only be made on highly dilute samples (concentrations of 0.01–0.1% w/w).

Infrared and Raman spectroscopy, on the other hand, have been used to monitor changes in protein solutions as they occur at higher concentrations. FTIR spectroscopy is best for monitoring changes in β-sheet, although solutions have to be prepared in D_2O, while Raman and CD are better at following changes involving α-helix. For example, Clark *et al.* (1981a, 1981b) used scanning infrared and Raman spectroscopy to monitor the aggregation of a range of globular proteins. Development of β-sheet during aggregation is indicated by the development of a 'shoulder' (around 1620 cm^{-1}) in the Amide I carbonyl stretching band. This is observed more or less generally, although the amount of sheet seems to vary from system to system.

Traditional microscopy methods (i.e. using light or electrons) can be used to probe aggregate and gel structures over longer distances. Since gelled and aggregated solution samples contain large amounts of water, extensive sample preparation is required before an image can be formed. However, one aspect which can be investigated in detail with conventional (transmission) electron microscopy is the degree of network homogeneity and the nature – transparent or opaque – of the gel. Our archetype – ovalbumin, the protein gel traditionally formed by boiling a hen's egg – is of course opaque, but this is not an intrinsic property. In principle, and indeed in practice, a transparent egg 'white' can be prepared by dialyzing away some of the naturally occurring salt, or alternatively treating in

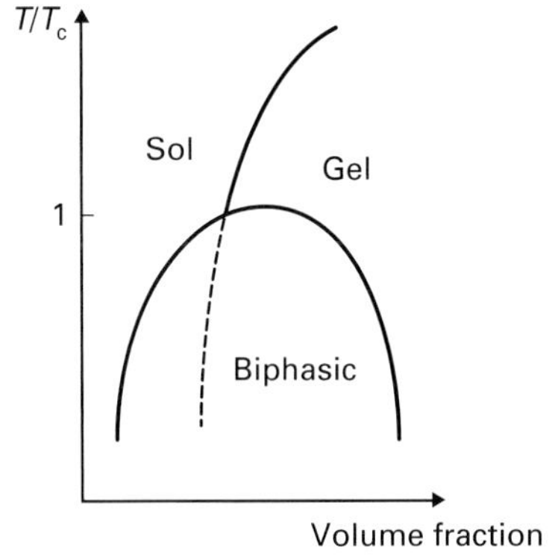

Figure 9.1 Sol–gel state diagram showing typical behaviour for protein gels prepared by heating at pH values away from pI.

the ancient Chinese fashion (Doi, 1993; Eiser *et al.*, 2009). One critical parameter in determining whether a transparent gel is formed is the protein isoelectric point pI.

For BSA, pI ~ 5.5 (although some sources give values as low as 4.7), while for β-Lg it is 5.1. Close to the pI, a coagulate is formed rather than a gel. By increasing the charge, increasing the ionic strength or altering the pH away from pI, increasingly transparent gels tend to be formed. Conversely, as pI is approached, a gel can still be formed but it tends to be opaque and to show significant syneresis (propensity for water release), and so enters a two-phase region. This, qualitatively at least, is consistent with predictions from the theory of Coniglio *et al.* (1982). Figure 9.1, not specific to any protein, shows typical behaviour for gels prepared at pH values away from pI. A clear illustration of these tendencies is given in the early work of Clark and co-workers (Clark *et al.*, 1981a). In Figure 9.2, TEM images of BSA under various conditions show changes in network heterogeneity. At or close to pI, a very dense globular structure is seen.

Much work has been performed on the microstructure of β-Lg gels, especially by the Hermansson group. From this work it was concluded that various gel structures occur, depending on pH and heating regime. For example, Langton and Hermansson (Langton and Hermansson, 1992) describe what they call 'particulate' networks, at pH values in the range 4–6, which are (a) 'regular' in the pH range 5–6, with uniform distributions of particles and pores; (b) 'irregular' at pH 4.5, where there are coarse, uneven aggregates; and (c) a mixture of network types at pH 4 with clusters of particles embedded in a fine-stranded network. The particles in these gels are almost spherical (probably formed by gelling liquid droplets) and their size depends on pH. Below pH 4 and above pH 6, fine-stranded networks (discussed later in this chapter) are formed, with shorter and apparently stiffer linear strands occurring at the lower pH. As the pH is lowered still further, the width and length of the strands increases, with a similar effect occurring above pH 6.

A recent EM study has indicated that at pH 7 short 'worm-like' rods of width ~3–5 nm (i.e. close to a single monomer in width) are formed, in agreement with earlier dynamic light scattering analysis (Griffin *et al.*, 1993; Kavanagh *et al.*, 2000a, 2000b, 2000c). Other microscopy data has suggested, however, that spherical particles of diameter

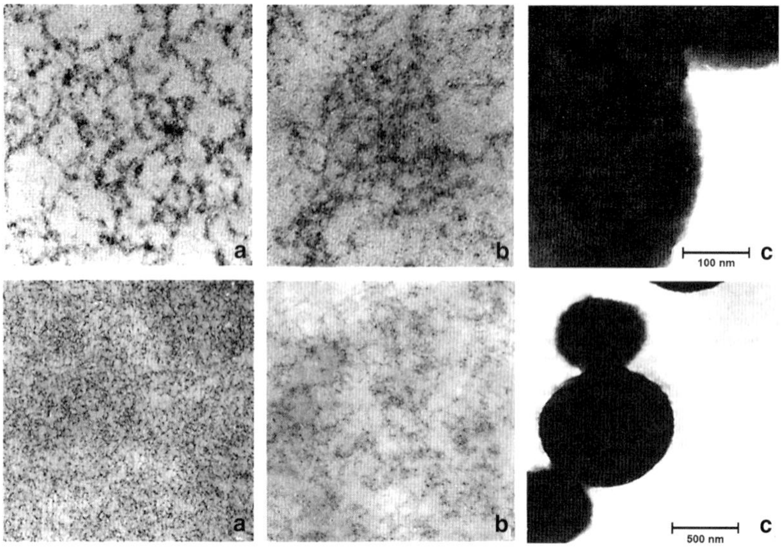

Figure 9.2 TEM images of BSA under various conditions: (a) pH 6.5, 10% w/w transparent gel in water; (b) pH 6.5, 10% w/w turbid gel in 125 mN NaCl; (c) pH 5.1, 10% w/w coagulate at the isoelectric point. The upper and lower images correspond to different magnifications, as indicated for (c). Reproduced with permission from Clark *et al.* (1981a) © 1981 Munksgaard International Publishers, a John Wiley & Sons subsidiary.

35–70 nm are created, although these structures are probably made up of rod-like particles, also with widths ~3–5 nm (Carrotta *et al.*, 2001). In agreement with this latter picture, neutron scattering results at high ionic strength suggest the formation of spherical structures ~20 nm in diameter (Aymard *et al.*, 1996b).

Most recently of all, probe microscopy methods have been very widely employed – perhaps reflecting the ready availability of such instrumentation – and atomic force microscopy (AFM) has proved of great value. However, most of the latter work has been applied to the class of fibrillar gels now designated as amyloid gels (Hughes and Dunstan, 2009), discussed in more detail in Section 9.5.

Nevertheless, Gosal and co-workers (Gosal *et al.*, 2004a) carried out an AFM analysis of aggregates formed when β-Lg was heated for 24 h at 80°C at pH 7. The concentration employed was 4% w/w, which is significantly below the gel concentration. In agreement with previous EM investigations, short rod-like particles were observed, with a typical height of ~1.8 nm. However, this differs from previous EM estimates, where widths of ~3–5 nm were found (Clark *et al.*, 1981a, 1981b).

A great deal of work, especially including light scattering studies, has concentrated on the specifics of the early aggregation stages. In some of these publications, a combination of static and dynamic light scattering has been used to determine the sizes and shapes (i.e. average molecular masses and radii of gyration) of the aggregates formed, including fractal dimensions (Griffin and Griffin, 1993; Griffin *et al.*, 1993; Gimel *et al.*, 1994; de Kruif *et al.*, 1995). From work of this kind it was concluded that the mode of aggregation of β-Lg at neutral pH and at a given ionic strength was essentially unaltered over a range of concentrations and heating times. In relation to this mechanism, Griffin and Griffin (Griffin and Griffin, 1993; Griffin *et al.*, 1993) and de Kruif *et al.* (1995) proposed that the aggregation process occurs in two stages. First, a primary linear aggregate is formed, followed by more general aggregation of these species. The work by Gimel (Gimel *et al.*, 1994), however, suggests that the primary aggregate is globular rather than linear, with a size independent of the initial protein concentration. This globule then forms fractal aggregates.

To summarize, a number of workers have tried to establish more specific assembly mechanisms, especially for β-Lg. However, the techniques employed, including neutron and light scattering, valuable though they are (Nicolai and Durand, 2007), can only measure such parameters as strand thickness, and cannot address the most significant aspect, particularly for this volume, which is why and how the linear segments branch or intertwine to give rise to the structure necessary for gel formation. Complications are due to a number of additional factors, and the precise balance of structural pre-gel intermediates appears controversial. In this respect, and for these systems, lower-resolution techniques such as TEM and optical microscopy have, in our view, proved more useful.

While the gelation of animal globular proteins, particularly those from milk, has been studied widely, until recently plant (particularly soybean) protein gelation has been less extensively investigated, although work carried out on glycinin (11S) and β-conglycinin (7S) soy proteins (Nagano *et al.*, 1994a, 1994b) found that the FTIR band at 1618 cm^{-1}, normally associated with β-pleated sheet, begins to increase around 80°C for 11S and around 65°C for 7S. In previous work, it was not possible to measure FTIR in aqueous solution because of strong interference from water absorption, but in this case the protein solution in D_2O was set between two CaF_2 discs using a spacer. These results were interpreted to suggest that heat-induced gels of glycinin and β-conglycinin are formed just like the milk proteins, by the formation of intermolecular β-sheet. In FTIR measurements for heated 7S in the gel state (Nagano *et al.*, 1994a), the β-sheet peak at 1620 cm^{-1} was found to increase with decreasing pH, consistent with findings of Clark *et al.* (1981a, 1981b) for the gelation of other globular proteins, including BSA, insulin and lysozyme.

More recently, Samoto *et al.* (2007) reported that earlier descriptions of soybean protein were insufficient and, in addition to the previously known 11S (glycinin) and 7S (β-conglycinin), they reported the existence of a new protein portion which they termed LP (lipophilic protein). They found that LP itself had no strong gelling ability but it interfered with the gelling behaviour of 7S and 11S.

9.2.2 Rheological measurements

Rheological properties of these systems have been carried out by a considerable number of workers. Before around 1980, such measurements were often made using home-designed, sometimes crude, equipment. Experiments reported the dependence of the gels' modulus on concentration (Clark and Ross-Murphy, 1987) or the properties of 'fully cured' gels under various conditions of pH, ionic strength (as NaCl concentration) and BSA concentration (Richardson and Ross-Murphy, 1981). The essential conclusions were that (a) for a given set of gelling conditions, a critical gel concentration could be established, although this varied considerably (for these systems, from ~3% to ~12% w/w) depending on pH and [NaCl]; (b) log G' showed only a slight increase with ω; (c) log G'' showed a slight minimum whose frequency increased with temperature; (d) both G' and G'' fell with increasing temperature, but this was essentially reversible on cooling; (e) linear viscoelastic behaviour was found up to quite large values (>10% strain), especially far from pI.

All of these essentially qualitative observations have been confirmed for other globular proteins under equivalent conditions. Many such data has been collected, particularly for β-Lg (Stading *et al.*, 1992, 1993a) but also for α-lactalbumin (Kavanagh *et al.*, 2000a, 2000c), ovalbumin (Koike *et al.*, 1996), soy glycinins (Kohyama *et al.*, 1995a, 1995b) and a number of other systems. Indeed it seems that, at one level, the species and type of globular protein is not significant. On the contrary, by adjusting pH and salt conditions, essential commonality of behaviour can be seen.

For example, rheological studies performed by Stading and Hermansson (Stading *et al.*, 1993b) determined the critical gelation concentration of β-Lg. The samples were pH adjusted over a range of pH, but the effect of additional salt was not pursued. Samples were heated to 95°C and held there for 1 h, and critical gelation concentration was defined as the concentration at which G' exceeded the noise level after heating. At intermediate pH levels (4.5–5.6) close to the protein isoelectric point, the critical concentration was as low as 1% w/w, while at lower pH values (2.5 and 3) the critical concentration increased to 5% w/w, and increased further to 10% w/w when the pH was 7.

Subsequently, Renard and Lefebvre (Renard and Lefebvre, 1992) examined the effects of pH and ionic strength. Various concentrations of β-Lg were prepared at a variety of pH values (2, 5, 6, 7 and 9) and ionic strengths (0–0.15 M NaCl). The samples were heated in sealed tubes and c_0 determined, using the simple method of ensuring the meniscus did not deform upon tilting the tube. At extreme pH values (2 and 9), high critical concentrations were found to be *c.*8% w/w. Increasing the salt content reduced c_0, while at pH values near the isoelectric point (pH 5, 6, 7) values were close to 1% w/w, regardless of the salt content. Because of differences in the samples and heating regimes employed, only semi-quantitative agreement would be expected between the results obtained by these two groups.

As mentioned above, the low pH systems (Section 9.5) are fibrillar in nature. One interesting observation in gelation kinetics measurements is that, somewhat unexpectedly, G' is already higher than G'', even from the beginning of the measurement, as pointed out by several workers (Matsumoto and Inoue, 1993; Inoue, 1994; Ikeda and Nishinari, 2000, 2001; Ikeda *et al.*, 2000). This tendency has been recognized for other protein concentrations and temperatures, and also at various frequencies and strains. One possibility is that the sample is already structured before the gelation point, in a way analogous to a charged colloid system (Nicolai and Durand, 2007).

Figure 9.3 shows typical results for the gelation kinetics of β-Lg, heated at pH 7 to 80°C. As can be seen just by inspection of this data, the critical concentration must be below 13% w/w: modelling the results gives a value around 12.1%. The corresponding frequency sweep follows the behaviour suggested above, and the minimum in G'' is centred around 0.3 rad s^{-1}.

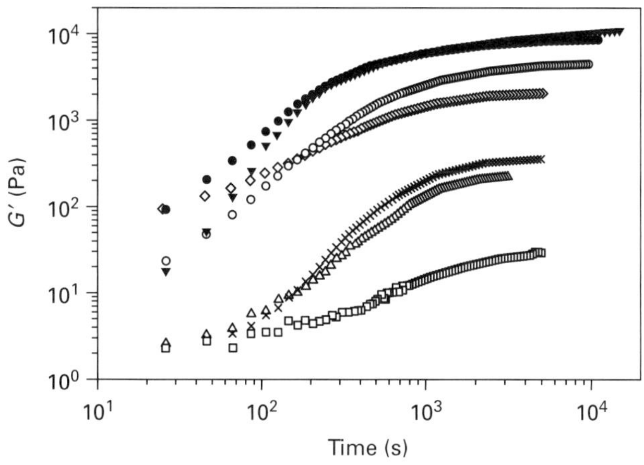

Figure 9.3 Typical results for the gelation kinetics (G') of β-Lg, heated at pH 7 to 80°C: 13.0% w/w (□), 13.2% w/w (Δ), 13.5% w/w (×), 14.0% w/w (◊), 14.8% w/w (○), 15.9% w/w (●), 16.7% w/w (▲). Reproduced with permission from Kavanagh *et al.* (2000a) © 2000 American Chemical Society.

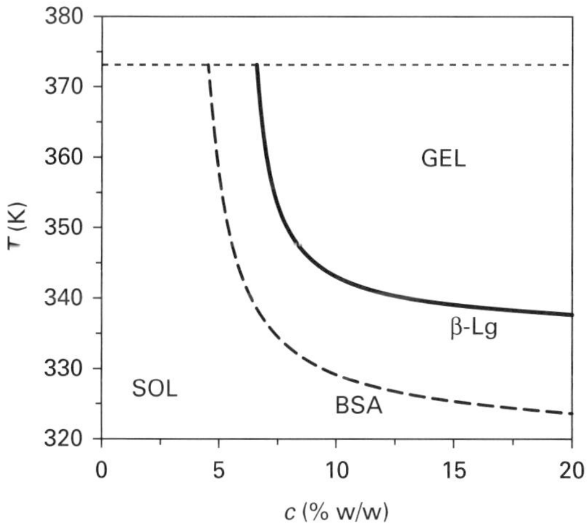

Figure 9.4 Tobitani diagram of infinite gelation time as a function of concentration and temperature. Reproduced with permission from Tobitani and Ross-Murphy (1997a) © 1997 American Chemical Society.

One approach which may have potential for further generalization is the pseudo-kinetic model suggested by Tobitani and Ross-Murphy (1997a, 1997b). In this model, the gelation time is the investigated parameter. A series of experiments is carried out, and the actual gelation time corresponding to a particular concentration and temperature established. By interpolation and extrapolation of the data set, a locus of nominally infinite gelation time is established, which is a useful form of state diagram (Figure 9.4). In the diagram the upper region represents the gel state and the lower the sol state. The boundary between the sol and the gel regions corresponds to the gelation curve, and with this diagram the possibility of gel formation for any given concentration and temperature can be predicted. Results are shown here for both β-Lg and BSA; the latter lies below and to the left, so the system gels more readily.

Experimental studies on this type of state diagram have been carried out on other systems, using a variety of techniques, by Kawanishi and Tan for synthetic polymers (Tan *et al.*, 1983; Kawanishi *et al.*, 1986) and by Tanaka for biopolymers (Tanaka *et al.*, 1979).

In most of these diagrams, the two-phase region, as well as the gel region, was located on the lower-temperature side, i.e. a two-phase region exists at lower temperatures, giving an upper critical solution temperature (UCST) diagram. By contrast, when the two-phase region is located on the higher-temperature side, as here, we have a lower critical solution temperature (LCST) phase diagram. Work by San Biagio and co-workers (San Biagio *et al.*, 1996) suggests that complex LCST type phase behaviour (there attributed to spinodal demixing) is seen for unfolded BSA, but does not appear to be present with the native protein.

Using small-deformation oscillatory measurements, Kohyama and Nishinari (1993) compared the gelation process of 11S and 7S soy proteins in the presence of the common acidifier glucono-delta-lactone (GDL). The gelation rate was greater and the gelation time shorter for 11S than for 7S. Nagano *et al.* (1994a) made small-deformation measurements on soy proteins and found that, by plotting log G' for glycinin (80°C) and β-conglycinin (65°C) against the log of FTIR absorption at 1618 cm^{-1} from their work described above, a very good correlation was seen.

Kohyama and Nishinari (1993) also examined the effect of GDL on large-deformation uniaxial compression of gels as a function of time and found a tendency similar to their small-deformation results. There are very few other reliable measurements in the large-deformation regime, except those from Foegeding and co-workers (Foegeding, 2006). More recently it has become possible to study (small amounts of) protein gels during fracture under the microscope, especially using the environmental confocal scanning method (Olsson *et al.*, 2002). By using a tension–compression stage, changes in stress and strain can be correlated with changes in microstructure, as demonstrated by Plucknett and co-workers (Plucknett *et al.*, 2001). Using essentially this method, Ohgren *et al.* (2004) were able to demonstrate that the fracture behaviour of particulate β-Lg gels changed with protein concentration. A ‘dense’ gel showed brittle behaviour when the clusters were rigid, and the crack propagated smoothly compared to a gel with an open network structure, which showed discontinuous crack growth. Differences in the extensibility of the aggregated β-Lg structure, induced by addition of a polysaccharide (amylopectin) solution, were shown and these were related to differences in stress–strain behaviour and crack propagation.

While all of the above involve heat-set gels, a recent innovation in this area is the discovery and characterization of globular protein gels that form on cooling. In effect, solutions are heated at concentrations below the critical value and then Ca^{2+} (or presumably other divalent cations) added. The work by Bolder and co-workers (Bolder *et al.*, 2006) has characterized a number of these, including from β-Lg and from mixtures of this with other whey proteins. Other work has examined the same effect for soy protein gels (Maltais *et al.*, 2005). At one level there is no great surprise in this finding: one might expect divalent ions to alter gelation characteristics, just as they do for polysaccharide gels (Chapter 5). At the same time, changing the ionic strength will also encourage gelation; this is an interesting area which may develop further in the future.

9.3 Gels from milk proteins

Of course β-Lg is a milk protein, but this section is limited to work on casein and whey protein isolate systems. Casein gels have long been associated with products such as cheese and yoghurt. Some are purely physical gels, but most involve the enzymatic action of rennet, the common name for the enzyme chymosin, usually extracted from the fourth stomach of young calves. Caseins remain the most important proteins in milk, and one of the four main caseins, κ- casein, located on the surface of spherical casein micelles, stabilizes casein micelles by steric repulsion. Chymosin cleaves the 'hairs' on the surface of the casein spherical particles, disrupting the steric stabilization, and then aggregation occurs.

9.3.1 Casein micelles

The phosphorylated caseins, including αS1-, αS2-, β- and κ- derivatives, make up a large part of the protein content of milk. The major component of the proteins (90%) occurs as casein micelles, and both κ-caseins and phosphate groups occur widely at the micelle surface, while α- and β-caseins dominate the inner part (Figure 9.5). Since the stability of casein micelles depends on both electrostatic and steric effects, at pH 6.7 electrostatic conditions alone are not sufficient to maintain the micelles in milk (Payens, 1979). Instead it is understood that the hydrophilic C-terminal part of

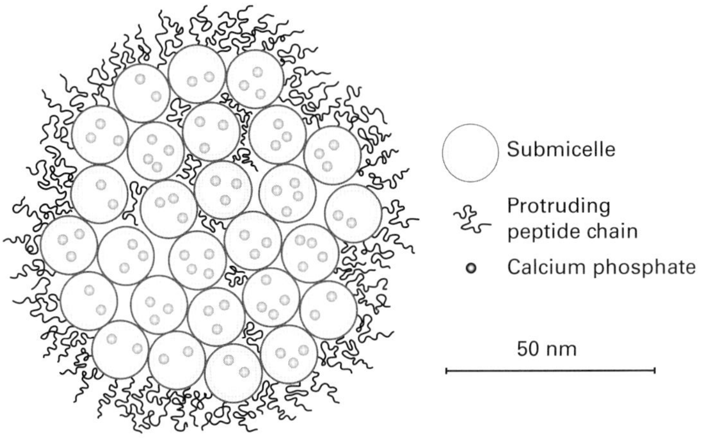

Figure 9.5 Schematic cross-section of a casein micelle. Reprinted with permission from Walstra (1999) © 1999 Elsevier.

κ-casein, the glycomacropeptide (commonly abbreviated GMP), mainly located at the outside of the casein micelles, protrudes from the casein micelle surface into the solution. This is the species that helps stabilize the casein micelles by the steric mechanism (Walstra, 1999).

Flocculation of casein micelles and subsequent gel formation can then be induced either at neutral pH and temperatures above 20°C by the addition of a proteolytic enzyme (rennet), which acts specifically to cleave off the glycomacropeptides, or by lowering the pH. After acidification of a milk dispersion to pH 4.6 (the isoelectric pH of the casein particles), a physically stable suspension of casein particles is produced at low temperatures. Gelation results from the aggregation of these casein particles at temperatures above 10°C, but these particles have a complex structure due to the association of numerous different casein molecules.

9.3.2 Gelation kinetics

κ-casein is a substrate for chymosin and plays the most important role in the gelation of casein micelles. The substrate specificity of chymosin is very high, and it attacks only one specific bond in κ-casein. Gel formation of casein micelles consists of three steps: (i) cleavage induced by chymosin upon the κ-casein within micelles, (ii) aggregation of the destroyed κ-caseins and (iii) consolidation by the aggregation of micelles. The various stages are identified by turbidity, and by yield stress measurements when the gel forms (Walstra *et al.*, 1985). Of these stages, only the first step, the proteolytic cleavage, can be monitored totally independently of the others by following the release of the glycomacropeptide, or the formation of para-κ-casein. The aggregation reaction of the destabilized micelles is a consequence of this proteolysis, but its rate cannot be separated easily from that of the proteolysis reaction, and aggregation overlaps the proteolysis reaction (Figure 9.6; Horne and Banks, 2004).

The early aggregation phase can be followed readily by turbidity or light scattering (Payens *et al.*, 1977; Dalgleish *et al.*, 1981), whereas gel formation and development is most easily monitored in the laboratory by rheometry (Tokita *et al.*, 1982).

The shear modulus appears suddenly at a so-called latent time and increases with reaction time. This latent time increases with decreasing enzyme concentration, while the curvature of the gelation curve, which reflects the rate constant of the gelation process, depends on the concentration of enzyme. However, gelation curves seem to converge to the same value of the shear modulus at an infinite reaction time, i.e. the final value of the elastic modulus G of a casein gel is independent of enzyme concentration. Niki and co-workers (Niki *et al.*, 1994a, 1994b) prepared casein micelles of various sizes from skim milk using differential centrifugation, and the observed gelation curve as a function of time, after rennet was added to the casein micelle solutions, was apparently well approximated as a first-order reaction.

Measurement of non-protein nitrogen (NPN) liberated from casein micelles has been conventionally used to follow the progress of the enzymatic phase of milk-clotting. Hooydonk and Olieman (1982) estimated the amount of the glycomacropeptide (GMP) using an HPLC system, and measured the extent of enzymatic reaction after the addition

(a)

(b)

(c)

(d)

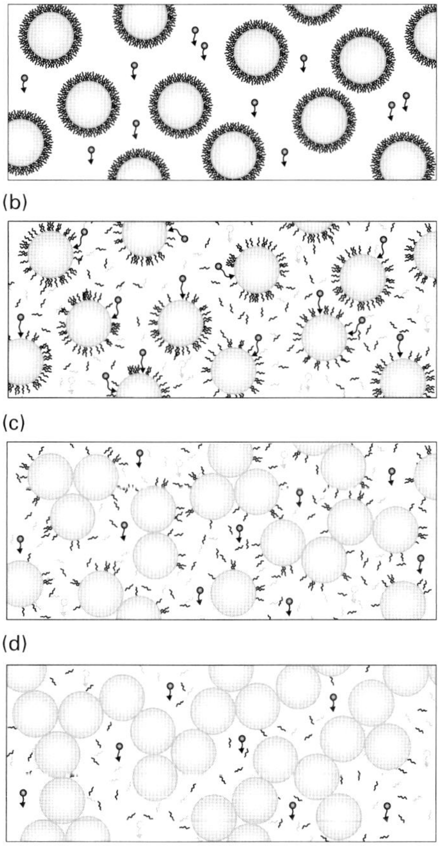

Figure 9.6 Schematic of the stages envisaged for enzymatic coagulation of milk, starting from the initial mixture of casein micelles and enzyme (a) and proceeding through proteolysis (b), initial aggregation into small clusters (c) and attainment of the gel point at percolation (d). Adapted from Horne and Banks (2004) © 2004 Elsevier.

of chymosin. Niki *et al.* used the method of Hooydonk and Oliemann to estimate GMP from κ-casein in casein micelles by rennet and to monitor the reaction of chymosin, and the liberated GMP as a function of time (Ekstrand *et al.*, 1978; McGann *et al.*, 1980; Niki *et al.*, 1994a, 1994b).

The gel time for small micelles was shorter than that for large micelles, in good agreement with previous finding (Ekstrand and Larsson-Raznikiewicz, 1980). The gelation time was found to be a decreasing function of temperature. The final value of G' was proportional to the square of concentration, and showed a maximum as a function of temperature. It is suggested that while hydrogen bonding is not negligible in the gelation of casein micelles, this event is governed mainly by hydrophobic interactions. Smaller micelles form a more solid-like gel than larger micelles.

9.3.3 Syneresis and permeability

Fractal analysis has been applied to confocal scanning laser microscopic (CSLM) images and permeation results of casein gels, and the fractal dimensional d_f, lower cut-off length r_0 (see Equation (2.37)) and apparent pore size of the linear scaling regime calculated from microscopy data. Values of d_f and apparent pore size were also calculated from permeability data (Mellema *et al.*, 2000). During ageing of the gels, a coarsening of the structure was observed; the pore size increased and the clusters became more compact. This was reflected in the fractal parameters, since both r_0 and pore size increased during gel ageing, although values were generally higher than those obtained by computer simulation. The d_f value was also high (~2.2–2.6), which was thought to be an indication of slow aggregation or rearrangements during aggregation. From these results, it was concluded that rearrangements such as particle fusion and strand fracture occurred during gel ageing, and were accelerated by increasing temperature and, even more pronouncedly, by decreasing pH.

Fractal aggregation theories have been applied to the flocculation of casein particles by Bremer *et al.* (1993). Because fractals ought to be scale invariant, such a gel should also be scale invariant. In other words, a gel formed from a system with a high volume fraction (concentration) of particles will resemble a gel with a low particle concentration, when examined at the appropriate shorter distance scale. Measurements of the permeability of gels of varying particle concentration can yield information about the effective fractal dimensionality. For example, the effective fractal dimensionality of one gel was estimated as 2.39 but by allowing time for the addition of rennet, the permeability was found to increase, an effect attributed to microsyneresis or rearrangement of the network (Dejimek and Walstra, 2004). The relative rate of change of permeability with time was found to depend on the casein concentration. At higher concentrations the permeability was found to change faster than at lower concentration, so Bremer *et al.* (1989) re-estimated the effective fractal dimensionality of the gel before significant microsyneresis as 2.23.

9.3.4 Elastic modulus

The enthalpic contribution to the elastic modulus for casein gels was discussed by van Vliet and Walstra (1985) and later by Roefs and van Vliet (1990). The simple power-law exponent in the concentration dependence of the storage modulus of acid casein gels is reported to be 2.6, from which Roefs *et al.* (1990) concluded that the network is very heterogeneous. Roefs and van Vliet (1990) examined the magnitude of the dynamic moduli G' and G'' of acid casein gels as a function of ageing time and temperature, measuring temperature, pH, ionic strength and composition, casein concentration and the time scale of the measurement.

By combining these with permeability results, Roefs and co-workers (Roefs and van Vliet, 1990; Roefs *et al.*, 1990) described acid casein gels as a collection of fractal clusters, but with different levels. At the first level, protein particles probably do not have a homogeneous structure with respect to the distribution of the different casein molecules throughout the particles. At the second level, the strands and small conglomerates are inhomogeneous, because of the size distribution of the particles and because of the mechanism of floc formation. Finally, the whole network is inhomogeneous, as it consists of large conglomerates and cavities (each up to 10 μm in size). Since the concept of fractal structure implies self-similarity over a range of length scales, this is an interesting variation of the concept.

de Kruif *et al.* (1995) tried to correlate the viscoelastic properties and structure of milk gels, and obtained $d_f = 2.1$ for acid casein gels and $d_f = 2.3$ for rennet casein gels, based on Bremer's theory (Bremer *et al.*, 1989). Structural observations by CSLM gave a similar value for d_f, and they concluded that these gels have similar structures, i.e. stretched stranded gels. However, although all three types of milk protein gels, chymosin treated casein gels, acid casein gels and whey protein gels (Section 9.2) showed fractal structures at some level, gel properties could not be described completely by the scaling laws proposed by Bremer *et al.* (1989) and Shih *et al.* (1990).

Finally, van Vliet *et al.* (1991) examined the relationship between syneresis and the rheological properties of particle gels of milk, and found that low fracture stresses and high tan δ correlated with an increased tendency to show syneresis. The relation between structure and rheological properties of gels of the protein mix known as whey protein isolate (WPI) – a commercial blend of β-Lg, BSA and α-lactalbumin – was studied by gel permeability, electron microscopy and dynamic viscoelasticity by Verheul and Roefs (1998). They found that gels became coarser (with larger pore size) with increasing [NaCl], as confirmed by an increase in gel permeability. Although the modulus continued to increase, no structural change was detected after gel formation, while only a part of the protein in the dispersion contributed to the gel network, again indicating that, without adjustment, the fractal concept cannot be applied to whey protein gels (Verheul and Roefs, 1998).

9.4 Fibrillar gels formed from partially denatured proteins

When proteins, particularly β-Lg and BSA, are heated below pH 4, fine-stranded or fibrillar networks are formed and the strands become shorter and apparently stiffer as the pH is lowered further (Stading *et al.*, 1993a; Walkenstrom and Hermansson, 1994). These materials appear to share many features in common with the so-called amyloid fibrils which occur in a number of pathological disease states (Gosal and Ross-Murphy, 2000; Dobson, 2001; Hughes and Dunstan, 2009).

9.4.1 Amyloid gels

'Amyloid' is a term that goes back to the nineteenth century and refers to a tendency of starch polymers to stain iodine – hence the starch polysaccharides are known as amylose and amylopectin. Nowadays, however, it refers almost exclusively to proteinaceous systems and, in a clinical context, describes an abnormal deposition of dense and insoluble fibrillar material – 'plaques' – which damage organs or tissues, including the brain. Amyloid deposits occur in the so-called prion diseases such as bovine spongiform encephalopathy (BSE) and new variant Creutzfeldt-Jakob disease (CJD), in dementias such as Alzheimer's and Huntington's chorea and in other pathologies such as primary

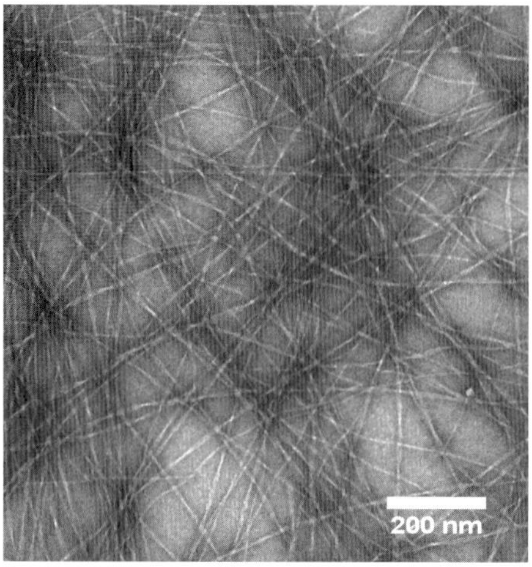

Figure 9.7 TEM image of hen lysozyme amyloid fibrils (scale bar = 200 nm). Reproduced from Corrigan *et al.* (2006) © American Chemical Society.

and dialysis-related amyloidoses, which affect patients on haemodialysis and cause severe discomfort. At the same time, protein amyloid fibrils have many features that are potentially of great value for the construction of functional materials. Natural examples provide a model for novel fibrous biomaterials for application in materials science and nanobiotechnology (MacPhee and Woolfson, 2004).

Figure 9.7 shows a TEM image of the very dilute amyloid fibrils forming randomly oriented long fibrils with a constant cross-section.

In view of the main focus of this volume, we concentrate on the ‘simple’ case of globular proteins heated as with the particulate or colloidal systems, but at these lower pHs. Early studies of temperature-induced denaturation, especially those with lysozyme (hen egg white), insulin and serum albumin (both human and bovine), revealed that fibril (and gel) formation could be induced by thermal denaturation, and that these form highly fibrillar networks.

Aggregates formed at pH 2 have been studied by a combination of light and neutron scattering (Aymard *et al.*, 1996a; Aymard *et al.*, 1996b). Rod-like structures formed at pH 2, rather than the globules and aggregates of globules formed at pH 7. The rods were thought to occur when disulphide exchange (implied in globule formation) was inhibited. Increasing the salt concentration increased the level of branching (or rod flexibility), since high salt screens out repulsive electrostatic interactions. The β-Lg dimer was suggested to be the fundamental repeat unit in these linear structures, although an unfolded monomer could not be excluded.

A number of studies have been carried out using TEM (Langton and Hermansson, 1992; Stading *et al.*, 1992), wide-angle X-ray diffraction and, most particularly, Fourier transform IR (Kavanagh *et al.*, 2000a) to try to understand what factors influence the relative propensity for fibrillar growth in the resultant gel. Microscopy showed long linear aggregates forming in solutions at pH 2 (and sometimes 2.5) after prolonged heating. Wide-angle X-ray diffraction (WAXD) showed a diffuse β-sheet halo in patterns for both dried native and aggregated protein (irrespective of pH) with only a small change (sharpening) of this feature on heat treatment. At all pH values studied, β-Lg showed only limited secondary and tertiary structural changes on aggregation (Gosal *et al.*, 2004b; Kavanagh *et al.*, 2000a), in contrast to previous studies of insulin aggregation where highly ordered crystalline fibrils were indicated (Clark *et al.*, 1981a).

Although electron microscopy and AFM have shown quite consistently that heating globular proteins at low pH gives rise to the fibrillar structures mentioned above, it is important to point out that almost all of these studies have been carried out well below the critical gel concentration c_0 (Chapter 3). Nevertheless some conclusions can be reached. In one such study, TEM revealed many long extended fibrils (~0.1–2 μm in length). The width distribution of these fibrils, calculated from negatively stained samples, was narrowly dispersed, the average being of order 10 nm. The absence of higher-order fibril structure within them was suggested by the absence of any regular helical twist. Some heterogeneity was revealed by the presence of much shorter, 'worm-like' fibrils, these being typically 100–200 nm in length but very similar in width (Gosal *et al.*, 2004b).

AFM results on the same materials revealed numerous fibrils, varying in length from ~0.1 to greater than 1 μm, similar to those observed in the EM images. However, the average height of the members of the major population of long fibrils was found to be ~3–4 nm. A very small number of thicker fibrils were also seen, apparently assembled from two of the smaller underlying components. Work by Donald and co-workers (Bromley *et al.*, 2005; Domike and Donald, 2009; Krebs *et al.*, 2009) has investigated other structural aspects of β-Lg amyloid formation and shown how a very large spherulitic aggregate consisting of radially arranged amyloid fibrils can be formed.

At this point it is worth considering the overall conclusions of all this structural work, especially studies on β-Lg, regarding the assembly of globular proteins. At pH $\gtrsim$ 7 β-Lg occurs as dimers, at pH < 3 as monomers and at intermediate pH in a more complex form, sometimes referred to as an 'octamer'. On being heating to moderate temperatures, the native structure is in equilibrium with a partially folded intermediate, the so-called molten globule form, formed prior to such transformation to the so-called 'H' state, which seems to be unstable and prone to aggregation (Hughes and Dunstan, 2009). In the model proposed by Dobson, protein aggregation/gelation is the reverse of protein folding: native monomer can interact to form native-like aggregates that can then go on to produce β-structured aggregates, which form fibrils (amyloid fibres) (Dobson, 2001, 2003). Gelation at or close to the isoelectric point is indicative of native-like aggregates, and particulate but non-coagulated gels contain features of both types. The whole field of structural aspects of protein unfolding and subsequent aggregation is very active; for example, we refer the reader to the work of Dobson and co-workers (Calamai *et al.*, 2005; Chiti and Dobson, 2006, 2009).

9.4.2 Rheological measurements

Rheological measurements on fibrillar protein gels have been carried out by a number of groups, including Hermansson and Clark and their respective co-workers. For example, Gosal *et al.* (2004a) measured β-Lg gels at a range of concentrations with pH at 2 and temperature at 75°C and 80°C. The results were qualitatively similar to those found around pH 7, but the critical concentrations were typically lower, with good cure data being measured around 8% w/w and extrapolated values of the critical concentration c_0 being found around 5–6%.

Gosal *et al.* analysed the results in terms of both a constant power law fractal approach (Bremer *et al.*, 1989, 1993) and a classical branching model (Clark and Ross-Murphy, 1987; Clark *et al.*, 2001). The range of concentration data was quite limited – by practical considerations – over the range 8% to 14%, so it could be described equally well by either of the models. It was argued, however, that the fractal description appeared inconsistent with the uniform fibrillar networks expected for such fine-stranded protein gels, these networks being unlikely to be self-similar over a significant length scale.

An analysis was carried out to evaluate critical exponents, using a method due to van der Linden and Sagis (2001). They suggested that the power-law exponent γ for gel formation (here defined as the exponent in $G' \sim (c/c_0 - 1)^{\gamma}$), for a variety of protein systems, belongs to a certain 'universality class' which gives $\gamma \sim 1.7$. This method requires knowledge of c_0, but the technique employed calculates this at the same time as the exponent γ.

Analysis of the data in the Gosal paper used a different method and produced a good fit, with the parameter γ lying in the range 2.2–2.8. Within this region there did not appear to be any obvious discontinuity in γ values, even though data was used from pH 2 to pH 7 corresponding to the structural transition from essentially fibrillar systems to increasingly colloidal or particulate ones. Other analyses in the same paper produced a variety of estimates, typically 2.7 ± 0.5. Unfortunately this covers almost all the expected range, including the classical value of 3. As pointed out in Chapter 3, such approaches are very testing – perhaps too testing for the experimental data currently available.

Another technique of considerable value is particle tracking microrheology (PTM) (Corrigan and Donald, 2009), introduced in Chapter 3. Using PTM, gels were observed to form at significantly lower concentrations than determined by bulk rheometry, where the authors claim oscillatory shear forces may disrupt either fibril or network formation. This result has important consequence, although, as we pointed out in Chapter 3, the strain and frequency regimes employed in PTM are rather different from those used in small-deformation oscillatory testing.

As far as large-deformation data is concerned, there is very little on 'pure' protein systems, simply because of the requirements for tens of grams of sample. In one study, WPI (a mixture of β-Lg and other milk proteins) was used to form translucent heat-induced gels. Those gels formed at pH 7.0 and 6.5 were classified as 'strong' (they had fracture stresses of ~60 kPa) and 'rubbery' (fracture strain > 1.2), whereas gels formed at low pH generally had lower fracture stresses (1–15 kPa) and fracture strains (0.3–0.7) (Errington and Foegeding, 1998). Although this behaviour is observed in a number of other studies, it is fair to say that comparison has not, in our perception, been made under comparable conditions. For example, if the materials had the same low deformation modulus, then fracture measurements might be easier to compare. Interestingly, few if any studies have examined the area under the failure stress/strain curve. Recent work by van den Berg and co-workers (van den Berg *et al.*, 2007a, 2007b, 2008) has tried to correlate failure properties with macroscopic and microscopic appearance for a range of gel structures.

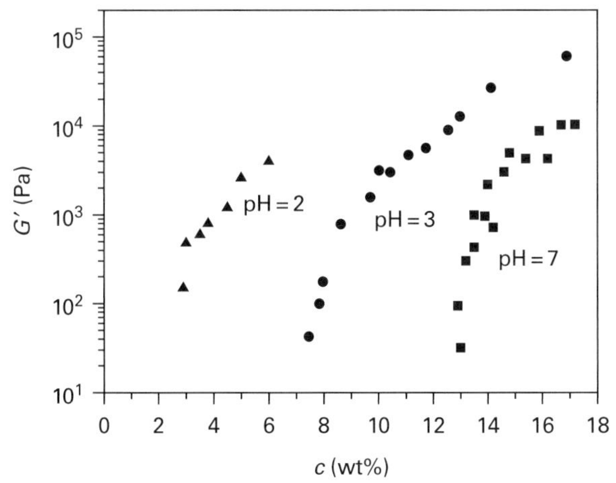

Figure 9.8 Observed log modulus versus log concentration for three β-Lg gels: at pH = 7 (particulate, ■) and at pH = 3 (fine-stranded, ●), adapted from Kavanagh *et al.* (2000c); and at pH = 2 (amyloid strands, ▲), adapted from Veerman *et al.* (2002a). © American Chemical Society.

As pointed out in Chapter 3 and elsewhere in this volume, two of the most fundamental parameters which need to be acquired for all the above systems – not least to establish whether or not we have a gel or a sol – are c_0 and the modulus concentration dependence exponent. Values of c_0 for these materials are typically 1–15%, and tend to decrease with increasing ionic strength or more extreme preparation conditions (i.e. heating to 80°C rather than 60°C) (Clark, 1998). What is clear is that, in general, c_0 values are higher by as much as two orders of magnitude than those found, for example, for fibrous polysaccharide gels, and higher even than typical values for gelatin gels. That said, as we mentioned above, the fundamental fibrillar structure is much larger – say, 3–5 nm in width – than for most of the polysaccharide systems – typically an order of magnitude lower.

Figure 9.8 shows a comparison between three types of β-Lg heat-set acid gels at three different pHs: pH 2 from Veerman *et al.* (2002b) and pH 3 and pH 7 from Kavanagh *et al.* (2000c). In the first case, solutions were heated at 80°C for 10 h and subsequently cooled to 20°C for 2 h. After this procedure a strain sweep was performed. In the latter two solutions, gelation was studied at 80°C for times of up to 6 h. Differences in the heating protocol and ionic strength are likely to influence any observed differences in c_0 and concentration exponent, and any correlation with the morphology and flexibility of the aggregated structures. This figure illustrates the challenge in interpreting the rheological properties of these protein gels, clearly showing that polymer concentration alone is not sufficient to determine the value of the shear modulus.

Reported values of the concentration exponent tend to vary widely, many clustering around 2–2.5 but both lower and higher values being reported (Clark and Ross-Murphy, 1987). We would not expect a single exponent to describe the concentration dependence over a wide range. It is true that at high concentrations – particularly when expressed as c/c_0, say $>10\ c/c_0$ – an apparent limit exponent is reached. Equally, as we approach c_0, higher and higher apparent exponents are seen, since here the log G' versus c plot tends to a pronounced curvature.

Sagis and co-workers (Sagis *et al.*, 2004) investigated the domain of a transparent gel formed at low β-Lg ovalbumin or BSA concentrations from homogeneous solutions by heating at low pH and for very extended times (up to 10 h). They excluded from their analysis conditions where non-homogeneous gels could appear (with liquid crystal inclusions; see below). For their amyloid fibrillar systems, Sagis *et al.* (2004) measured the frequency-independent G' values in the linear regime as a function of protein concentration, and then fitted the data versus $(c - c_0)$ using a power law with the exponent γ, as a function of ionic strength. The range of concentrations investigated covered $c/c_0 = 0.01–0.1$. For all proteins, the authors found a decreasing c_0 with increasing ionic strength or decreasing Debye length, and for almost all samples they found $\gamma \approx 2$, which tends to suggest isotropic force percolation and a homogeneous network, even though others have reported a range of values from 1.8 to 2.5. However, as pointed out above, this form of fitting tends to produce a correlation between γ and c_0.

They concluded, in contradiction to much other work, that there is no cross-linking of the fibrils and that the results can be interpreted simply in terms of these very extended and stiff strands, whose properties can be interpreted in terms of mesoscopic (liquid crystalline) behaviour, as discussed in the next section. Assuming amyloid fibrils are semi-flexible rods, the critical percolation mass fraction was expressed according to the model developed for charged semi-flexible fibres by Khokhlov and Semenov (1981) and adapted from the Onsager treatment (Section 3.11) . The model requires an assumption that we replace the chain contour length L_c by its persistence length l_p, so that the volume fraction of objects at the isotropic–nematic (I-N) transition is then written

$$\phi_{\text{iso}} \approx \langle \alpha \rangle \frac{d_{\text{eff}}}{l_p}, \quad l_p >> d_{\text{eff}}, \tag{9.1}$$

where d_{eff} is the effective diameter of the rods, which depends on the repulsion interactions through the Debye length (lower salt levels yield less screening and therefore a larger effective diameter), and $\langle \alpha \rangle$ is a coefficient close to 1. The model suggests that the persistence length is a control parameter for this critical concentration when $L_c >> l_p$: no effect of the fibril length is expected for their estimates of c_0. The persistence length was deduced from TEM images while the effective diameter was calculated independently using a theoretical model.

Figure 9.9 shows that this estimate of critical volume fraction increases with the effective diameter of the semi-flexible fibres. The contour length L_c was derived from TEM images. Sagis and co-workers found a range of values for contour length from 0.2 to 5 μm and persistence lengths from 0.02 to 2 μm, and they suggested that the persistence lengths in these systems were independent of ionic strength. Their value for persistence length l_p for β-Lg is very large ($l_p = 1.6$ μm), while for ovalbumin it is smaller ($l_p = 300$ nm). The contour lengths of fibrils at pH 2 near their percolation threshold vary between 2.5 and 4.5 μm for β-Lg and between 50 and 190 nm for ovalbumin. The difference is somewhat surprising, but may reflect the particular heating regime adopted.

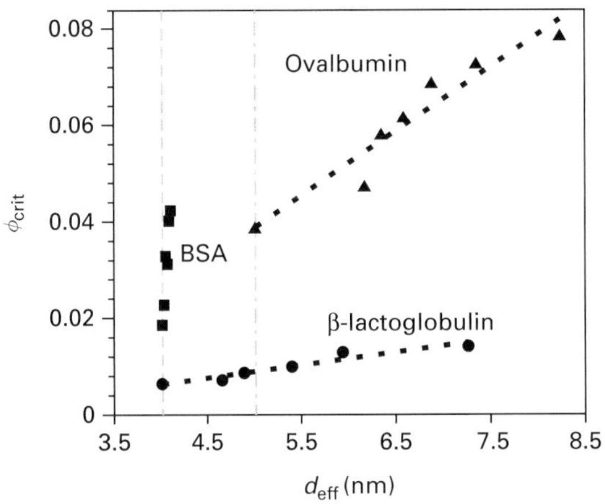

Figure 9.9 Critical percolation volume fraction (ϕ_{crit}) versus effective diameter for semi-flexible amyloid fibrils of β-Lg, BSA and ovalbumin. The vertical lines indicate the diameter of the rods for ovalbumin (4 nm) and BSA or β-Lg (5 nm). Adapted from Sagis *et al.* (2004) © 2004 American Chemical Society.

9.4.3 Phase behaviour

According to Sagis *et al.* (2004), their isotropic solution becomes an isotropic gel at the percolation transition, before reaching the biphasic region predicted by the Onsager model. At larger concentrations, a two-phase gel is observed, consisting of nematic droplets dispersed in an isotropic gel. At even higher concentrations (>10 wt%) a single nematic phase develops. What this does suggest is that some solutions containing amyloid fibrils have interesting phase behaviour, which resembles that expected for an assembly of rod-like systems (Chapter 3).

Incubation of proteins at relatively high concentrations (10–15% w/v) results in substantial yields of amyloid fibrils, and according to some reports these solutions gel after several days. Polarized optical microscopy images reveal that the gels contain large numbers of small birefringent domains, indicating the presence of liquid crystal inclusions which do not allow the fibrils to align over long distances. Sagis *et al.* (2004) have suggested the term 'liquid crystal glass' to describe the appearance of many small liquid crystal domains.

Adamcik *et al.* (2010) re-examined the morphology of β-Lg fibrils. They performed AFM measurements of β-Lg fibrils produced by heat denaturation at pH 2: the images show semi-flexible fibrils several micrometres in length, in agreement with previous literature reports such as those mentioned above. Maximum height measurements were performed to characterize the various fibril cross-sections. Hierarchical helical structures were observed. Periodic fluctuations in cross-section, persistence length and height along the contour length of the fibrils allowed them to distinguish single-filament, double-, triple- and multi-stranded left-handed helical aggregates, the most frequently occurring population being the double-stranded helix. In particular, the authors showed that a strong tendency to inter-fibril aggregation exists, in particular for a positive large linear charge density. They suggested that a strong energetic driving force must exist to promote aggregation under these conditions. Once aggregation has occurred, the electrostatic repulsive forces induce a twisting of the ribbons. The attractive force leading to aggregation might arise from the amphoteric nature of the protein fibrils, which can promote strong attractive 'hydrophobic' interactions. The ribbon-like cross-section may arise from the balance of hydrophobic short-range attraction and electrostatic long-range repulsion.

Jung and Mezzenga (2010) determined a phase diagram for β-Lg by optical observations of flasks. Each flask contained a magnetic bar to stir the solution during the heating process, and it is reported that stirring enhanced the conversion rate and the birefringence, and reduced the formation of spherulites. This method was also used by Adamcik *et al.* (2010) for sample preparation before their AFM observations. The resulting systems are fundamentally different from the ones previously investigated in quiescent conditions, because they contain virtually no spherulites. Consequently, the structure of the final fibres is highly dependent on the exact heat denaturation process.

Sagis *et al.* (2004) reported on the coexistence of nematic droplets (spherulites) in the isotropic phase around 0.5 wt% and pH 2, whereas in the work of Jung and Mezzenga (2010) the transition was purely an isotropic–nematic one. For Jung and Mezzenga, solutions were liquid and transparent up to 5 wt%, where a gel was obtained. Under polarized light, the sample at 0.3 wt% was not birefringent, whereas at 0.4 wt% the sample started to be very slightly birefringent. The isotropic–nematic phase transition occurred at 0.4 wt% at pH 2. The sample became progressively more birefringent as the concentration was increased. At 2 wt% and above, the sample was birefringent with bright colours, suggesting the possible presence of cholesteric liquid crystalline phases (chiral helical conformations of the fibres). The gel revealed a microscopic phase separation, with coexisting isotropic (non-birefringent) and nematic (birefringent) phases, visible under polarized light. According to this work, the entire region of the phase diagram below the gelation threshold is presented as reversible to pH and concentration changes, in agreement with the thermodynamic character of the phase diagram.

For the gel, however, dissolution of the nematic phase did not occur upon dilution, and the system retained birefringent spots both between cross-polarizers and under polarized light microscopy, indicating that the aggregation in the gel region does not follow thermodynamic behaviour, but rather a concentration-induced irreversible aggregation of the fibres, similar to the formation of spherulites. A gel was obtained by concentrating the pure nematic region, which induced irreversible aggregation within the nematic phase, and phase separation into a fibre-enriched region (the aggregates) and a fibre-impoverished region (the isotropic continuous phase). This scheme was different from the one suggested by the work of Sagis *et al.* (2004), who based their analysis on a percolation transition in the isotropic phase. As a consequence of this complexity, direct application of Onsager theory based on excluded-volume interactions appears to be inappropriate to describe these systems. Flory's theory, accounting for hydrophobic interactions of rod suspensions in water, seems to produce theoretical predictions which are more consistent with experimental observations.

A challenging issue in the self-assembly of complex supramolecular structures is therefore understanding how kinetically efficient pathways emerge from the multitude of possible transition states and routes (Pouget *et al.*, 2010). In general, *in vitro* reassembly of proteins produces architectures that exhibit the same molecular and supramolecular structures as their *in vivo* counterparts, indicating that self-assembly is driven by thermodynamic equilibrium determined by external conditions such as ionic strength, temperature and pH. Self-assembly kinetics is typically very slow or very fast depending on the growth phase, making it difficult to observe the processes in equilibrium conditions. The *in vitro* formation of such fibres results from a very slow nucleation stage followed by a very fast growth stage. These inherent difficulties are further complicated by the existence of simultaneous pathways leading to various morphologies that have been experimentally observed or predicted by molecular dynamics.

The application of model ‘amyloids’, both as ways of understanding disease states and as interesting modern materials, is a very active area. Useful recent reviews and articles include Chiti and Dobson (2006) and Pedersen and Otzen (2008). Their suggested use as nanomaterials is discussed by Cherny and Gazit (2008).

9.5 Specific assemblies from peptides and proteins

There are many examples of fibril formation involving more specific assembly mechanisms, including the formation of actin and tubulin fibrils, which play a role in the properties of the cytoskeleton or fibrin gel in blood clots (Shah and Janmey, 1997; Janmey *et al.*, 1998). Fibrinogen may also be considered to fall into this category. Although its assembly into fibrin (the mechanism of blood clotting) requires chemical cross-linking (‘ligation’), there is work on unligated fibrin gels, and we discuss this briefly below. Although all of these systems are critical to *in vivo* function, we restrict ourselves to *in vitro* work on pure components and/or very simple mixtures.

9.5.1 Insulin and lysozyme gels

Insulin is the pancreatic hormone involved in glucose transport, and the failure of this mechanism gives rise to the various forms of diabetes. Insulin is a small globular protein, but for many years it has been known that, when heated above its unfolding temperature, particularly at acid pH, an opalescent gel can be formed, with critical concentration around 1%. Electron microscope studies have showed that the gel consists of long, large persistence length fibrils, with minimum strand thicknesses around 5–10 nm (Clark *et al.*, 1981a). More recent studies have confirmed this using AFM, although estimates for fibril thickness are lower by a factor 2 (Gosal *et al.*, 2004b). Secondary structure measurements have shown an increase in β-sheet, and an early paper established this to be of parallel cross-β structure. Growth of the fibrils seems to follow the spherulitic pattern reported by Donald and co-workers for other amyloid protein gels (Domike and Donald, 2009); consequently it can be regarded as the prototype of these. Images tend to suggest that the ‘gel’ structure is made up of intertwined fibrils, and the ‘pasty’ gel appearance reported would suggest that they are somewhat different from the other heat-set gels discussed in this chapter, and more akin to the hydrophobic gels discussed in Chapter 6.

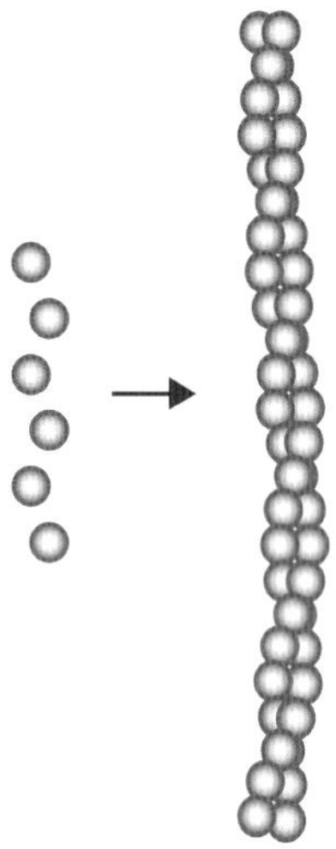

Figure 9.10 Self-assembly from G actin to F actin.

Following pioneering work by Clark and co-workers (Clark *et al.*, 1981a), who examined lysozyme gels at pH 2 and found very extended thin fibrils, these protein gels have been investigated by a number of other groups. Many of their properties seem to follow those of the amyloid gels discussed previously.

9.5.2 G and F actin

The globular protein actin – along with myosin, tropomyosin and troponin – is an intrinsic component of muscle, but is also a major constituent of the cytoskeleton of cells, in particular in association with tubulin. G actin, as the prefix suggests, is the globular subunit extracted from muscle, but when treated with salts (e.g. 0.1 M KCl) this readily undergoes self-assembly into long rod-like fibrils of F actin (see Figure 9.10). These have been known for many years to be formed from two helical strands, although, unlike *in vivo*, polymerization *in vitro* produces a length distribution (Berg *et al.*, 2007a, 2007b).

Actin polymers in aqueous solution have been studied rheologically by a number of groups. The 'thixotropic' character of actin solutions has long been known, and an early quantitative study pursued this aspect via measurements of viscosities and rigidities of appropriate preparations (Kasai *et al.*, 1960). A transient network was inferred from the results. Some years later Maruyama and co-workers (Maruyama *et al.*, 1974) carried out a similar investigation. They believed that the samples used in the Kasai study could have contained impurities, which could have introduced cross-links into F actin preparations. Studies using steady shear viscosity were performed on highly purified F actin solutions near neutral pH, and at various concentrations, and these indicated strong shear-thinning behaviour above a critical concentration. Below this, the F actin depolymerized. The viscosity at low shear rates was very high, and it was suggested that the system possessed a finite yield stress. Measurements in oscillatory shear also allowed G' and G'' to be established and, at the low concentrations studied, these were very small indeed.

However, plots of equilibrium values for G' against concentration suggested a c^2 dependence. The fall in modulus with increasing strain was also recorded, and the viscosity behaviour led these workers to conclude that the actin system was a temporary network, easily destroyed by external forces. It was not clear, however, whether the temporary cross-links could be ascribed to entanglement or to non-covalent interactions, or indeed to residual impurities still present in the samples. Subsequent studies of F actin have led to similar conclusions, although some workers have concluded that an entanglement network was involved, and that modulus–frequency data agreed well with a prediction for a solution of rigid rods.

The mesh size of the gel also decreased, albeit entirely as expected, as the concentration of actin increased, but some current thinking suggests that the elasticity of the gels results entirely from the thermal fluctuations of the filaments (Janmey and Weitz, 2004). In other words, because they are so long and thin, they possess inherent thermal fluctuations that reduce their end-to-end length. When a strain is applied, these fluctuations will be reduced, a reduction which costs energy and is entropic (Higgs and Ball, 1989; Storm *et al.*, 2005), similar to classical rubber elasticity. The magnitude of the elastic response depends on both actin and cross-link concentration, but gives rise to strain hardening; but at larger deformations enthalpic effects begin to contribute (Storm *et al.*, 2005). Such a strain hardening effect is also seen in an early theory for rods, in which there are no cross-links (Doi and Kuzuu, 1980).

9.5.3 Tubulin

Microtubules, derived from the globular protein tubulin, are highly organized, hollow, rod-like assemblies, commonly found in the cytoplasmic material of cells, where they can be assembled and disassembled by processes involving associated proteins, which influence the nucleation step which is part of the cooperative assembly process (Berg *et al.*, 2007a). Like F actin, the microtubules are part of the filamentous structural components of cell cytoplasm, and are involved in several processes including cell mitosis.

The microtubules are large structures, around 50 nm in diameter and with persistence lengths believed to be as great as 1 mm, and thought to grow from a nucleation ring of tubulin dimer molecules by addition of further dimers to either surface. The assembly process has been studied by a variety of techniques including turbidity measurements and small-angle X-ray scattering.

Lin and co-workers (Lin *et al.*, 2007) examined the linear and non-linear viscoelastic properties of isotropic solutions of purified microtubules, as well as networks permanently cross-linked with a specific reagent. In the linear regime, both solutions and networks were soft elastic materials with small G' and G''. For the networks, G' was approximately proportional to $c^{1.7}$, with lower exponents for the solutions, although these show a concentration-dependent apparent yield stress, which suggests the presence of effective cross-linking even in these pure microtubule solutions. Such results show that, like F actin, microtubule assemblies do not generate high-modulus networks on their own. Consequently, *in vivo* they must depend on their participation in architectures involving other components.

9.5.4 Non-ligated fibrin gels

The precursor of the blood clotting fibrin gels is fibrinogen, a small rod-like protein. The usual fibrin gels are produced by enzymatic treatment of fibrin with the peptide fibrinoligase – known to experts in the field as clotting factor XIIIa (Berg *et al.*, 2007b) – to produce a 'chemical' cross-linked structure, the so-called ligated gels. Nevertheless, gels can be produced without this treatment and so they are discussed here. They can be prepared by altering pH and ionic strength, and show typical fibrillar strands under microscopic examination. Another reason for including them here is that the rheological work carried out in the mid-1970s is a model study from what was then the world's leading laboratory in this area, led by the late J. D. Ferry.

Roberts and co-workers (Gerth *et al.*, 1974; Roberts *et al.*, 1974) described a series of measurements of such fibrin clots, both ligated and unligated, using the specialized Birnboim apparatus. Various preparation regimes gave rise to gels described as 'fine' and 'coarse clots'. Unligated clots showed little G' frequency dependence, and a concentration dependence around $c^{1.5}$. Creep measurements showed some flow, perhaps slip induced in the coarse clots, but little for the finer materials. A later study found a slightly higher concentration exponent, around 1.9. It was suggested that the elasticity did not arise simply from steric effects, which would give higher exponents still, but that there was some branching, as seen in later EM pictures (Muller *et al.*, 1984), and recent work has examined strain stiffening. Both the similarities and the differences between the rheology of these materials and gels formed from tubulin and actin have been discussed (Janmey *et al.*, 1991; Rammensee *et al.*, 2007).

9.5.5 'Amyloid' protein gels formed in non-aqueous media

The self-assembly of proteins and peptides can be induced by a variety of treatments other than heating, most of which initiate partial unfolding of the native monomer, so that such methods as solvent or chemically induced unfolding of the protein are possible. The denaturation of proteins in alcohol–water mixtures, in particular, is a well-documented feature of protein chemistry, leading to the formation (in most cases, and at appropriate molar concentrations of solvent) of an expanded helical conformation often referred to as the 'H-state' (Tanford *et al.*, 1962; Buck, 1998).

For β-Lg, numerous studies (for example, using ORD and CD spectroscopy) have shown that the largely β-sheeted secondary structure of this protein can be converted to a mainly α-helical form upon addition of various alcohols, and at the same time the protein expands (Tanford *et al.*, 1960; Townend *et al.*, 1967; Kamatari *et al.*, 1996). In addition, the partially folded intermediate 'molten globule' is formed prior to such transformation to the H-state. The H-state itself, however, seems to be unstable and prone to aggregation. Hence, in some cases, fibrillar aggregates have been shown to occur in alcohol–water mixtures – for example, lysozyme in high molar concentrations of ethanol (Goda *et al.*, 2000). Aggregation of the β-Lg H-state has been investigated mostly at neutral pH, and in ethanol–water mixtures.

Amyloid fibrils can be stained with the dye Congo Red (Hughes and Dunstan, 2009), and show a distinct green birefringence when viewed under polarized light. An alternative stain is Thioflavin T (Krebs *et al.*, 2005). Under appropriate microscopy all show uniform, smooth and persistent fibrils with a strand thickness of the order of 8 nm, and all show a distinct form of β-sheet structure. Similar properties have already been noted above for the low-pH gels, although amyloid structures *in vivo* often involve very specific proteins, which can convert spontaneously from soluble protein to insoluble plaques (Dobson, 2001).

Gosal and co-workers (Gosal *et al.*, 2002) studied the gelation and amyloid formation of β-Lg by dissolving the protein in aqueous alcohols. Images of the systems (made at sub-gelling concentrations) show the expected fibrils, and rheological measurements of the gels showed some features to be very similar to those of the heat-set pH 2 systems, although, compared to these, G'' sometimes showed a minimum, perhaps suggesting that these were more flexible networks.

9.5.6 Gels from peptide self-assembly

Over the last decade or so there has been a great deal of interest in developing fibrillar self-assembled, so-called biomimetic, systems from specially synthesized peptide sequences. Among the pioneering work, that by Boden and co-workers is particularly noteworthy. They were able to design anti-parallel β-sheet 'tapes' (Aggeli *et al.*, 1997a, 1997b) which formed highly entangled gel-like systems, and their rheological properties could be dictated by either shear history or chemistry, for example by altering solvent polarity or hydrogen bonding ability. Altering the response to such influences could be achieved by modifying the properties of the β-sheet. A vast array of fibrillar structures (protofilaments, fibrils and ribbons) were obtained by changing the pH or ionic conditions. Yet another useful peptide 'motif' is the coiled coil (Petka *et al.*, 1998; Wang *et al.*, 1999).

Both pH and temperature control assembly, making the resultant structures thermo-reversible, and the response can be encoded into the peptide sequence; even conformational switching from a coiled-coil to β-sheet fibril has been achieved by introducing acyl chains (Takahashi *et al.*, 1999). This introduces a hydrophobic defect, which controls assembly. Recent reviews which cover this topic in more detail are those by Hamley (2007), Jung *et al.* (2010) and Ho *et al.* (2011).

Of particular note here is the recent themed journal issue on peptide- and protein-based materials, which summarizes both systems and applications (Ulijn and Woolfson, 2010). Many of the articles in this issue are of relevance, but we concentrate on the review of rheological measurements on peptide gels (Yan and Pochan, 2010). Pochan and co-workers have established a very sound reputation in this area, and their work on the MAX1 and MAX8 peptides, described there, is particularly significant.

MAX1 is a 20 amino acid containing a central β-turn sequence, with adjacent residues of the hairpin containing alternating hydrophobic (valine) and hydrophilic (lysine) residues. In aqueous solution this forms a short piece of 'random' coil because the positively charged lysines prevent folding. However, in electrolyte solution the positive charges are screened out, and then the peptides fold into hairpins, which in turn self-assemble to give a 'rigid' fibrillar network. MAX8 has one of the valines replaced by a glutamic acid residue, which allows gel formation both to be faster and to produce a higher modulus G' under the same conditions of solvent and peptide concentration. By changing these parameters, both gels can be tuned. Using a small (10%) strain gelation could be followed in terms of G' and G'' by oscillatory strain and also by dynamic light scattering methods, and a gelation time identified using the Winter–Chambon method (Chapter 3). However, both MAX1 and MAX8 are highly strain sensitive. Under shear flow they have a low viscosity, but can 'heal' very fast. Rheologically, then, they show more similarity with so-called 'structured liquids' such as xanthan solutions (Chapter 5).

Other peptide systems discussed by Yan and Pochan (2010) include the decapeptides of Ramachandran *et al.* (2005) and the block co-polypeptides of Aulisa *et al.* (2009). These show quite similar behaviour, with relatively high G' (typically >100 Pa at 1% w/w) but with a linear viscoelastic limit extending only out to, say, 1% strain. Consistent with this, it seems all the systems can be readily disrupted by applying a steady shear flow, but then can recover fast, say, in a few minutes. Most experiments seem to be of the form: (a) measure values of G', (b) shear thoroughly and (c) monitor G' recovery as a function of time (Kohyama and Nishinari, 1993). It seems that, so far, no group has performed rigorous shear overshoot/recovery measurements of the type discussed for xanthan 'gels' in Chapter 5. As we discuss in Chapter 11, such fast shear-reversibility is of great potential value for *in vivo* delivery of pharmaceutical materials by syringe injection, and this remains the long-term aim of much of this work.

9.6 Conclusions

Much early work on protein gels was driven by applications in the food, particularly dairy, industries. This tended to be qualitative in its approach, but work since the 1980s seems to have improved the quality of this work. The recent huge interest in this area is two-fold: first, investigation of the apparent similarities between globular protein gel formation at low pH and amyloid diseases states; and, second, biomimetic work on peptide gels, driven originally by the availability of specifically designed and synthesized peptides. However, one of the problems with the former work reflects the very nature of protein based studies. In many recent papers there seems a disappointing tendency to neglect necessary sample details in the search for exciting and novel results. This may be the reason for the lack of agreement between various groups; consequently we do implore workers in this area to be more specific about sample treatments, thermal histories etc., so progress can continue to be made.

References

Adamcik, J., Jung, J.-M., Flakowski, J. *et al.*, 2010. *Nat. Nanotechnol.* **5**, 423–428.
Aggeli, A., Bell, M., Boden, N. *et al.*, 1997a. *Nature* **386**, 259–262.
Aggeli, A., Bell, M., Boden, N. *et al.*, 1997b. *J. Mater. Chem.* **7**, 1135–1145.

Aulisa, L., Dong, H., Hartgerink, J. D., 2009. *Biomacromolecules* **10**, 2694–2698.
Aymard, P., Durand, D., Nicolai, T., 1996a. *Int. J. Biol. Macromol.* **19**, 213–221.
Aymard, P., Gimel, J. C., Nicolai, T., Durand, D., 1996b. *J. Chim. Phys. Phys.-Chim. Biol.* **93**, 987–997.
Berg, J. M., Tymoczko, J. L., Stryer, L., 2007a. Blood clotting is accomplished by a cascade of zymogen activations. In *Biochemistry*, 6th edn. W. H. Freeman, New York, pp. 293–296.
Berg, J. M., Tymoczko, J. L., Stryer, L., 2007b. Actin is a polar, self-assembling, dynamic polymer. In *Biochemistry*, 6th edn. W. H. Freeman, New York, pp. 985–990.
Bolder, S. G., Hendrickx, H., Sagis, L. M. C., van der Linden, E., 2006. *Appl. Rheol.* **16**, 258–264.
Bremer, L. G., van Vliet, T., Walstra, P., 1989. *J. Chem. Soc., Faraday Trans. 1* **85**, 3359–3372.
Bremer, L. G. B., Bijsterbosch, B. H., Walstra, P., van Vliet, T., 1993. *Adv. Colloid Interface Sci.* **46**, 117–128.
Bromley, E. H. C., Krebs, M. R. H., Donald, A. M., 2005. *Faraday Discuss.* **128**, 13–27.
Buck, M., 1998. *Q. Rev. Biophys.* **31**, 297–355.
Calamai, M., Chiti, F., Dobson, C. M., 2005. *Biophys. J.* **89**, 4201–4210.
Carrotta, R., Bauer, R., Waninge, R., Rischel, C., 2001. *Protein Sci.* **10**, 1312–1318.
Cherny, I., Gazit, E., 2008. *Angew. Chem., Int. Ed.* **47**, 4062–4069.
Chiti, F., Dobson, C. M., 2006. *Annu. Rev. Biochem.* **75**, 333–366.
Chiti, F., Dobson, C. M., 2009. *Nat. Chem. Biol.* **5**, 15–22.
Clark, A. H., 1998. Gelation of globular proteins. In Hill, S. E., Ledward, D. A., Mitchell, J. R. (eds), *Functional Properties of Food Macromolecules*. Aspen Publishers, Gaithersburg, MD, pp. 77–142.
Clark, A. H., Ross-Murphy, S. B., 1987. *Adv. Polym. Sci.* **83**, 57–192.
Clark, A. H., Ross-Murphy, S. B., 2009. Biopolymer network assembly: Measurement and theory. In *Modern Biopolymer Science*. Elsevier, San Diego, pp. 1–27.
Clark, A. H., Judge, F., Richards, J. B., Stubbs, J. M., Suggett, A., 1981a. *Int. J. Pept. Protein Res.* **17**, 380–392.
Clark, A. H., Saunderson, D. H. P., Suggett, A., 1981b. *Int. J. Pept. Protein Res.* **17**, 353–364.
Clark, A. H., Kavanagh, G. M., Ross-Murphy, S. B., 2001. *Food Hydrocolloids* **15**, 383–400.
Coniglio, A., Stanley, H. E., Klein, W., 1982. *Phys. Rev. B* **25**, 6805–6821.
Corrigan, A. M., Donald, A. M., 2009. *Eur. Phys. J. E* **28**, 457–462.
Corrigan, A. M., Müller, C., Krebs, M. R., 2006. *J. Am. Chem. Soc.* **128**, 14740–14741.
Dalgleish, D. G., Paterson, E., Horne, D. S., 1981. *Biophys. Chem.* **13**, 307–314.
Dejimek, P., Walstra, P., 2004. The syneresis of rennet-coagulated curd. In Fox, P. F., McSweeney, P. L. H., Cogan, T. M., Guinee, T. P. (eds), *Cheese: Chemistry, Physics and Microbiology Vol. 1*, 3rd edn. Elsevier, Amsterdam, pp. 71–103.
de Kruif, K. G., Hoffmann, M. A. M., van Marle, M. E. *et al.*, 1995. *Faraday Discuss. Chem. Soc.* **101**, 185–200.
Dobson, C. M., 2001. *Philos. Trans. R. Soc. London B* **356**, 133–145.
Dobson, C. M., 2003. *Nature* **426**, 884–890.
Doi, E., 1993. *Trends Food Sci. Technol.* **4**, 1–5.
Doi, M., Kuzuu, N. Y., 1980. *J. Polym. Sci., Part B: Polym. Phys.* **18**, 409–419.
Domike, K. R., Donald, A. M., 2009. *Int. J. Biol. Macromol.* **44**, 301–310.
Donald, A. M., 2008. *Soft Matter* **4**, 1147–1150.
Eiser, E., Miles, C. S., Geerts, N., Verschuren, P., MacPhee, C. E., 2009. *Soft Matter* **5**, 2725–2730.
Ekstrand, B., Larsson-Raznikiewicz, M., 1978. *Biochim. Biophys. Acta* **536**, 1–9.
Ekstrand, B., Larsson-Raznikiewicz, M., Perlmann, C., 1980. *Biochim. Biophys. Acta* **630**, 361–366.

Errington, A. D., Foegeding, E. A., 1998. *J. Agric. Food Chem.* **46**, 2963–2967.
Foegeding, E. A., 2006. *Food Biophys.* **1**, 41–50.
Gerth, C., Roberts, W. W., Ferry, J. D., 1974. *Biophys. Chem.* **2**, 208–217.
Gimel, J. C., Durand, D., Nicolai, T., 1994. *Macromolecules* **257**, 583–589.
Goda, S., Takano, K., Yamagata, Y. *et al.*, 2000. *Protein Sci.* **9**, 369–375.
Gosal, W. S., Ross-Murphy, S. B., 2000. *Curr. Opin. Colloid Interface Sci.* **5**, 209–215.
Gosal, W. S., Clark, A. H., Pudney, P. D. A., Ross-Murphy, S. B., 2002. *Langmuir* **18**, 7174–7181.
Gosal, W. S., Clark, A. H., Ross-Murphy, S. B., 2004a. *Biomacromolecules* **5**, 2420–2429.
Gosal, W. S., Clark, A. H., Ross-Murphy, S. B., 2004b. *Biomacromolecules* **5**, 2408–2419.
Griffin, W. G., Griffin, M. C. A., 1993. *J. Chem. Soc., Faraday Trans. 1* **89**, 2879–2889.
Griffin, W. G., Griffin, M. C. A., Martin, S. R., Price, J., 1993. *J. Chem. Soc., Faraday Trans. 1* **89**, 3395–3406.
Hamley, I. W., 2007. *Angew. Chem., Int. Ed.* **46**, 8128–8147.
Higgs, P. G., Ball, R. C., 1989. *Macromolecules* **22**, 2432–2437.
Ho, R.-M., Chiang, Y.-W., Lin, S.-C., Chen, C.-K., 2011. *Prog. Polym. Sci.* **36**, 376–453.
Hooydonk, A. C. M., Olieman, C., 1982. *Neth. Milk Dairy J.* **36**, 153.
Horne, D. S., Banks, J. M., 2004. Rennet-induced coagulation of milk. In Fox, P. F., McSweeney, P. L. H., Cogan, T. M., Guinee, T. P. (eds), *Cheese: Chemistry, Physics and Microbiology*. Elsevier, Amsterdam, pp. 47–70.
Hughes, V. A., Dunstan, D. E., 2009. Amyloid fibrils: Self-assembling proteins. In *Modern Biopolymer Science*. Elsevier, San Diego, pp. 559–594.
Ikeda, S., Nishinari, K., 2000. *Biomacromolecules* **1**, 757–763.
Ikeda, S., Nishinari, K., 2001. *Int. J. Biol. Macromol.* **28**, 315–320.
Ikeda, S., Nishinari, K., Foegeding, E. A., 2000. *Biopolymers* **56**, 109–119.
Inoue, H., 1994. *J. Rheol.* **38**, 973.
Janmey, P. A., Weitz, D. A., 2004. *Trends Biochem. Sci.* **29**, 364–370.
Janmey, P. A., Shah, J. V., Janssen, K. P., Schliwa, M., 1998. *Subcell. Biochem.* **31**, 381–397.
Jung, J.-M., Mezzenga, R., 2010. *Langmuir* **26**, 504–514.
Jung, J. P., Gasiorowski, J. Z., Collier, J. H., 2010. *Biopolymers* **94**, 49–59.
Kamatari, Y. O., Konno, T., Kataoka, M., Akasaka, K., 1996. *J. Mol. Biol.* **259**, 512–523.
Kasai, M., Kawashima, H., Oosawa, F., 1960. *J. Polym. Sci.* **44**, 51–69.
Kavanagh, G. M., Clark, A. H., Gosal, W. S., Ross-Murphy, S. B., 2000a. *Macromolecules* **33**, 7029–7037.
Kavanagh, G. M., Clark, A. H., Ross-Murphy, S. B., 2000b. *Int. J. Biol. Macromol.* **28**, 41–50.
Kavanagh, G. M., Clark, A. H., Ross-Murphy, S. B., 2000c. *Langmuir* **16**, 9584–9594.
Kawanishi, K., Takeda, Y., Inoue, T., 1986. *Polym. J.* **18**, 411–416.
Khokhlov, A. R., Semenov, A. N., 1981. *Phys. A (Amsterdam)* **108**, 546–556.
Kohyama, K., Nishinari, K., 1993. *J. Agric. Food Chem.* **41**, 8–14.
Kohyama, K., Murata, M., Tani, F., Sano, Y., Doi, E., 1995a. *Biosci., Biotechnol., Biochem.* **59**, 240–245.
Kohyama, K., Sano, Y., Doi, E., 1995b. *J. Agric. Food Chem.* **43**, 1808–1812.
Koike, A., Nemoto, N., Doi, E., 1996. *Polymer* **37**, 587–593.
Koltun, W. L., Waugh, D. F., Bear, R. S., 1954. *J. Am. Chem. Soc.* **76**, 413–417.
Krebs, M. R. H., Bromley, E. H. C., Donald, A. M., 2005. *J. Struct. Biol.* **149**, 30–37.
Krebs, M. R. H., Domike, K. R., Donald, A. M., 2009. *Biochem. Soc. Trans.* **37**, 682–686.
Langton, M., Hermansson, A. M., 1992. *Food Hydrocolloids* **5**, 523–539.
Lin, Y. C., Koenderink, G. H., MacKintosh, F. C., Weitz, D. A., 2007. *Macromolecules* **40**, 7714–7720.

MacPhee, C. E., Woolfson, D. N., 2004. *Curr. Opin. Solid State Mater. Sci.* **8**, 141–149.
Maltais, A., Remondetto, G. E., Gonzalez, R., Subirade, M., 2005. *J. Food. Sci.* **70**, C67–C73.
Maruyama, K., Kaibara, M., Fukada, E., 1974. *Biochim. Biophys. Acta* **371**, 20–29.
Matsumoto, T., Inoue, H., 1993. *J Appl. Phys.* **74**, 2415–2419.
McGann, T. C. A., Donnery, W. J., Kearney, R. D., Buchheim, W., 1980. *Biochim. Biophys. Acta* **630**, 261–270.
Mellema, M., Heesakkers, J. W. M., van Opheusden, J. H. J., van Vliet, T., 2000. *Langmuir* **16**, 6847–6854.
Nagano, T., Akasaka, T., Nishinari, K., 1994a. *Biopolymers* **34**, 1303–1309.
Nagano, T., Mori, H., Nishinari, K., 1994b. *Biopolymers* **34**, 293–298.
Nicolai, T., Durand, D., 2007. *Curr. Opin. Colloid Interface Sci.* **12**, 23–28.
Niki, R., Kim, G. Y., Kimura, T. *et al.*, 1994a. Milchwissenschaft (Milk Sci. Int.) **49**, 325–329.
Niki, R., Kohyama, K., Sano, Y., Nishinari, K., 1994b. *Polym. Gels Networks* **2**, 105–118.
Ohgren, C., Langton, M., Hermansson, A. M., 2004. *J. Mater. Sci.* **39**, 6473–6482.
Olsson, C., Langton, M., Hermansson, A. M., 2002. *Food Hydrocolloids* **16**, 477–488.
Payens, T. A. J., 1979. *J. Dairy Res.* **46**, 291–306.
Payens, T. A. J., Wiersma, A. K., Brinkhuis, J., 1977. *Biophys. Chem.* **6**, 253–261.
Pedersen, J. S., Otzen, D. E., 2008. *Protein Sci.* **17**, 2–10.
Petka, W. A., Harden, J. L., McGrath, K. P., Wirtz, D., Tirrell, D. A., 1998. *Science* **281**, 389–392.
Plucknett, K. P., Baker, F. S., Normand, V., Donald, A. M., 2001. *J. Mater. Sci. Lett.* **20**, 1553–1557.
Pouget, E., Fay, N., Dujardin, E. *et al.*, 2010. *J. Am. Chem. Soc.* **132**, 4230–4241.
Ramachandran, S., Tseng, Y., Yu, Y. B., 2005. *Biomacromolecules* **6**, 1316–1321.
Renard, D., Lefebvre, J., 1992. *Int. J. Biol. Macromol.* **14**, 287–291.
Richardson, R. K., Ross-Murphy, S. B., 1981. *Int. J. Biol. Macromol.* **3**, 315–322.
Roberts, W. W., Kramer, O., Rosser, R. W., Nestler, F. H., Ferry, J. D., 1974. *Biophys. Chem.* **1**, 152 160.
Roefs, S. P. F. M., van Vliet, T., 1990. *Colloids Surf.* **50**, 161–175.
Roefs, S. P. F. M., de Groot-Mostert, A. E. A., van Vliet, T., 1990. *Colloids Surf.* **50**, 141–159.
Sagis, L. M. C., Veerman, C., van der Linden, E., 2004. *Langmuir* **20**, 924–927.
Samoto, M., Maebuchi, M., Miyazaki, C. *et al.*, 2007. *Food Chem.* **102**, 317–322.
San Biagio, P. L., Bulone, D., Emanuele, A., Palma, M. U., 1996. *Biophys. J.* **70**, 494–499.
Shah, J. V., Janmey, P. A., 1997. *Rheol. Acta* **36**, 262–268.
Shih, W.-H., Shih, W. Y., Kim, S.-I., Liu, J., Aksay, I. A., 1990. *Phys. Rev. A* **42**, 4772–4779.
Stading, M., Hermansson, A. M., 1991. *Food Hydrocolloids* **5**, 339–352.
Stading, M., Langton, M., Hermansson, A. M., 1992. *Food Hydrocolloids* **6**, 455–470.
Stading, M., Langton, M., Hermansson, A. M., 1993a. *Food Hydrocolloids* **7**, 195–212.
Stading, M., Langton, M., Hermansson, A. M., 1993b. *Makromol. Chem.* **76**, 283–290.
Storm, C., Pastore, J. J., MacKintosh, F. C., Lubensky, T. C., Janmey, P. A., 2005. *Nature* **435**, 191–194.
Takahashi, Y., Ueno, A., Mihara, H., 1999. *Bioorg. Med. Chem.* **7**, 177–185.
Tan, H. M., Moet, A., Hiltner, A., Baer, E., 1983. *Macromolecules* **16**, 28–34.
Tanaka, T., Swislow, G., Ohmine, I., 1979. *Phys. Rev. Lett.* **42**, 1556–1559.
Tanford, C., De, P. K., Taggart, V. G., 1960. *J. Am. Chem. Soc.* **82**, 6028–6034.
Tanford, C., Lively, E. P., Buckley, C. E., De, P. K., 1962. *J. Biol Chem.* **237**, 1168–1171.
Tobitani, A., Ross-Murphy, S. B., 1997a. *Macromolecules* **30**, 4845–4854.
Tobitani, A., Ross-Murphy, S. B., 1997b. *Macromolecules* **30**, 4855–4862.
Tokita, M., Hikichi, K., Niki, R., Arima, S., 1982. *Biorheology* **19**, 695–705.

Townend, R., Kumosins, T. F., Timashef, S. N., 1967. *J. Biol Chem.* **242**, 4538–4545.

Ulijn, R. V., Woolfson, D. N., 2010. *Chem. Soc. Rev.* **39**, 3349–3350.

van den Berg, L., van Vliet, T., van der Linden, E., van Boekel, M. A. J. S., van de Velde, F., 2007a. *Food Hydrocolloids* **21**, 420–432.

van den Berg, L., van Vliet, T., van der Linden, E., van Boekel, M. A. J. S., van de Velde, F., 2007b. *Food Hydrocolloids* **21**, 961–976.

van den Berg, L., van Vliet, T., van der Linden, E., van Boekel, M. A. J. S., van de Velde, F., 2008. *Food Biophys.* **3**, 198–206.

van der Linden, E., Foegeding, E. A., 2009. Gelation: Principles, models and applications to proteins. In *Modern Biopolymer Science*. Elsevier, San Diego, pp. 29–91.

van der Linden, E., Sagis, L. M. C., 2001. *Langmuir* **17**, 5821–5824.

van Kleef, F. S. M., Boskamp, J. V., van den Tempel, M., 1978. *Biopolymers* **17**, 225–235.

van Vliet, T., Walstra, P., 1985. *Neth. Milk Dairy J.* **39**, 115.

van Vliet, T., Dijk, H. J. M., Zoon, P., Walstra, P., 1991. *Colloid Polym. Sci.* **269**, 620–627.

Veerman, C., Ruis, H., Leonard, M., van der Linden, E., 2002a. *Biomacromolecules* **3**, 869–873.

Veerman, C., Ruis, H., Sagis, L. M. C., van der Linden, E., 2002b. *Biomacromolecules* **3**, 869–873.

Verheul, M., Roefs, S. P. F. M., 1998. *Food Hydrocolloids* **12**, 17–24.

Walkenstrom, P., Hermansson, A. M., 1994. *Food Hydrocolloids* **8**, 589–607.

Walstra, P., 1999. *Int. Dairy J.* **9**, 189–192.

Walstra, P., van Dijik, H. J. M., Geurts, T. J., 1985. *Neth. Milk Dairy J.* **39**, 209–246.

Wang, C., Stewart, R. J., Kopecek, J., 1999. *Nature* **397**, 417–420.

Yan, C. Q., Pochan, D. J., 2010. *Chem. Soc. Rev.* **39**, 3528–3540.

10 Mixed gels

10.1 Introduction

It has been well known since the 1940s that two high polymeric species in the melt state tend not to mix. Consequently only some combinations lead to miscible blends, and the same applies for solutions of the two polymers. Whether two polymeric species are miscible or tend to phase separate is explained in the next section.

The compositional range, especially where the two polymers separate into two phases, depends on the temperature, and also on the respective molecular masses of the two components. Generally, when the temperature of the solution of two polymers is raised, that range narrows, and above a certain temperature that range of immiscibility will disappear. This temperature is called the upper critical solution temperature (UCST) (Figure 10.1a). Along the curve, the two phases coexist and therefore it is referred to as the co-existence or binodal curve. Within this curve lies the spinodal curve or limit of absolute stability, and between the binodal and spinodal curves is a region of metastability. According to the Flory–Huggins theory and its subsequent modifications (Koningsveld *et al.*, 2001), as the molecular mass increases, the compatibility range tends to shrink. This is why gelation – which can be considered, at one level at least, a very large increase in molecular mass – tends to drive compatible pre-gel solutions into phase separation.

However, there are cases where the miscibility *decreases* with increasing temperature and the phase diagram is the inverse of that shown in Figure 10.1. For such a pair of polymers, the range where the species separate into two phases becomes larger as temperature increases. When the temperature is raised enough, the range of miscibility disappears, at a temperature now known as the lower critical solution temperature (LCST). Since the entropy of mixing is smaller for polymers with higher molar mass and/or especially those which have been hydrophobically modified, such a mixture can show LCST behaviour. For lower molar mass polymers, a UCST may be seen instead.

Mixed gels consisting of more than two polymers are classified into two groups: gels formed by 'simple' phase separation and those formed by specific (in the literature commonly, though not always accurately, referred to as 'synergistic') molecular interactions, where the enthalpy of the specific interactions is sufficient to overcome the normal tendency for phase separation.

In the incompatible case, gels formed by phase separation tend to be produced by the so-called 'exclusion effect' between the two polymers, which then leads to the increase in the effective concentration of each component within the mixture.

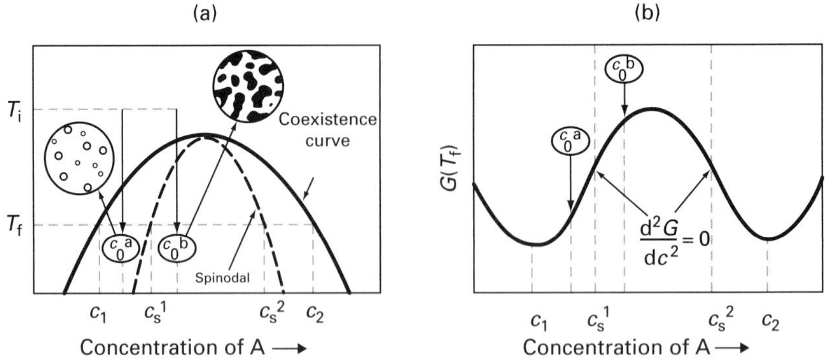

Figure 10.1 (a) Phase separation in a binary mixture of polymers A and B as a function of concentration of A, showing the coexistence (binodal) and spinodal curves. Below the binodal curve the mixture is separated into two phases, and above it is miscible. The temperature where the binodal and spinodal touch is the upper critical solution temperature (UCST). Starting at high temperature T_i and quenching the mixture at the final temperature T_f, we can see a phase transformation through nucleation and growth in the case where the initial concentration is c_0^a, and via spinodal decomposition when it is c_0^b. c_1 and c_2 are the equilibrium concentrations of the two phases at T_f, and c_s^1 and c_s^2 are the concentrations on the spinodal line. (b) The corresponding free energy of mixing versus concentration of A in the mixture A + B, at temperature T_f determining the boundaries of the phase diagram.

For example, if a solution of nominal concentration around 0.5% separates into a phase with a volume fraction ϕ of 0.4, its *effective* concentration is now $(c/\phi) = 1.25\%$.

On the other hand, gels formed by 'synergistic' interactions can be divided into two types: coupled networks, formed by specific binding between two different polymers; and interpenetrating polymer networks (IPNs). The latter comprise two networks, both of which completely fill the available space and so mutually interpenetrate but, structurally at least, appear not to interact or influence one another. For example, if images are made, what will appear is a superimposition of the images of the two separate components. In modern parlance they can be regarded as nanocomposites. According to Lipatov and Alekseeva (2007), the IPN results from incomplete phase separation of networks, where the kinetics of phase separation and of gelation proceed in non-equilibrium conditions.

A number of excellent reviews and chapters on mixed gels have been published (Morris, E. R., 1990; Morris, V. J., 1998; Tolstoguzov, 1998, 2006; Morris, E. R. *et al.*, 2009). The aim of the present chapter is to compliment these publications rather than to present a complete catalogue of systems (which would require several volumes rather than a single chapter). It is important to stress that comparison between workers is particularly difficult because of variability in samples and in preparative and experimental conditions. There are very few 'complete' studies because most workers have pursued only a few techniques. For this reason there is still much scope for future research in this area.

In many applications, gelling systems contain mixtures of several biopolymers, and we shall concentrate almost exclusively on these. Water is generally a good solvent for the biopolymer species of the blend composition (proteins, polysaccharides), but phase separation for such systems is still a very common phenomenon (Harding *et al.*, 1995).

Indeed miscibility remains an exception because of the high molecular mass of the polymers, as explained above. Among polymer properties, the type of polymer, its molecular mass, the presence of certain functional groups (e.g. ionized or hydrophobic groups) and the ionic composition are all of crucial importance. The thermodynamics of such mixtures, in particular the phase diagrams, plays a key role but in general is far from sufficient to predict the final state of the systems for which they were designed. The morphologies observed are never in equilibrium: the mechanisms of transport in systems containing polymers are very slow, but during physical gelation of one of the constituents, phase separation becomes even slower and the structures that result are the result of a process which is limited kinetically. This also occurs if one of the phases becomes glassy, as discussed in Chapter 8.

The theoretical background for phase separations, devised for small-molecule mixtures, was extended to polymer mixtures by the Flory–Huggins lattice model. Physical gelation has a profound effect on the kinetics of phase separation, and offers great *potential* control of the phase morphology. In this chapter we first discuss the mechanisms which affect the kinetics of phase separation, in particular those incipient stages known to affect the morphologies of the mixture, i.e. nucleation and growth versus spinodal decomposition, and their influence on structure. Next we consider the structure and properties of mixed-gel systems, first where phase separation follows the usual routes described here, then where more specific molecular interactions begin to influence the phase behaviour, as is the case in many polysaccharide–polysaccharide mixtures. In the latter case we give only selected examples, stressing those where research or applications are of current interest.

10.2 Equilibrium thermodynamics

The equilibrium thermodynamics of any mixture is determined by the free energy of mixing (Gibbs energy) as a function of composition and temperature (Ragone, 1995), and the phase diagram can be derived by minimizing the free energy function. The simplest way to approach the problem is to consider a two-component (A and B) mixture: at a given temperature, the minima of the free energy diagram define the limits of coexistence of two phases. The locus of such points is the binodal curve, and the symmetric form of Figure 10.1a is valid for simple mixtures or regular solutions. More generally, the limit of solubility is determined by the common tangent to the free energy curve as a function of composition (Cahn *et al.*, 1991). The points below the curve of Figure 10.1a are metastable to local fluctuations in concentration.

Gibbs derived a necessary condition for the stability or metastability of a 'fluid' phase. He showed that a two-component solution will transform spontaneously if $(\partial^2 G/\partial c^2)_{T,P} < 0$, where G is the Gibbs free energy per mole of solution and c is the concentration at constant temperature T and pressure P. On a phase diagram, the boundary of the unstable region is defined by the locus$(\partial^2 G/\partial c^2)_{T,P} = 0$, and is called the spinodal. The spinodal and co-existence curves meet at a single point, the critical point, and this defines the limit temperature where phase separation takes place. For ternary systems the phase diagram is three-dimensional, with stability limits defined as surfaces.

Starting at a high temperature T_i in the one-phase region, and cooling down to T_f in the two-phase region, an initially clear solution becomes turbid when the temperature is lowered below the binodal. In Figure 10.1b we can see that the first derivative with respect to concentration is zero at the two free-energy minima, corresponding to the two miscible phases, and the second derivative is zero at the points of inflexion. (The above treatment considers the case of upper critical solution temperatures or UCSTs.)

10.3 Phase dynamics: nucleation and growth versus spinodal decomposition

10.3.1 Nucleation and growth

For this case, consider the dynamics of the phase transformations induced by an instantaneous quench from the initial temperature T_i to a final temperature T_f. If a system is quenched from an equilibrium state in the one-phase region to a state below the coexistence line, phase separation tends to occur, and in the final equilibrium state both phases coexist, with equilibrium concentrations c_1 and c_2.

However, if the system is quenched to the final temperature T_f in the metastable region between the spinodal and the binodal, phase separation starts with the nucleation of small droplets whose structure and properties vary from that of the parent phase material, but approach those of the more stable phase. A finite fluctuation is required to render the solution unstable. This fluctuation results in the formation of a nucleus, and the work of forming such a nucleus is a measure of the metastability of the phase.

For a nucleus to form through thermal composition fluctuations, the nucleation barrier, typically of the order of several $k_B T$, has to be overcome. Therefore, there is a critical droplet size r^* which is in unstable equilibrium with the exterior phase. With any infinitesimal increase in size, such a droplet (or critical nucleus) can continue to grow without further external intervention, because the free energy decreases. In the metastable region below the binodal, the solution is supersaturated and nucleation is homogeneous. As shown in Figure 10.2, the nucleus rich in component A is surrounded by a concentration gradient of A which provides the driving force for solute diffusion and gives rise to its growth. The growth rate is controlled by the rate at which molecules A diffuse towards the interface or by the rate at which they cross the interface. Once the molecules are incorporated into the growing interface, the solution is depleted of these molecules and diffusion is slower.

The work required to create a nucleus is strongly influenced by surfaces such as inclusions (seed particles) which may catalyze the nucleation of the new phase. For example, solidification in supercooled solutions allows the solid to start growing from an existing surface. The critical radius of the nucleus is unchanged, because it depends mainly on the temperature, but the free energy for the nucleus to overcome is decreased by so-called heterogeneous nucleation.

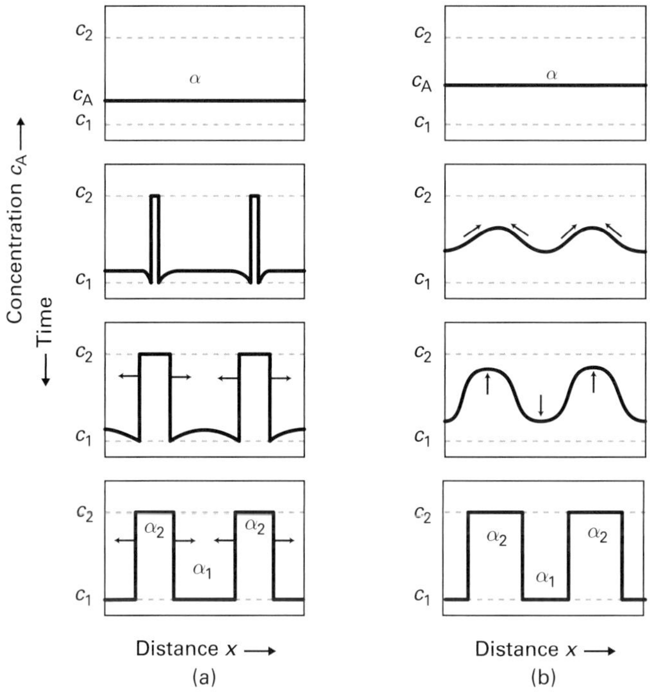

Figure 10.2 Nucleation and growth (a) and spinodal decomposition (b) mechanisms. Two phases α_1 and α_2 form in the mixture A + B when temperature is lowered below the coexistence line, at T_f. If the initial value of c_A (the concentration of constituent A) is c_0^a, phase separation proceeds through nucleation and growth (a). The nucleus rich in constituent A has an initial diameter $2r^*$. As time increases, the size of the nuclei increases but their composition is stable. If the initial value of c_A is c_0^b, spinodal decomposition takes place (b) and concentration fluctuations increase versus time. These have a characteristic wavelength which also increases with time. In both cases, at equilibrium, the concentrations of the two phases are c_1 and c_2, but the volume fractions are different.

The rate of growth of the transformed phase rate depends on two factors: the nucleation barrier and the activation energy for the movement of the nuclei. When the temperature is lowered, these terms work in opposite directions, the critical nuclei become smaller and smaller (inversely proportional to the supercooling temperature) and their nucleation rate increases, but the diffusion rate of the nuclei decreases and thus the factor related to the activation energy becomes weaker. The overall balance gives rise to an optimum temperature, which gives a maximum rate to the phase transformation.

10.3.2 Spinodal decomposition

Figure 10.2b shows a binary mixture with a concentration c_0^b of the constituent A in the homogeneous phase. The mixture is homogeneous at time $t = 0$. If the mixture is quenched at a temperature below the spinodal, locally spontaneous fluctuations create inhomogeneities in composition: the solution undergoes local concentration fluctuations in concentration, so that in some regions it is $c_0 + \delta c$ and in others $c_0 - \delta c$. A Taylor series expansion of the change in the free energy ΔG around the composition c_0 shows that it is related to the second derivative of the free energy with respect to concentration. If this is positive, ΔG accompanying the fluctuation is positive, so the fluctuations will die out and the phase is stable. Conversely, in the unstable region of the phase diagram, any small fluctuations in concentration that occur spontaneously because of thermal fluctuations will tend to grow. Since there is no thermodynamic barrier to reaction inside the spinodal region, decomposition is determined solely by diffusion. If diffusion takes place in a stable solution, the molecules will diffuse 'downhill' from the high- to the low-concentration domain and the system will return to equilibrium with a uniform density (single phase). However, if the system is unstable, the mutual diffusion coefficient is negative and material is transported from regions of low concentration to regions of high concentration ('uphill diffusion'), a process called spinodal decomposition (SD). Such phase separation processes have been observed, for instance, in the solidification of metal alloys and in block copolymer blends (Hashimoto *et al.*, 1984; Sasaki and Hashimoto, 1984), and they differ qualitatively from those in the nucleation and growth process. For a range of compositions that would give approximately equal volume fractions, and close to the critical point composition, the two phases form an interconnected, often bicontinuous structure.

The theoretical formulation of SD was first proposed by Cahn and Hilliard in a series of papers (Cahn and Hilliard, 1958, 1959; Cahn, 1965). Below the spinodal, the free energy can be lowered ($\Delta G < 0$) if the amplitude of the fluctuations is large enough, and so these fluctuations will tend to increase. Then the inhomogeneous phase is no longer stable, and the so-called demixing with SD occurs. Consequently, if the system is quenched rapidly at a composition close to the critical composition, the spinodal mechanism is likely to take place, while systems quenched slowly through the metastable region will decompose by nucleation and growth steps, allowing critical nuclei to be formed (Papon *et al.*, 2002).

Consider an inhomogeneous solution whose composition everywhere differs only slightly from the average composition, and with small composition gradients. The difference between the free energy of inhomogeneous and homogeneous parts of the system corresponds to the free energy of creation of the new phase, written

$$\Delta G = \int \left[\frac{1}{2} \left(\frac{\partial^2 f}{\partial c^2} \right)_{c_0} (c - c_0)^2 + K |\nabla c|^2 \right] dV, \tag{10.1}$$

where c_0 is the overall composition and $f(c)$ is the free energy density of a mixture of a homogeneous material with composition c. The additional term $K|\nabla c|^2$ accounts for the contribution to the free energy density of small concentration gradients.

In the development of the Cahn–Hilliard theory, in order to predict the time and spatial dependence of fluctuations (or the fluctuations in the reciprocal space), it is convenient to recast the theory in terms of Fourier components of the composition, rather than the composition itself. If $f(c)$ is the free energy density of a mixture of homogeneous material, the approach shows that, in the unstable region, the term $(\partial^2 f/\partial c^2) < 0$, so that fluctuations with a short wavelength (large $\boldsymbol{q}$ vectors) tend to increase the free energy and therefore stabilize the homogeneous phase. On the other hand, fluctuations with large wavelengths (small $\boldsymbol{q}$ vectors) stabilize the inhomogeneous phase and work in favour of transformation into separate phases. K, the energy gradient coefficient, is always positive and a characteristic of the system, so demixing will take place when $\Delta G_q < 0$, and this determines a critical wave vector q_c:

$$\Delta G_q \sim \left[\left(\frac{\partial^2 f}{\partial c^2} \right)_{c_0} + 2Kq^2 \right] \tag{10.2}$$

$$\left(\frac{\partial^2 f}{\partial c^2} \right)_{c_0} + 2Kq^2 < 0 \quad \text{when} \quad q^2 < q_c^2,$$

with

$$q_c^2 = -\frac{\left(\frac{\partial^2 f}{\partial c^2} \right)_{c_0}}{2K}. \tag{10.3}$$

Equation (10.2) states that

- when $q > q_c$ the solution is stable with respect to concentration fluctuations, so spinodal fluctuations will decay
- when $q < q_c$ the solution becomes inhomogeneous and spinodal decomposition occurs.

10.3.3 Kinetics of demixing

The difference in chemical potentials between the components is obtained from the derivative of ΔG with respect to composition. The Fourier components of the concentration fluctuation $\delta c(\boldsymbol{q},t)$ decay with a characteristic rate ω_q depending on the wave vector $\boldsymbol{q}$:

$$\omega_q = D_{\text{eff}} q^2 \left[\frac{q^2}{q_c^2} - 1 \right], \tag{10.4}$$

where D_{eff} is the effective diffusion coefficient that changes sign at the spinodal line.

In the thermodynamically unstable region for wave vectors ($q < q_c$) the characteristic decay rates become negative ($\omega_q < 0$) and the concentration fluctuations are not dampened, so they grow with time. If the material is represented in one dimension (Figure 10.2), fluctuations with large wavelengths are amplified with time and lead to SD within the blend. This wavelength has a maximum at $q_m = q_c/\sqrt{2}$ and it becomes an 'amplification factor'. The theory predicts that one particular wavelength of fluctuations in the composition q_m grows most rapidly, leading to a well-defined peak in the scattering pattern. Thus, at some time after phase separation starts, a description of the composition in the solution will be a superposition of sine waves of fixed wavelength $q = q_m$, but random in orientation, phase and amplitude. A periodic distribution of higher and lower concentrations increases, with a characteristic periodicity length $\Lambda_m = 2\pi/q_m$ predicted.

Overall, then, the Cahn–Hilliard linearized model suggests:

(a) The logarithm of the scattered intensity plotted against time should increase linearly.
(b) A plot of ω_q/q^2 versus q^2 (Cahn–Hilliard plot) is a straight line if the effective diffusion coefficient D_{eff} is independent of q. From the slope and the intercept, D_{eff} and q_m can be calculated.
(c) In order to compare the coarsening of phase separating systems at various temperatures and test for universal behaviour, the processes are considered on a reduced time scale, which is the time it takes to build up an inhomogeneity of characteristic size $1/q_m$. The characteristic time τ for the early stages of SD can be defined by

$$\tau = \frac{1}{D_{eff} q_m^2}. \tag{10.5}$$

10.3.4 Spinodal decomposition in polymer mixtures

A specific model was developed for polymer mixtures by de Gennes (1980) for incompressible binary liquids composed of macromolecules, which he called the 'metallurgy of polymer blends'. de Gennes discussed the qualitative features of the concentration fluctuations of binary polymer melts of two species A and B, with individual molecular masses (or weights) N_A and N_B, starting with a description of the free energy density of homogeneous mixtures using the Flory–Huggins formula.

The interaction parameter or Flory parameter χ is often positive in polymer blends and favours segregation. If the two species are not too different, χ is small. Therefore the de Gennes model mainly concerns the case of small positive χ. Provided that both N_A and N_B are large, the static properties of the blends are supposed to be well described by the mean field approach. For instance, in the symmetric case ($N_A = N_B = N$) there is a critical value χ_s (spinodal point or critical point of the polymer blend) defined by

$$\frac{1}{\chi_s} = 2Nc(1 - c), \tag{10.6}$$

where c is the concentration of polymer A. The χ parameter is temperature-dependent. For $\chi < \chi_s$ the system is entirely miscible, while for $\chi > \chi_s$ phase separation may occur.

As far as the early stages of demixing kinetics are concerned, by comparison with other systems, demixing in polymer mixtures proceeds very slowly, so that the initial stages should be studied more easily. Among others, Hashimoto *et al.* (1983) showed by time-resolved light scattering studies that binary polymer mixtures of poly(styrene) (PS) and poly(vinylmethylether) (PVME) obeyed de Gennes's predictions for the early stage of SD.

10.3.5 Spinodal decomposition of ternary systems

A more complex situation is encountered with a ternary polymer system, composed of polymer A, polymer B and a solvent. However, ternary systems can be treated as pseudo-binary systems so that, with good approximation, theories developed for binary systems can be applied when a neutrally 'good solvent' for both polymers is used. The kinetics of the phase separation of ternary PS–PB–Tol and PS–SB–Tol mixtures (PS, PB, SB and Tol represent respectively poly(styrene), poly(butadiene), poly(styrene)–poly(butadiene) diblock polymer and toluene, a good solvent for each of these polymers) were studied by Sasaki and Hashimoto (1984) using time-resolved light scattering. Their results were in good agreement with the predictions of Cahn's linearized theory of SD, modified by de Gennes. The phase separation of the ternary systems could be treated as that between PS solution and PB (or SB) solution, in which the solvent essentially weakens the effective interaction between the two polymers.

Figure 10.3, from the work of Sasaki and Hashimoto (1984), shows a typical change of scattering profiles with time, after a temperature jump, for a nearly critical PS–PB–Tol mixture. The scattering profiles were reconstructed from plots of scattered intensity versus time at various scattering vectors q. After initiation of phase separation, a clear scattering maximum is seen, with a characteristic length $2\pi/q_m \sim 3\ \mu$m. Up to 150 s, the scattering intensity increases, whereas the scattering vector q_m of maximum intensity does not significantly change with time, qualitatively in agreement with the theoretical prediction given by Cahn's linearized theory of SD.

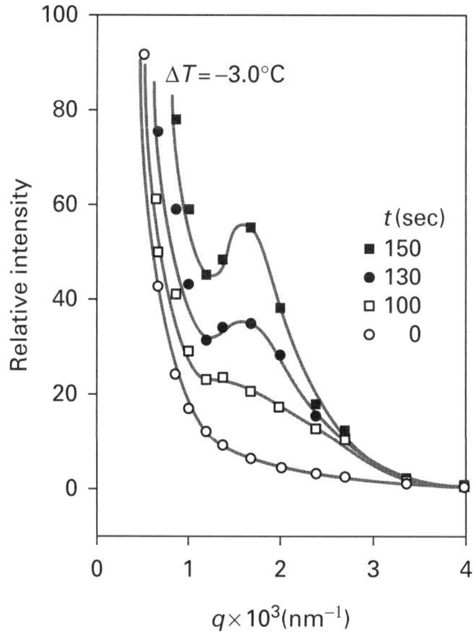

Figure 10.3 Change of the light scattering profile $I(q)$ with time for the ternary solution PS–PB–Tol at solvent volume fraction 0.925 at $\Delta T = -3$°C below the cloud point, for periods of time of 0 to 150 s after temperature equilibration for a nearly critical mixture. From Sasaki and Hashimoto (1984) © 1984 American Chemical Society.

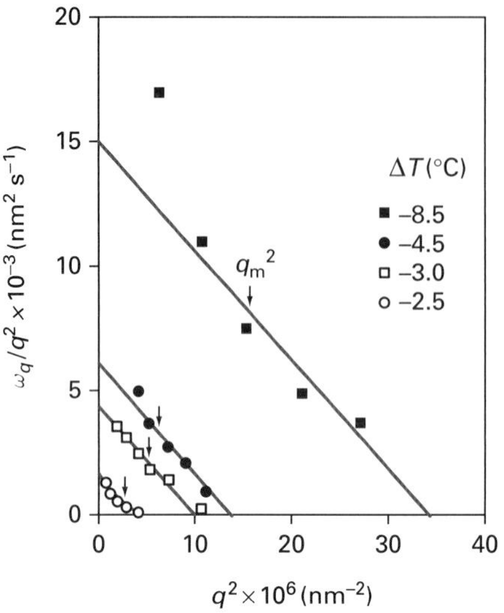

Figure 10.4 Cahn–Hilliard plots of the early stages of phase separation of the ternary mixture PS–PB–Tol at solvent volume fraction 0.925 (nearly critical mixture) quenched to various depths ΔT from −2.5°C to −8.5°C below the cloud point temperature. The arrows indicate the position of the maximum q_m. From Sasaki and Hashimoto (1984) © 1984 American Chemical Society.

In the early stage of phase separation ($t \leq 100$ s), the relaxation (or growth) rate of the intensity or of concentration can be measured from the slope of scattered intensity versus q. Later ($t \geq 150$ s), the intensity change with time deviates from exponential, owing to the onset of coarsening processes. As phase separation progresses further, the peak position starts to shift towards small angles, indicating that the periodicity of fluctuations, as well as the amplitude of fluctuations, increases further with time.

Cahn–Hilliard plots of ω_q/q^2 versus q^2 are shown in Figure 10.4, and different temperature jumps at the early stages help to determine D_{eff}. This is found to be independent of ΔT up to 10°C, for various proportions of these ternary systems. In measurements performed at large quench depths, deviations from a straight line appear in the plots of ω_q/q^2 versus q^2, suggesting the limitations of linearized theory.

10.3.6 Kinetics of demixing and gelation in ternary systems

In both biological and application systems, polysaccharides and proteins are often present together. When a certain concentration of these biopolymers in the mixture is exceeded, phase separation often results (Tolstoguzov *et al.*, 1985; Zasypkin *et al.*, 1997; Doublier *et al.*, 2000; Scholten *et al.*, 2002). A very large number of ternary systems containing proteins and polysaccharides in aqueous medium are described as thermodynamically incompatible. When the Flory–Huggins interaction parameter for biopolymer A–biopolymer B interactions is positive, indicating a net repulsion between the biopolymers, segregation occurs, so that the system finally demixes into two phases, each enriched with one of the biopolymers. This occurs for most such biopolymer systems, whether charged or uncharged. Another type of phase separation phenomenon is associative phase separation or complexation. This occurs when interactions between the two biopolymers are favoured (negative Flory parameter), for example when the polymers carry opposite charges, for instance at a pH slightly lower than the isoelectric point of the protein while the polysaccharide carries a negative charge. One example of this is the mixture of milk proteins and carrageenans in certain food products.

Consequently the main parameters involved in the phase separation mechanism of aqueous mixed polymer solutions are pH and ionic strength. During the past three decades, a significant amount of work has been performed on the stability of polysaccharide–protein mixtures. These studies have contributed to an understanding of the concentrations and conditions under which phase separation processes are expected to occur. Substantial richness and complexity are introduced to the problem when there is competition between phase separation and other processes, such as gelation. A similar case was noted in Chapter 8 for gelation of synthetic polymer solutions via phase transformations, either in organic solvents or in water. When competition between phase separation and gelation occurs, stable but *non-equilibrium morphologies* can be generated whose characteristics may be strongly dependent on the history of the sample. The aim of such investigations is often to design systems with pre-specified properties, of value in particular applications, by varying phase morphology.

10.3.7 Rheology of mixed systems

As pointed out in earlier sections, segregative phase separation tends to produce a structure with one phase dispersed in another. What are the rheological implications of this? Of course work on the rheology of dispersed systems has a long and detailed history. The use of fibre-reinforced polymeric materials, for example, has revolutionized modern construction work. In evaluating the properties of such materials, a number of assumptions are made, at least initially. These include (1) that the structure remains constant; (2) the so-called 'rule of mixtures', that the properties of a composite material can be estimated from the volume fractions and, say, moduli of the two phases; and (3) that there is perfect adhesion between the phases, so stress and strain are not disrupted across the phase boundary.

Allowing for the above assumptions, observed values of elastic modulus can, in principle, be compared with those calculated by a formula for composite systems such as that proposed by Okano (1962), which in turn is a generalized form of Kerner's formula (Kerner, 1956). This was suggested by Watase and Nishinari (1980) for mixed gels. There are a number of other models in the literature, but one simple but effective approach is the so-called polymer blending law, proposed by Takayanagi *et al.* (1963). Although this treatment does not give the elastic modulus of the mixture for each polymer volume fraction explicitly, it does give upper and lower bounds for the moduli under so-called isostrain and isostress conditions, respectively:

$$G_c = \phi_x G_x + \phi_y G_y \quad \text{(upper bound)} \tag{10.7}$$

and

$$\frac{1}{G_c} = \frac{\phi_x}{G_x} + \frac{\phi_y}{G_y} \quad \text{(lower bound)}. \tag{10.8}$$

In other words, knowing the moduli G_x and G_y of the two components and their respective volume fractions ϕ_x and ϕ_y, it is possible to predict the modulus G_c of the composite. Of course there are other formulations for the composite modulus, but these tend to be more sophisticated.

For example, Piculell and co-workers (Piculell *et al.*, 1992) reintroduced a phenomenological model for a granulated composite essentially analogous to that for random conductive percolation. This moderates the upper bound of (10.7) but at the expense of another parameter, the exponent 1/5:

$$G_c{}^{1/5} = \phi_x G_x{}^{1/5} + \phi_y G_y{}^{1/5}. \tag{10.9}$$

It has been suggested that this form, apparently appropriate for nanocomposites, may also be apposite for IPN systems. This assumption does not appear to have been tested, or even to be testable, since the determination of the two volume fractions would seem to be quite arbitrary.

However, in using any of these formulae as they stand, there is a fundamental weakness. As a mixed gel is formed, the polymeric components tend to be associated with a volume fraction of solvent, but this is common to both. Consequently both the concentration of each phase (which tends to control the modulus of each phase) and the volume fraction of solvent associated with that phase remain essentially unknown. However, a common approach is that due to Clark *et al.* (1983), in which a so-called solvent distribution parameter p_a, assumed to be independent of x and y and intended to measure the relative affinities of networks x and y for solvent, would modify these equations. In this way both the new 'effective concentration' of each phase, which is assumed to govern the modulus of each phase, and their respective volume fractions can be related just to p_a. Using this approach, large-deformation material properties can also be related back to phase composition using, for example, the BST equation approach introduced in Chapter 7 in studies of agarose and gelatin.

At this level the treatment is purely phenomenological, but p_a might be determinable by experimental methods such as measurement of the appropriate Flory–Huggins χ values for agarose and gelatin gels, measurement of the volume fractions ϕ_x and ϕ_y by microscopy, or direct determination of p_a using water NMR relaxation or FTIR microscopy.

10.4 Gels involving segregative phase separation

10.4.1 Protein–polysaccharide mixtures

Mixtures of protein and polysaccharide have attracted much attention, and the Russian school led by Tolstoguzov (1998, 2006) accumulated data on a wide range of mixtures, particularly by determining phase boundary conditions. One of the major techniques for studying incompatible mixed-gel systems is microscopy, and both electron and optical microscopy have been used by a number of groups, including early work on BSA and agarose gels (Clark *et al.*, 1982). Since many microscopy stains are taken up preferentially by the protein component, it is relatively simple to study the phase composition simply by examining images showing these contrasting areas.

10.4.1.1 Gelatin–dextran

In recent years several groups have examined the gelatin–dextran system. Gelatin is considered in detail in Chapter 7; the co-system dextran is a very soluble, and usually long-chain branched, polyglucan produced by lactic acid bacteria. Strictly speaking, the combination does not form a mixed gel since dextran itself never forms a gel under normal conditions. On the other hand, as a model system for investigating phase behaviour it has advantages since both components can be obtained 'off the shelf' quite easily. Because of this the system has been studied quite widely over recent years. As expected, close to the cooling conditions where gelatin begins to gel, phase separation begins, but it can be trapped (or to use the parlance common in this area, 'pinned') in various non-equilibrium states.

The kinetics of segregative phase separation was investigated in dextran–gelatin system by Tromp *et al.* (1995) using time-resolved small-angle light scattering and phase contrast microscopy. After a quench into the two-phase region of the phase diagram, at or below the gelling temperature, the resultant two-dimensional scattering pattern was recorded. This showed a ring corresponding to a peak in radially averaged scattered intensity, and is characteristic of spinodal decomposition (SD). For the samples quenched to the higher temperature, the logarithm of the scattered intensity plotted versus time showed an initial linear increase, as expected from linearized SD theory. At lower temperatures the initial linear growth regime was of very short duration, making it hard to observe. When the final quench temperature was fixed above or below the gelation temperature, a difference in behaviour was clearly observed by phase contrast microscopy as the morphology coarsened.

As shown in Figure 10.5, the position of the lower quench temperature scattering peak moves to lower q values as time progresses. This indicated that for q values close to q_m the system was outside the range of validity of the linear theory of SD. The initial time measurements in Figure 10.5 showed a maximum at $q_m \sim 1\ \mu m^{-1}$, corresponding to a characteristic length of 6 μm. Cahn–Hilliard plots were also drawn and they are seen to be far from linear, for both gelling and non-gelling systems. A possible explanation for the overall curvature of the Cahn–Hilliard plots is that there is more than one time scale governing the phase separation kinetics in this case (e.g. polydispersity effects) or that the very early stages were not captured because of experimental limitations.

Direct microscopic observation of this system gave images consistent with the scattering data. Phase contrast micrographs at two temperatures showed that, for both high- and low-temperature quenches, the pattern has the same network-like character, consistent with a common mechanism of phase separation in the early stages. Qualitative differences appeared at later times: at the higher temperature, droplet-like domains appeared after about 10 min, whereas such a transition did not take place for the low-temperature

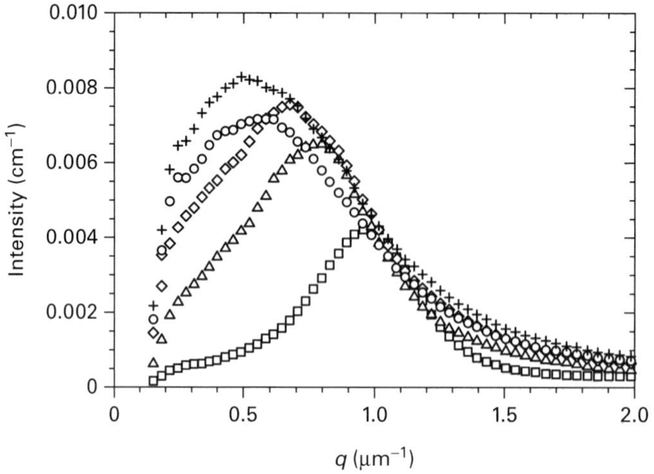

Figure 10.5 Change of light scattering profile $I(q)$ versus time for a ternary solution of 4.2% gelatin, 4.2% dextran and water quenched to $T = 21°C$ at 126 s (□), 378 s (Δ), 560 s (◊), 1160 s (○) and 1382 s (+). From Tromp *et al.* (1995) © 1995 American Chemical Society.

quench. Subsequently, phase separation above the gelation temperature proceeded via formation of large gelatin-rich spheres that grew by coalescence. For the gelling system the morphology was 'frozen' after about 20 min, when it was still a fine network.

10.4.1.2 Gelatin–maltodextrin

Another example of segregative phase separation and gelation is provided by aqueous gelatin–maltodextrin mixtures quenched to various end temperatures (Lorén *et al.*, 2001, 2002). The co-system, maltodextrin, represents one of a series of commercially degraded starches. The authors determined the time evolution of the bicontinuous microstructures by confocal laser scanning microscopy (CLSM) of a fluorescently labelled maltodextrin, and the growth of the bicontinuous microstructures was quantified by Fourier image analysis. The mixtures were quenched from 60°C to various end temperatures both above and below the gelatin gelation temperature, and the gelatin concentration was chosen so the continuous phase was gelatin-rich. Measurements showed that the mixture phase separated through SD at all temperatures. The authors noted that the time of residence in the region where phase separation competes with gel formation was of critical importance in influencing the morphology of the resulting system, gelation of the continuous phase having a strong influence on the structure evolution. Phase separation finally became trapped kinetically ('pinned') by gelation (Lorén *et al.*, 2002).

A series of six CLSM micrographs of the ternary solution recorded at various times during a temperature quench is shown in Figure 10.6. The series of micrographs clearly show the mottled pattern attributed to the SD phase separation mechanism. The onset of phase separation is shown in micrograph (a) as fluctuations in concentration with a characteristic wavelength, i.e. a characteristic distance between the maltodextrin domains (bright phase). The micrograph series also shows that the distance between

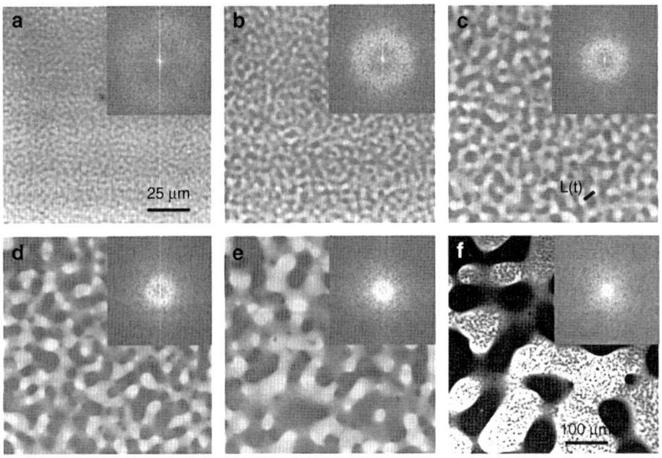

Figure 10.6 CLSM micrographs for a ternary solution of 4.2% gelatin, 7.9% maltodextrin and water quenched to $T = 20°C$: (a) 47.3 s after quench start; (b) 52.9 s; (c) 58.5 s; (d) 64.1 s; (e) 69.7 s; (f) 40 min. The bright phase is maltodextrin-rich and the dark phase is gelatin-rich. The corresponding two-dimensional fast Fourier transform (2D-FFT) is shown as an inset for each micrograph. After Lorén *et al.* (2002) © 2002 American Institute of Physics.

the centres of the maltodextrin bicontinuous microstructure grows with time. The difference in intensity between the phases increases and the interfacial width decreases with time. However, since the end temperature is below the gelation temperature of this gelatin, phase separation is pinned. The arrested microstructure is shown in micrograph (f), where the size of maltodextrin domains is about 100 μm. Secondary phase separation appears in micrograph (f) as dark gelatin inclusions in the maltodextrin phase. The formation of small, bright maltodextrin inclusions in the dark gelatin phase during secondary phase separation is perturbed by gelation.

The two-dimensional fast Fourier transforms (2D-FFT) corresponding to the CLSM micrographs are included as insets in Figure 10.6. The rings reflect the strong correlations in the local maltodextrin concentration that are directly visible in the micrographs. The rings intensify and collapse towards the centre, which indicates that the correlated maltodextrin domains grow larger with time, in agreement with the results of Tromp *et al.* (1995), and the characteristic wavelength derived from FFT increases with time following power-law behaviour. On the other hand, the experimental values of the exponent reported by Lorén *et al.* (2002) in the gelling system depart from theoretical predictions; in particular, the exponent is temperature-dependent and increases with the quench depth. The reason for this is not clear: it could be related to the rate of helix formation in gelatin gels, which could accelerate the phase separation. It is also important to mention that maltodextrins have a complex chemical structure and this may interfere with the observations and interpretations.

10.4.1.3 Gelatin–agarose

Both components here are the subject of Chapter 7. The gelatin–agar(ose) mixed-gel system has been studied by Watase and Nishinari (1980) and by Clark *et al.* (1983).

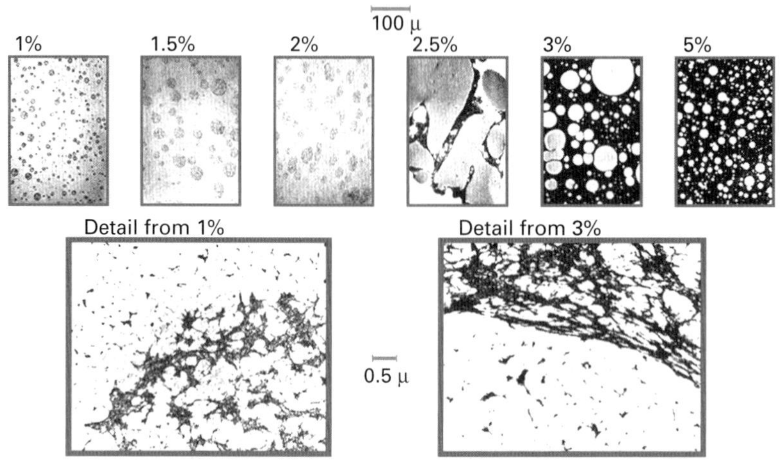

Figure 10.7 Light microscope (top row) and electron microscopy (bottom row) results for a series of agarose–gelatin gels (1% w/w agarose). Gelatin appears as the darkly staining material, the maximum concentration being 5% w/w. From Clark *et al.* (1983b) © 1983 American Chemical Society.

Stress relaxation measurements were made on aqueous gels (Watase and Nishinari, 1980) and the observed values of elastic modulus were compared with those calculated using a formula for composite systems proposed by Okano (1962), as mentioned above. Since the observed values were much smaller than the calculated values, the gelling process of the agarose was considered to be not a simple dispersion of agarose in a gelatin medium, but rather the hindering of agarose gelation by gelatin. In calculating the elastic moduli of the mixed gels, no allowance was made for the redistribution of water after mixing the solutions of agarose and gelatin, whereas this was allowed for in the work of Clark *et al.* (1983).

These workers also examined the same co-gels by electron and light microscopy (Figure 10.7) and mechanical spectroscopy. The microscopy evidence showed phase separation of the two polymer networks, with phase inversion occurring at specific mixture compositions – for example in Figure 10.7 at around 1% agarose, 2.5% w/w gelatin. Mechanical characterization of the gels was achieved by measurement of the shear modulus at small deformations, introduction of the solvent partition parameter p_a into the two phases and use of the polymer blending law proposed by Takayanagi *et al.* (1963); see Equations (10.7) and (10.8).

This approach and variations thereof have been employed for a wide range of phase separating mixed-gel systems over the succeeding years, although its basis is still essentially empirical. Applications of this approach have included papers on gelatin–low methoxy pectin, sodium caseinate–barley β-glucan and gelatin–maltodextrin (Kasapis, 2008). Morris (1992) has even suggested a way to determine self-consistent values of p_a by analysing modulus behaviour.

As far as large-deformation measurements on mixed gels are concerned, the vast majority of workers have used compression measurements, which remain difficult to interpret in terms of fundamental parameters. However, McEvoy and co-workers (McEvoy *et al.*, 1985) extended their work on the failure of agarose and gelatin gels, discussed in Chapter 7, to include mixtures of the two. Interpretations were again made in terms of failure envelopes and the use of the BST equation.

Finally it is worth noting that microscopy images of phase separated mixed gels often appear quite complex, just as they do in Figures 10.6 and 10.7. Images sometimes seem to consist not of two pure, discretely separated phases but of a hierarchy of inclusions of mini-phases within phase separated regions, in a self-similar pattern. This is actually what is predicted if phase separation proceeds by spinodal decomposition rather than by nucleation and growth, and may ultimately lead to the anticipated bicontinuous structure. Much work has been performed on such effects in the phase separation of synthetic polymer blends, but such observations have not been widely reported for mixed gels, although observational evidence in the form of microscopy images has been reported.

10.4.1.4 Casein–dextran

Casein (Chapter 9) is a major milk protein. At constant temperature and pH, the stability limit of the system water–casein–neutral polysaccharide is determined by changing the concentration of the polymers and salt. Figure 10.8 shows the phase diagram of the water–casein–dextran system for various NaCl concentrations, plotted not in the traditional triangular form but in the Tolstoguzov form, showing c_2 (concentration of the protein, casein) and c_3 (concentration of the polysaccharide, dextran) for fixed salt concentrations at pH 6.5 and 25°C. As the salt concentration increases from 0.15 M, the binodal curve shifts to lower concentrations.

To find out how temperature affects the stability of the water–casein–dextran–NaCl system, a two-phase system was prepared at 25°C, one of the phases (in most cases that enriched with a polysaccharide) was selected and its behaviour studied under heating. In

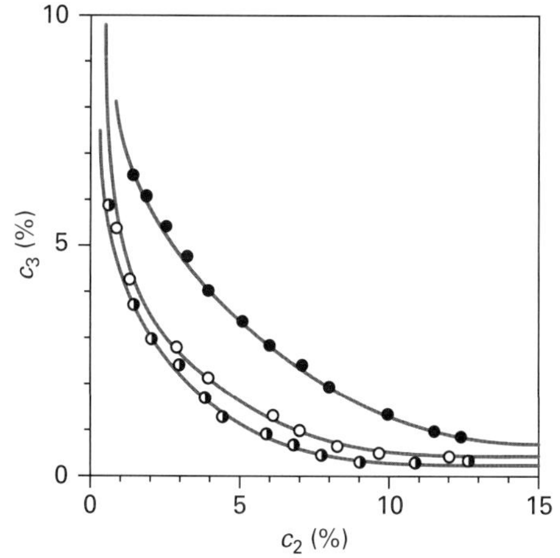

Figure 10.8 Phase diagram of water–casein–dextran in the presence of NaCl with various concentrations determined by cloud point observation at pH 6.5 and 25°C; c_2 is the concentration of casein and c_3 is that of dextran. The two-phase state is above the curves and the single phase is below the curves. Concentration of NaCl: 0.15 M (●), 0.3 M (○), 0.5 M (◑). Reprinted with permission from Antonov *et al.* (1977) © 1977 Springer.

all cases, turbidity was observed at sufficiently elevated temperatures. Cloud point curves are well reproduced even after repeated heating and cooling cycles.

10.4.2 Depletion interaction

A different type of phase separation, associated with an attractive interaction between polymers, commonly referred to as depletion interaction, was reported by Tuinier *et al.* (2000). The solution had a more complex composition, containing whey protein particles (Chapter 9) of colloidal size, mixed with exocellular polysaccharides from lactic acid bacterium.

The native whey proteins were subjected to a preliminary heating step and became denatured. Denatured whey proteins are mutually attractive and the proteins are partially aggregated. Phase separation kinetics was evidenced by small-angle light scattering (SALS) in this system. It was found that the overall value of the scattered intensity increases with time. All curves went through a maximum in the scattered intensity as a function of q. The first recorded value of the characteristic length was about 9 μm.

The solutions eventually phase separated because the partial aggregation of the whey proteins did not prevent sedimentation of the phases by gravity, as was the case for mixtures containing gelatin.

10.4.3 Protein–protein mixed-gel systems

A number of these, particularly those formed from dairy products, are of great commercial importance, although specific details are not discussed here since whole volumes (Fox and McSweeney, 2003), and indeed whole journal runs, are dedicated to them. Model systems of, for example, BSA with the whey protein components β-lactoglobulin (Lg) and α-lactabumin (La) (Chapter 9) have been studied by a number of groups, following the pioneering work by Hermansson and co-workers (Langton and Hermansson, 1992). They studied microscopy of these mixtures using a variety of techniques, and in later work extended the mixtures to include those with added gelatin, and performed small-deformation measurements on the same mixtures (Walkenstrom and Hermansson, 1994). They concluded from their gel formation studies that the components gel individually, suggesting phase separation. They detected a change in overall properties above *c.*10% whey protein and 3% gelatin; gel formation of the whey protein mixture was independent of the presence of gelatin, while that of gelatin depended on the whey proteins. At concentrations below these values, the mixed gels were thermoreversible, so the system was gelatin-continuous. At concentrations above this, the microstructure of the mixed gels suggested that a phase separated, bicontinuous system was formed.

When a mixture of globular proteins is heated above the denaturation temperatures of the components it appears that mutual aggregation allows mixed networks to form. Many years ago it was suggested that immuno-staining would allow this effect to be studied in more detail. Work by Comfort and Howell (2002) has applied this method to samples, albeit rather crude, of soy and whey protein mixtures.

Mixed protein gel rheology has been carried out by a number of workers, including the work mentioned above. Kavanagh and co-workers (Kavanagh *et al.*, 2000) studied β-Lg and α-La mixtures at two different pHs (3 and 7), keeping the total protein concentration at 15% w/w. Their results tended to confirm the idea that the two protein species, when partially unfolded into the molten globule state, gave results consistent with non-specific mixed network formation between protein species, rather than a synergistic interaction between the individual protein gel systems. However, without immuno-staining it was not possible to establish whether this mixed network was formed by cross-interaction of fibrils of the two segregated species or by formation of mixed fibrils from the two unfolded species in the molten globule state. The modulus G' of 15% α-La was greater than for the same concentration of β-Lg. Results for G' for the mixed gels lay between the two bounds (10.7) and (10.8), and much closer to the upper bound for the pH 3 systems. Perhaps as expected from the structural viewpoint, the bicontinuous model (10.9) did not produce any improvement. Large-deformation measurements for mixed pure protein systems are scarce, reflecting the need for (and high cost of) large amounts of sample.

More recently, Loveday and co-workers (Loveday *et al.*, 2009) have reviewed experimental and theoretical work, particularly on the fibrillar mixed gels formed at low pH. Their overall conclusion was that fibrillar gels from α-La seem to be of higher modulus than those from β-Lg, most probably because of the higher density of fibrils and the different fibril properties. They concluded that more data on the rheological properties of the gels is needed, using a wide range of protein concentrations, and that systematic studies should include strain sweep data to determine whether strain-stiffening or strain-softening behaviour is exhibited, and the persistence and contour lengths of any fibrils. The same was true for fibril density (or fibril content) data. In addition, they asserted that more studies would be desirable on the stress–strain behaviour of the fibrils and on fibril-network models to predict modulus–strain relationships.

10.4.4 Polysaccharide mixtures

10.4.4.1 Gellan and κ-carrageenan

Mixtures of gellan and κ-carrageenan have been examined by Nishinari *et al.* (1996). At low polymer concentrations, a cooperative interaction between their ordered conformations was demonstrated from viscosity measurements. At higher polymer concentrations, a two-phase system is formed, as shown by differential scanning calorimetry and rheological measurements.

Figure 10.9 shows heating differential scanning calorimetry (DSC) curves for mixed gels of gellan and κ-carrageenan with mixing ratios 3/1 and 1/1, for various concentrations. All the DSC curves clearly show separated endothermic peaks from helix–coil transitions for κ-carrageenan (lower temperature peak) and for gellan (higher temperature peak), with peaks shifted to higher temperatures with increasing concentration. These DSC peaks should not be attributed to a gel–sol transition because the gels still show quite a large elastic modulus at these temperatures, as shown in Figure 10.10.

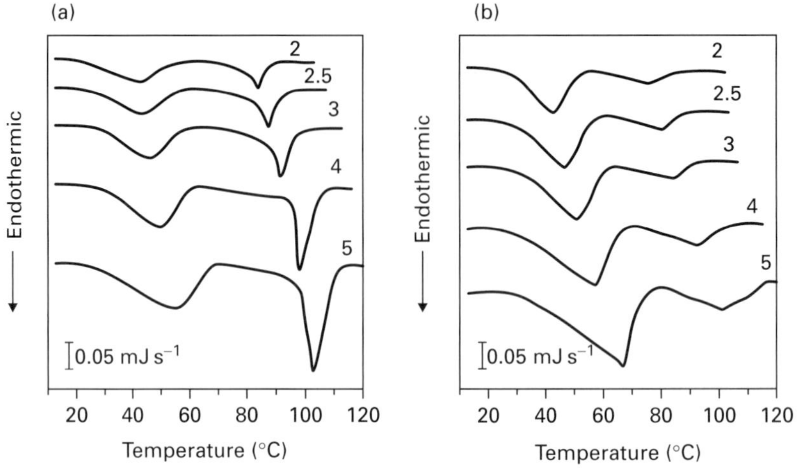

Figure 10.9 Heating differential scanning calorimetry (DSC) curves of mixed gels of gellan and κ-carrageenan with mixing ratio 3/1 (a) and 1/1 (b). Figures beside each curve represent the total polysaccharide concentration (% w/w). Reprinted with permission from Nishinari *et al.* (1996) © 1996 Elsevier.

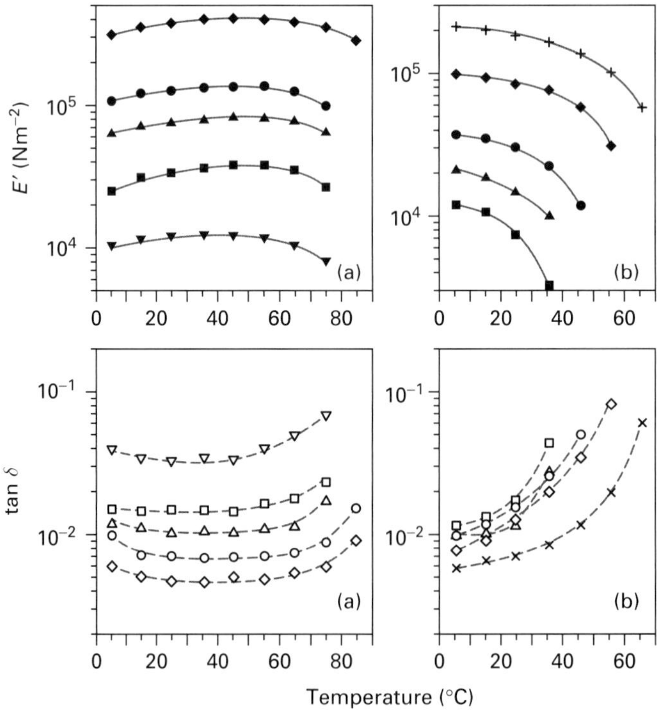

Figure 10.10 Temperature dependence of storage Young's modulus E' and mechanical loss tangent (tan δ) for a mixed gellan–κ-carrageenan gel with mixing ratio 3/1 (a) and 1/1 (b). Total polysaccharide concentration (% w/w) of gels: 1.5 (▼,▽); 2 (■,□); 2.5 (▲,△); 3 (●,○); 4(♦,◇); 5 (+,×). Reprinted with permission from Nishinari *et al.* (1996) © 1996 Elsevier.

The value of E' for single-component gels with the same concentration is significantly higher for gellan than for κ-carrageenan. The increase in E' with temperature observed for gellan or gellan-rich mixtures has also been found for agarose (Watase *et al.*, 1989) and polyvinyl alcohol systems (Nishinari *et al.*, 1985), and was interpreted by the reel-chain model discussed in Chapter 4. From this data, Nishinari *et al.* (1996) concluded that the gellan and κ-carrageenan mixed gel is a phase separated gel.

10.4.4.2 Gellan–agarose

Mixtures of gellan and agarose also show separated endothermic peaks in heating DSC curves (Nishinari *et al.*, 1993). A lower-temperature peak is mainly due to the agarose component and is shifted to still lower temperatures with increasing concentration, while a higher-temperature peak is caused mainly by gellan and is shifted to higher temperatures with increasing concentration. This is consistent with rheological observation: again, E' for agarose gels is larger than that of gellan gels in the concentration region 1–2% w/v, but this effect is reversed at higher concentrations of 3–4%. This suggests that, at concentrations higher than *c.*2%, the additional agarose does not contribute to the network structure effectively, and so that the mixture is a phase separated gel.

Amici *et al.* (2000, 2001) studied the same gellan–agarose mixture by rheology, DSC, transmission electron microscopy (TEM) and turbidity measurements. They observed two separated exothermic peaks in cooling DSC, originating from gellan and agarose, but they attributed these not to the usual phase separation – which, judging from their images, they could safely exclude – but to individual networks of each component. They concluded that this mixture forms a molecular interpenetrating network (IPN). Hitherto the evidence for IPN physical gels was equivocal, but here it is more convincing, at least from the structural evidence. For example, they did not observe any phase separation in TEM; instead images, including those employing the immuno-staining technique to label one component (in this case agarose), appear to support the 'separate network superposition' IPN hypothesis. It may be impossible to eliminate the possibility of cross-network interactions.

Deducing IPN behaviour from either small- or large-deformation rheology is much more complex. Amici *et al.* (2000, 2001) have argued this in terms of the mutual effective concentrations of the two components and their contribution to the modulus, but this remains an indirect approach. In compression measurements, adding agarose to gellan increased the stress at failure, although the strain at failure shifted to lower values. Although the TEM evidence appears very convincing, it is fair to say that there is still no consensus on this system. Allowing for sample differences, it is still difficult to establish the detailed structure of gellan–agarose mixed gels, although it is hoped this will be established by further studies.

10.4.4.3 κ-carrageenan–agarose

Interestingly, Amici *et al.* (2002) also showed evidence for IPN formation in the κ-carrageenan–agarose system. Again they were able to exclude substantial phase separation, although they conceded that a process of limited microphase separation during gelation had to be considered more seriously. However, the density fluctuations seen in images of the κ-carrageenan in the mixed system appeared very similar to those in the single-component system. From the large-deformation results, stress–strain curves for all of the gels studied were quite similar, and the constancy of, for example, failure strain was taken as support for their molecular IPN description.

10.4.4.4 Starch: a mixture of amylose and amylopectin

Starch lies behind only cellulose and chitin as the most common occurring polysaccharide, and whole volumes have been written about its properties and structure. Within the scope of this chapter, we will consider it only in terms of its properties as a mixed gel of its two components, the linear polyglucan amylose and the highly branched amylopectin. The plant source of the starch – e.g. pea, rice or maize – governs the relative amounts of amylose and amylopectin and the amount of crystallinity in the starch granule. When such a mixture or starch dispersion is heated at a given temperature, amylose is leached out of the granules. The viscosity begins to increase steeply, a phenomenon called gelatinization, and the result is that the partially crystalline portion is destroyed. When the gelatinized starch is kept at low temperature, the crystalline structure is partially recovered, a process called retrogradation.

It is well established that retrogradation is triggered by gelation of the amylose component (Miles *et al.*, 1985). Kalichevsky and Ring (1987) obtained a phase diagram for pea amylose (extracted, or leached, at 70°C) and waxy-maize amylopectin leached at both 70° C and 90°C (Figure 10.11), and the binodal was drawn through the experimental points. As the same binodal could be drawn through all of these, within experimental error, the results show no observable leaching temperature effect on the phase diagram. The concentration for the onset of coil overlap, obtained by viscometry, was 0.9% for amylopectin and 1.2% for pea amylose. From this they concluded that phase separation occurs well within the entangled regime of both polymers. However, it was not possible to investigate a broader temperature range because of the high gelling temperature of amylose.

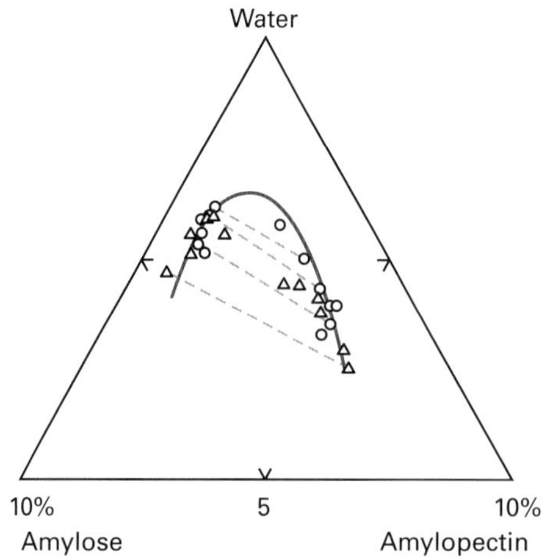

Figure 10.11 Phase diagram for the amylose–amylopectin–water system (using pea amylose leached at 70°C), obtained at 70°C (○) and at 90°C (Δ). Reprinted with permission from Kalichevsky and Ring (1987) © 1987 Elsevier.

Funami *et al.* (2005) reported that the addition of small amounts of konjac glucomannan (KGM;, Section 10.6.1.1) increased the peak viscosity for the starch system (13%) during the process of gelatinization. During short-term retrogradation, adding KGM (0.5%) increased tan δ for the starch system (5%) after storage at 4°C for 24 h. During long-term retrogradation (4°C for 14 days), addition of KGM (0.5%) decreased the rate constant expressing the relationship between storage time and creep compliance for the starch system (15%). They hypothesized that there are structural compatibility and molecular interactions between KGM and the starch components, amylose and amylopectin.

10.5 Filled gels

A range of systems closely related to mixed gels are those formed by adding an inert filler to a gelling system, in a manner quite analogous to the reinforcing of synthetic polymer resins by, for example, glass or carbon fibres. Clearly the results obtained, for example in mechanical measurements, depend on the size, shape and volume fraction of filler particles, just as they do for the resin.

One early publication is that by Richardson and co-workers (Richardson *et al.*, 1981). In this work, small-deformation measurements were performed on the model system of glass-filled gelatin gels. By using well-characterized samples of glass spheres, rods and irregular pieces of intermediate shape, e.g. plates at varying phase volume of filler, the effect of size, shape and phase volume on the shear storage and loss moduli could be investigated. The main effect was obviously to increase the small-deformation modulus, and in a predictable manner. The relative 'reinforcement factor' G_f/G_u for filled and unfilled particles was plotted against volume fraction relative to the maximum packing (volume) fraction ϕ_m . This was measured independently and found to be *c.*0.64 for the spheres and ~0.27 for the rods (length typically ~150 μm). These values produced quite satisfactory superposition, so that the reinforcement factor depended only on the ratio ϕ/ϕ_m.

Measurements were then carried out in the large-deformation regime using precast dumbbell-shaped replicates and extending the samples until they failed (Ross-Murphy and Todd, 1983). The effect of reinforcement could again be rationalized in relatively simple terms using Smith's failure envelope approach (Smith, 1963). In his review, Kasapis (2008) described his own data on gelatin filled with microcrystalline cellulose (MCC) fibrils. The corresponding value of ϕ_m is 0.52, which, in one model, corresponds to random orientation of fibres, although this seems high compared to the results for glass rods reported above.

A number of papers have subsequently been published in which alternative 'soft fillers' have been used, including microemulsion and gas-filled systems. Such systems have applications in the food and pharmaceutical sectors. Dickinson and his co-workers (Dickinson, 2012) have extended the work on filled gels by characterizing so-called emulsion gels. Here, an emulsion is first prepared by conventional routes but using a protein as emulsifier. By then heating or acidifying the system, the protein component can be induced to gel (Chapter 9), producing a material in which the included (filler) phase consists of oil droplets. Matsumura *et al.* (1993) examined the filler effects of oil droplets on the viscoelastic properties of emulsion gels by small-deformation mechanical measurements. They found that gels made from a fine emulsion (containing smaller oil droplets) exhibited higher G' and G'' values than the corresponding gels from a coarse emulsion.

Work on this topic has recently been reviewed by Dickinson himself (Dickinson, 2012). His work is noteworthy since it combines experimental and simulation work. For model systems of variable oil content and containing various proteins, almost all of which are food-relevant, he describes general trends of behaviour at small and large deformations, much of which is consistent with the work described above. Experimental measures such as G' and 'strain to fracture' are considered in relation to reinforcement effects for both active (reinforcing) and inactive filler particles. The former behave quite similarly to the glass-filled systems above, while the latter behave almost like voids. Brownian dynamics simulations (Chapter 2) of aggregated particle networks containing bonded and non-bonded particles are also described.

Of course if a polymeric viscous solution is prepared and then filled with particulates to a high enough degree, it will tend to develop an apparent yield stress, i.e. no Newtonian low shear rate plateau, an indication of solid-like behaviour, and so, by some definitions, it becomes a filled gel. Work by Rayment and co-workers (Rayment *et al.*, 1995, 2000) examined viscous solutions of guar galactomannan (see below) filled with non-gelatinized (small polyhedral shaped) rice starch and rod-like MCC particles. The apparent yield stress could be calculated by simply adding a stress term to the usual Cross equation (Cross, 1965) for fluids showing a Newtonian plateau. ϕ_m for the rice starch was found to be around 0.46, significantly below the value for spheres, reflecting the polydispersity of size and shape; the MCC rods gave $\phi_m \sim 0.28$.

Work by Kitano *et al.* (1981) has shown that there is a good (negative) linear correlation between the average aspect (length to diameter) ratio of rods and ϕ_m. Of course the area of reinforced polymer melts is a huge one, and outside the scope of the present volume. Nevertheless its importance in the area of modern structural materials remains paramount.

10.6 Gels involving molecular ('synergistic') interactions

Perhaps driven by the interests of the food and pharmaceutical industries, many of the systems involved in published mixed-gel work, particularly the so-called synergistic gels, involve mixtures of specific plant or microbial polysaccharides. This has been a very active field for researchers, with a large number of publications. For this reason we devote the rest of this chapter to these systems. We also introduce some polymers appearing for the first time in this chapter.

10.6.1 Polysaccharide systems

Some of these, including gellan, the carrageenans, xanthan and the galactomannans, have already been introduced (Chapters 5 and 8). However, several other non-gelling plant polysaccharides are important in mixed polysaccharide gel systems, and so are briefly introduced here. Some of these polysaccharides can be trapped into a gel-like state, either by applying freeze–thaw cycles (Chapter 8) or by alkali treatment, but under most normal conditions they form only viscous solutions. These mixed systems have a number of industrial applications, particularly in foods (dessert jellies and spreads) and pharmaceuticals (capsule additives).

10.6.1.1 Glucomannans

The commonest of these, konjac glucomannan (KGM), is a neutral polysaccharide that consists of β (1→4)-linked D-mannose and D-glucose with about 1 in 20 units being acetylated. The ratio of mannose to glucose is reported as 3/2. On its own KGM can form a gel when heated in the presence of alkali agents, and these alkali-induced gels are not thermoreversible. Mixed gels of KGM with other polysaccharides have been used extensively in industry (Nishinari *et al.*, 2007).

10.6.1.2 Xyloglucans

These are major structural polysaccharides which occur in the primary cell walls of higher plants, and include those from detarium (Wang *et al.*, 1996) and tamarind sources (Nishinari *et al.*, 2009). The seeds of the tamarind tree (*Tamarindus indica*) contain xyloglucan as a storage polysaccharide, tamarind seed xyloglucan (TSX), which has been widely used in the food industry. TSX has a β (1→4)-linked D-glucan backbone that is partially substituted at the O-6 position of its glucopyranosyl residues with α-D-xylopyranose. Some of the xylose residues are β-D-galactosylated at O-2, although the fine structure depends on the source of the plant. Alone it can form a thermoreversible gel by interacting with catechins or dyes, but it also forms gels in combination with other polysaccharides (Nishinari *et al.*, 2007) .

10.6.2 Polysaccharide mixed-gel interactions

It has long been known that galactomannans interact with certain gelling polysaccharides such as carrageenan and agar(ose) to improve gelling ability. Only a small amount of added locust bean gum (LBG) makes the gels 'firmer', i.e. of higher modulus, less brittle and more elastic (with higher failure to break). Since LBG and, particularly, guar gum are much less expensive than carrageenan and agar, it is an advantage to use a mixture of a galactomannan and less of the other component.

The interaction of galactomannans with xanthan has also attracted much attention. It was found that at, say, 1% concentration, both locust-bean gum and xanthan form only viscous solutions, but heating and cooling a mixture comprising 0.5% of each forms a firm gel. The experimental finding that these non-gelling polysaccharides can increase the modulus of agar and carrageenan gels, and cause gel-formation in the presence of xanthan, led researchers to suggest that galactomannans interact 'synergistically' with agar, carrageenan and xanthan, and in doing so form mixed junction zones (Dea and Morrison, 1975) .

The term 'synergism' literally means 'acting together' so that the total effect is greater than the sum of the two individually. In quite a few reported cases of synergism, the effect can be attributed simply to a failure to appreciate the 'effective concentration' contribution, together with the assumption that, for example, mechanical properties are simply additive. However, in the case of xanthan and the galactomannans it could be said that the change is qualitative rather than just quantitative, since neither component forms true gels at any practical concentration, whereas they can be formed simply by heating the two solutions together. In future sections we will continue to use the term, bearing in mind the above caveats.

10.6.2.1 Molecular models for polysaccharide gel interactions

A number of molecular models have been employed to understand the molecular interactions of such 'synergistic' gels, and these are now introduced. All are idealized, but over the years they have been refined. In later sections we introduce the evidence for each, and the areas of controversy.

The first is that of Dea *et al.* (1972) for the interaction of κ-carrageenan and galactomannan. They discovered that segmented κ-carrageenan – which does not gel alone at any concentration – forms firm, rubbery gels in mixtures with a galactomannan with a limited amount of galactose substituent, whereas the effect decreased with increasing galactose content; and that the galactomannans having highest content of galactose residues showed no interaction at all. From this Dea *et al.* proposed a picture of a molecular ribbon that is alternately 'hairy' and 'smooth', the 'hairy' regions corresponding to the presence of galactose substituents and the 'smooth' regions their absence. The 'smooth' regions are presumed to be able to bind to κ-carrageenan, whereas the 'hairy' regions cannot (Figure 10.12).

A similar explanation was applied to xanthan–galactomannan (X-GM) mixed gels, as illustrated in Figure 10.13, the interaction between xanthan and the galactomannan occurring only where the galactomannan is not side-chain branched (i.e. it is 'smooth'). Subsequently this area has become associated with such molecular diagrams – strictly speaking they are not models at all – and Figure 10.13 represents an alternate description of the binding of xanthan and galactomannans which no longer requires the galactomannan to be perfectly smooth.

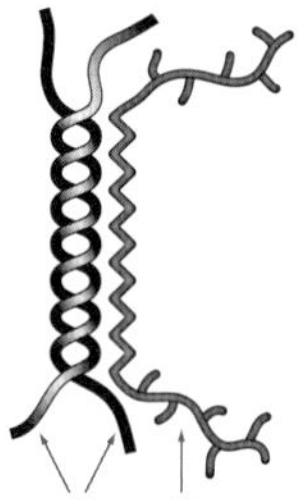

Figure 10.12 Model proposed for the interaction between chains of κ-carrageenan (C) and galactomannan (G). The double helix of κ-carrageenan binds with the non-substituted region of galactomannan. The zigzag line represents the so-called 'smooth' region of galactomannan. Reprinted with permission from Dea *et al.* (1972) © 1972 Elsevier.

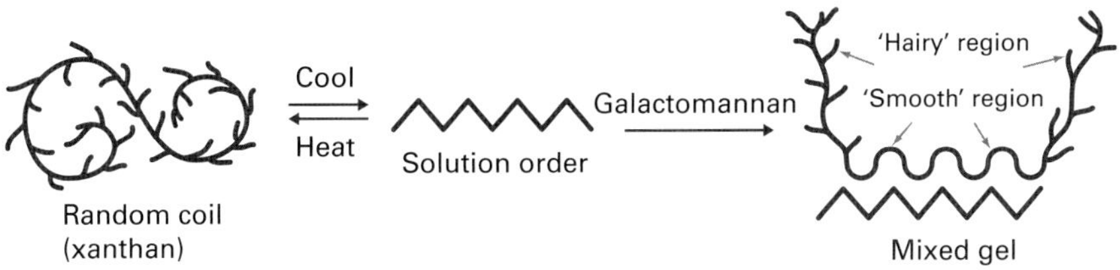

Figure 10.13 Molecular origin of xanthan interaction with galactomannan: the mixed junctions may be formed by cooperative association of mannan-backbone regions with the xanthan native structure, as indicated schematically. In this figure the zigzag line represents a single helix of xanthan. Reprinted with permission from Dea *et al.* (1977) © 1977 Elsevier.

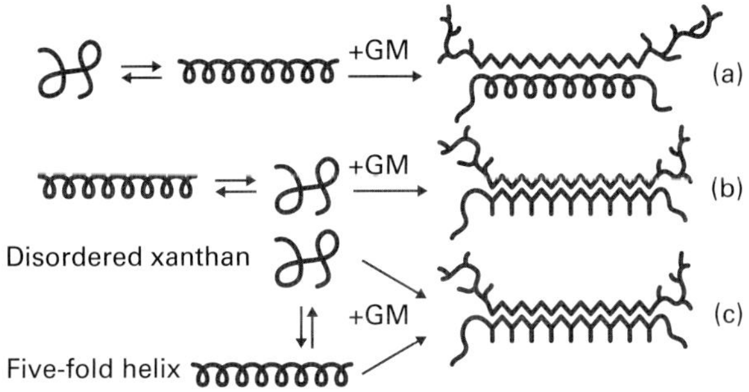

Figure 10.14 Proposed models for the interaction of xanthan with galactomannan or glucomannan (GM) chains: (a) 'Unilever' model; (b) 'Norwich' model; (c) proposal from Goycoolea *et al.* (1995). In each, the upper zigzag line on the right-hand side represents the smooth region of the galactomannan or glucomannan. Reprinted with permission from Goycoolea *et al.* (1995) ©1995 American Chemical Society.

The controversy over the fine structure of these interactions continues, and Figure 10.14 illustrates three models, the first being the original, and the other two being versions adapted to different experimental results. That said, we need to make two further comments. First, all of these figures show the xanthan macromolecule as a single ordered chain, although almost all current evidence suggests that the ordered structure involves two chains. Second, the evidence for all of these figures is somewhat indirect: there seems to be no data available to unequivocally establish *any* of these macromolecular structures; instead, these are deduced from spectroscopy or other 'local' probes. As we shall see, the evidence for some such binding is well established, for example by DSC, but the exact form of this is not settled. In our view, the ideal techniques to be used would seem to be either SAXS or, even better because of the potential to deuterium-label parts of one or the other molecule, SANS, but so far no such experiments appear to have been performed.

10.6.2.2 'Synergistic' mixed polysaccharide gels

Succeeding sections describe current experimental observations, and cover a range of other techniques. The sections gather the data collected on the most investigated mixtures containing κ-carrageenans, gellan and xanthan.

κ-carrageenan mixtures

κ-carrageenan–galactomannan

Dea *et al.* (1972) proposed their model for the junction zone (Figure 10.12) formed by the interaction between κ-carrageenan and galactomannan mainly on the basis of optical rotation measurements, supported by determination of the 'gel point' by a falling ball method, syneresis studies, X-ray diffraction and computer model building. They found thermal hysteresis in the temperature dependence of optical rotation for native (undegraded) and so-called segmented κ-carrageenan (in which the chain is split chemically to produce only the helix-forming parts), in both the absence and the presence of galactomannans with various mannose/galactose ratios. It was already known that a solution of native κ-carrageenan, diluted until it would not gel when cooled alone, could gel if mixed with a galactomannan with M/G = 3.35, but the same was true even for segmented κ-carrageenan, which would not gel alone at any concentration but would gel with this galactomannan. They found that both effects decreased with increasing galactose content, and that the galactomannans having the highest content of galactose residues (M/G = 1.08) showed no interaction at all.

Structural studies, including X-ray, infrared and enzymatic degradation, and other chemical analyses had already suggested that any ordered conformation of the galactomannan would be fairly fully extended and alternately rather densely and rather lightly substituted with galactose residues. From this evidence Dea *et al.* (1972) proposed their picture of the 'hairy' and 'smooth' molecular ribbon (Figure 10.12), where ordered binding occurs only between the carrageenan helix and parts of the galactomannan backbone that contain contiguous unsubstituted mannose residues.

κ-carrageenan–konjac glucomannan

Dea (1981) reported that konjac glucomannan (KGM) induced κ-carrageenan at a non-gelling concentration to form a thermoreversible gel. Based on optical rotation measurements, he concluded that KGM gelled with κ-carrageenan by inducing double helix formation. In 1987, Cairns *et al.* (1987) used X-ray fibre diffraction to study the interactions between KGM and κ-carrageenan. The diffraction patterns obtained for a film prepared from KGM and κ-carrageenan, mixed at the weight ratio of 1:1 in aqueous solution, was characteristic of pure κ-carrageenan, the unit cell dimensions were unchanged and there were no perturbations from KGM. They pointed out that if intermolecular binding had occurred between the two polymers, the mixed junction zones would have given rise to a diffraction pattern distinct from either of these two pure polymers. Since this did not occur in this test, they took issue with the model suggested by Dea, and proposed an alternative model. In this model the KGM molecules are not directly involved in the three-dimensional structure, but only distributed within the gel network of κ-carrageenan.

Williams *et al.* (1993) used DSC, electron spin resonance (ESR) and small-deformation rheology to study the interactions between KGM and κ-carrageenan. Figure 10.15 shows DSC cooling curves obtained for κ-carrageenan and KGM mixtures at a total polysaccharide concentration of 0.6% in 0.05 M KCl solutions. In the systems where carrageenan was in excess, a second peak appeared at 40°C. They concluded that

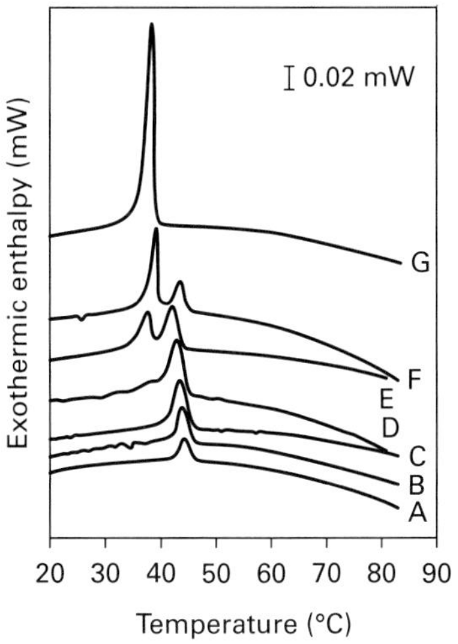

Figure 10.15 DSC cooling curves for mixtures with various ratios of κ carrageenan and glucomannan (0.6% total polysaccharide concentration) in 0.05 M KCl. Scanning rate was 0.1°C min^{-1}. Ratio of κ-carrageenan concentration to konjac glucomannan concentration: (A) 0.1/0.5; (B) 0.2/0.4; (C) 0.3/0.3; (D) 0.4/0.2; (E) 0.45/0.15; (F) 0.5/0.1; (G) 0.6/0. Reprinted with permission from Williams *et al.* (1993) © 1993 American Chemical Society.

the peak at higher temperature indicated an interaction between carrageenan and glucomannan, and the one at lower temperature was due to gelation of the excess carrageenan. ESR results showed that as the temperature is reduced the mobilities of the spin labels decrease more rapidly than for konjac glucomannan alone, signifying a greater reduction in segmental motion, presumably brought about through interaction of konjac glucomannan and κ-carrageenan molecules.

Using ^{133}Cs NMR measurements, Piculell *et al.* (1994) found enhanced chemical shifts and line broadening in κ-carrageenan gels containing KGM or LBG. They attributed this to the association between mannans and (aggregated) κ-carrageenan helices in the gel mixtures. They also pointed out that the ordered conformation of κ-carrageenan and its degree of specific ion binding was the same with or without the addition of galactomannans. In agreement with early findings by Dea *et al.* (1972), they also found that the conformational ordering of κ-carrageenan on cooling occurred at a higher temperature in the mixture with LBG, but otherwise the shift results for the two systems were quite similar. They concluded that the Cs line broadening was suggestive of a binding isotherm, where the available 'sites' for mannan association were saturated at high concentrations, and that the additional line broadening seen in mixtures of κ-carrageenan with LBG was indeed due to the formation of mixed aggregates.

Gellan mixtures

Gellan–konjac glucomannan

Miyoshi *et al.* (1996) used rheometry and DSC to study the interaction between gellan gum and KGM with and without NaCl and $CaCl_2$. As seen in Figure 10.16, a mixture of gellan and KGM at a ratio of 0.3/0.5 showed gel-like behaviour at 0°C, while no other

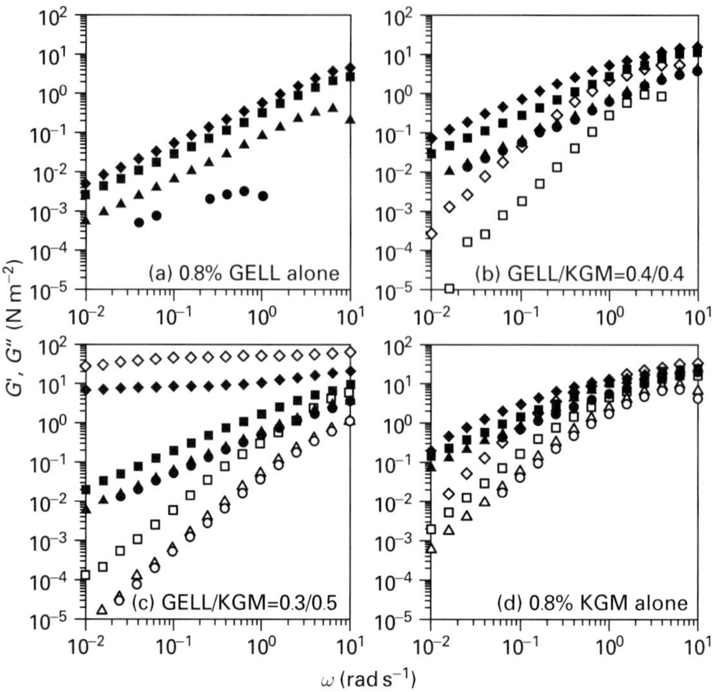

Figure 10.16 Frequency dependence of storage and loss shear moduli, G' (open symbols) and G'' (closed symbols), for various mixing ratios of gellan ('GELL') and KGM (total polysaccharide concentration 0.8 wt%) at various temperatures, in the absence of cations: 30°C (○, ●); 25°C (Δ, ▲); 15°C (□, ■); 0°C (◇, ♦). Reprinted with permission from Miyoshi *et al.* (1996) © 1996 Elsevier.

solutions formed a gel at any temperature. The authors concluded that KGM was adsorbed on to the surface of large aggregates of gellan gum helices, and the interactions between KGM and gellan gum were promoted with increasing concentration of cations. Nishinari *et al.* (1996) suggested that in KGM–gellan gum mixtures, the main ordered structures are formed by gellan molecules and that KGM molecules inhibit gel formation by gellan alone.

Mixtures of gellan–KGM showed a maximum synergy at the mixing ratio 3/5 (Miyoshi *et al.*, 1996). On adding monovalent cations, the storage and loss moduli of mixtures increased and the gelation temperature shifted to higher temperatures. Divalent cations also promoted the gelation of the mixtures, but excessive addition of divalent cations decreased the moduli, which was attributed to phase separation.

Gellan–xyloglucan

The mixture of gellan and xyloglucan was found by Nitta *et al.* (2003) to form a gel in a concentration range where the individual polysaccharides do not. Both TSX and (Na) GG (sodium form gellan gum) solutions behave as viscoelastic fluids at low concentrations or high temperatures. In the case of TSX solutions, G' was smaller than G'', and GG

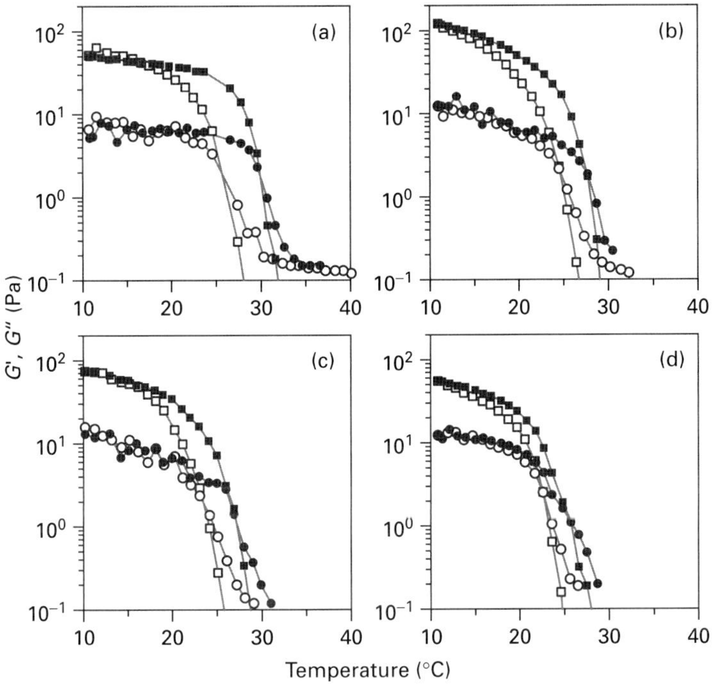

Figure 10.17 Temperature dependence of G' (□, ■) and G''(○, •) for 1 wt% mixed solutions of xyloglucan and Na-gellan on cooling (open symbols) and on subsequent heating (closed symbols). Frequency 0.2 Hz; stress 0.05 Pa. TSX content: (a) 0.9, (b) 0.6, (c) 0.5, (d) 0.4. Cooling and heating rates 0.5°C min^{-1}. Reprinted with permission from Nitta *et al.* (2003) © 2003 American Chemical Society.

solutions showed a solution-like behaviour. It is known that GG aqueous solutions undergo a helix–coil conformational transition, so at lower temperatures, when gellan gum is in the helix state, the values of the moduli increased even though the GG aqueous solution was still in the sol state.

Figure 10.17 shows a steep increase in storage and loss moduli on cooling. Further, G' exceeded G'', and G' became much larger than G'' at low temperatures. This data suggests that gel formation occurs by the mixing of TSX and GG. On subsequent heating, G' and G'' decreased gradually, and then more steeply around 30°C. That is, the characteristic temperatures at which the moduli increased or decreased were different, and thermal hysteresis was found even at slow scan rates. Both G' and G'' for the mixture of TSX and Na-GG showed a plateau at low temperatures and smaller values of $\tan\delta$. These results suggest that gel formation was induced by some form of interaction, as also revealed by DSC and circular dichroism. The appearance of a first DSC peak at a different temperature from the coil–helix transition of gellan gum, and the fact that this peak was observed even at very low gellan concentrations, suggest that the first peak is probably caused by intermolecular binding. However, images obtained by atomic force microscopy (AFM) showed no indication of heterotype junction zones (Ikeda *et al.*, 2004), which may suggest phase separation. Here, however, images were obtained from dried samples, not from aqueous solutions, so junction zone formation in solution cannot be ruled out.

Gellan–hyaluronan

A mixture of gellan and the important anionic glycosaminoglycan hyaluronan at non-gelling concentrations was found to form a gel in the presence of calcium ions (Mo *et al.*, 2000). At pH 2.5, G' and G'' for mixtures of calcium hyaluronate and sodium type gellan, each at non-gelling concentrations, showed a rapid increase after mixing and, after time, formed a typical high-modulus gel. Although gellan solutions form a phase separated gel because of lowered solubility, and become turbid at acidic pH (Moritaka *et al.*, 1995), a mixture of calcium hyaluronate and sodium type gellan gave more transparent gels at pH 2.5 than at pH 7.0.

Xanthan mixtures

Xanthan–galactomannans

When hot 1:1 solutions of locust bean gum and xanthan gum were mixed together and then cooled, the mixture formed what was reported to be a firm, rubbery, thermoreversible gel. The effect, which cannot be achieved using either hydrocolloid alone, was presumed to be the result of some form of junction zone formation. This was specified by Dea and co-workers to be between the galactose-deficient segments of the locust bean gum (LBG) and the xanthan gum (Dea and Morrison, 1975; Dea *et al.*, 1977; Morris *et al.*, 1977); see Figure 10.12.

Experimental findings obtained using HPLC (Cheetham *et al.*, 1986), NMR (Lazaridou *et al.*, 2001; Vittadini *et al.*, 2002), X-ray diffraction (Cairns *et al.*, 1987) and rheology (Shatwell *et al.*, 1990a, 1990b, 1990c) continue to support such an intermolecular binding process. The most commonly accepted model is that the binding region for galactomannan is an unsubstituted mannan region, the so-called ‘smooth region’, since locust bean gum, a galactomannan with a lower galactose content, has higher gelling abilities than guar gum, a galactomannan with higher galactose content. However, as mentioned above, this is still controversial (Figure 10.14). Indeed, yet another proposal is that galactomannan sections with all the galactose residues on the same side of the chain would form junction zones by interacting with xanthan (McCleary, 1979).

Shatwell *et al.* (1990a, 1990c) studied the effect of acetyl and pyruvate groups in xanthan on the interaction with LBG and KGM based on G', $\tan\delta$ and the minimum gelation concentration. Most xanthan samples formed a relatively good gel with LBG, but a mutant xanthan sample with high acetyl and low pyruvate content was an exception. Deacetylation of a xanthan sample resulted in a significant change, while depyruvylation made little difference. They suggested that the acetyl group has an inhibitory effect upon gelation but that the pyruvate group plays no role in the interaction. Pai and Khan (2002) examined the rheology of blends of xanthan and enzymically modified guar – EMG, a material structurally similar to LBG – and found that the elastic modulus increases with increasing extent of enzymatic modification, i.e. with decreasing galactose content (Figure 10.18). This again suggests that galactose side chains inhibit the association with xanthan.

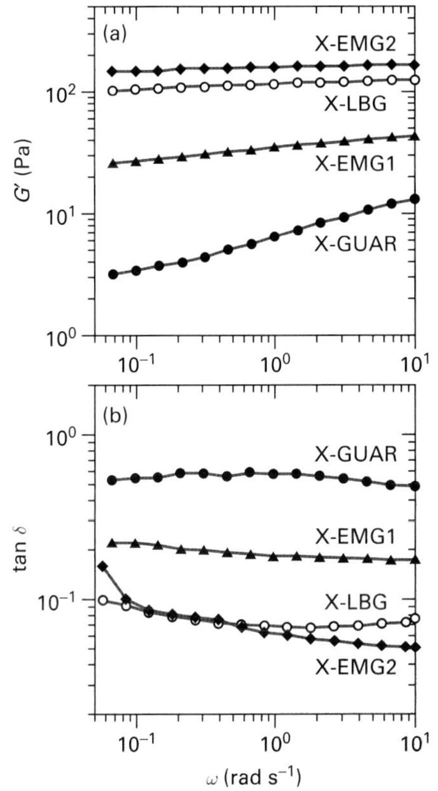

Figure 10.18 Shear modulus G' (a) and loss tan δ (b) as a function of frequency for blends of xanthan (X) with native guar (Man/Gal = 1.48), EMG1(Man/Gal = 1.85), EMG2(Man/Gal = 2.86) and locust bean gum (LBG) (Man/Gal = 2.82). Reprinted from Pai and Khan (2002) © 2002 Elsevier.

However, these explanations, although totally consistent, neglect any possible thermodynamic mechanism. For example, removing galactose side chains may lower the 'solvent quality' for the EGM-modified samples, making them more susceptible to both intramolecular and intermolecular aggregation *without* requiring any specific molecular mechanism. Richardson and Norton (1998) have induced 'gelation' in LBG by using the 'freeze/thaw' mechanism, which all suggests that polymer–polymer self-interactions are favoured, and that a conventional 'thermodynamic' approach, measuring, for example, the second virial coefficient by light scattering, might be fruitful.

Xanthan–konjac glucomannan

The above caveat being noted, X-ray fibre diffraction measurements by Cairns *et al.* (1986) helped to establish that intermolecular binding does occur in mixed KGM–xanthan gels. Dea *et al.* (1977) concluded that the interaction between KGM and xanthan was stronger than that of LBG and xanthan, based on the observation that the melt temperature of KGM–xanthan was nearly 20°C higher than that of LBG–xanthan. But Shatwell *et al.* (1990c) indicated that some thermoreversible gels were marginally weaker (of lower modulus) than for corresponding xanthan–LBG mixtures. They also found that the moduli were dependent on the degree of acetylation of xanthan, in agreement with Dea and co-workers.

Cairns *et al.* (1987) and Brownsey *et al.* (1988) suggested from X-ray diffraction and creep measurements that the KGM or galactomannan and xanthan intermolecular bindings occurred only when the xanthan helix had been disordered. Addition of 0.5 M $CaCl_2$ raises the denaturation temperature of xanthan so it is in an ordered conformation at high temperatures, which suggests that gelation does not occur between ordered xanthan and konjac glucomannan. The proposal that the junction zone is formed only between disordered xanthan and galactomannans with lower galactose content is supported by conformational studies using chiroptical methods (Cheetham and Mashimba, 1990; Bresolin *et al.*, 1998) and work by Williams *et al.* (1991).

Bordi *et al.* (2002) studied the interaction between KGM and xanthan using CD and suggested that, in the ordered conformation, the side chains of xanthan lie along the chain axis. Taking into account that the chromophore groups of xanthan are situated in the side chains, they concluded that the interaction involves this part of the xanthan moiety, in line with conclusions of Annable *et al.* (1994).

Williams *et al.* (1991) studied the interaction between xanthan and KGM by ESR, DSC and rheology. They reported the mid-point temperature for gelation to be higher in water than in NaCl solution, and concluded that, in NaCl solution, the interaction occurred between ordered xanthan chains and KGM, while in water the interaction is believed to be predominantly between KGM and the disordered xanthan chains, leading to a more stable gel.

Annable *et al.* (1994) used DSC and ESR methods to show that the interaction between KGM and xanthan occurred in water immediately following xanthan side chain–backbone association. In the presence of electrolyte, the conformational change of xanthan shifted to higher temperatures and depended on the nature of the cation. Their results also showed that the temperature for KGM–xanthan interaction was much lower than that for the xanthan conformational transition, and that electrolyte promoted xanthan self-association at the expense of KGM–xanthan interaction.

Goycoolea *et al.* (1995) suggested that there was a structural rearrangement after the initial stage of gel formation in xanthan–KGM, but this was not observed in xanthan–LBG gels. Figure 10.14c shows their proposition. Using X-ray diffraction, they suggested that the initial gelation involves so-called heterotypic junctions between KGM or LBG (i.e. different from those formed by either individual component) and xanthan or deacetylated xanthan, but that junctions involving KGM convert to a more compact six-fold arrangement at a lower temperature. Comparison with DSC traces for xanthan alone led the authors to conclude that conformational ordering of the xanthan component completed at 60°C. This was thought to be consistent with the concept of binding to the xanthan helix, as in the alternative model (Figure 10.14a), but inconsistent with the necessary involvement of disordered xanthan in synergistic gelation, as in Figure 10.14b. For a deacetylated xanthan, by contrast, they found that gel formation occurs with the xanthan in the disordered state, which is entirely consistent with Figure 10.14b but contradicts Figure 10.14a.

Xanthan–xyloglucans

Most of the work on xanthan–xyloglucan co-gelation has been carried out with xyloglucan from tamarind seeds (TSX), in which a mixture with xanthan or with gellan gum forms a transparent gel. Kim *et al.* (2006) showed that gelation occurs in a mixture of TSX and xanthan gum and that the system exhibits thermoreversible gel-like properties at low temperatures. A steep increase in both storage and loss moduli was observed on cooling, and a conspicuous exothermic DSC peak appeared at the same temperature, although these characteristic behaviours did not appear for the aqueous solution of either component alone. The effect on this interaction of acetyl and pyruvate groups in the xanthan side chain was also investigated. The mixture of xyloglucan and acetate-free xanthan showed essentially the same interaction as that with native xanthan, but the lack of a pyruvate group inhibits the interaction, indicating that the pyruvate group substituted at the terminal mannose of the xanthan side chain plays a key role. The linewidth from proton NMR showed a drastic restriction in the xanthan side chain, not observed even in the helix state of xanthan alone. This suggests that, in the mixture, the xanthan side chain adopts a new state different from either the coil or ordered form.

Gels from other polysaccharide mixtures

Dea *et al.* (1977) reported that only a small amount of locust bean gum was sufficient to cause a low concentration of agarose to form a cohesive gel, while a higher concentration of KGM was required. In a later paper, Dea (1981) reported that KGM induced agarose to form a thermoreversible gel at an otherwise non-gelling concentration. He pointed out that the temperature dependence of the optical rotation for KGM–agarose mixtures differed significantly from that of gelling agarose–locust bean gum mixtures, which he attributed to the difference of the backbone conformations of KGM and galactomannan.

Ridout *et al.* (1998) used rheometry to establish interactions between KGM and the bacterial polysaccharide acetan. They found that there was an enhancement of viscosity in dilute solution, whereas at higher concentrations gelation occurred, even though the interaction between KGM and acetan was weaker than that of LBG–acetan. Deacetylation of acetan particularly enhanced the interaction between KGM and acetan. They also used X-ray fibre diffraction to study the interaction. Figure 10.19 shows the X-ray fibre diffraction pattern of a deacetylated acetan–KGM mixed gel. The new structure corresponding to 'heterotypic' junction zones in the gels firmly established the existence of intermolecular binding between these two polysaccharides. The deacetylated acetan–KGM mixture X-ray diffraction pattern resembles that of acetan but the junction zones appear to be six-fold helices, whereas acetan alone forms five-fold helices. Chandrasekaran *et al.* (2003) extended this work and proposed a molecular model for binary 'weak' gelation systems. The model revealed that a double helix in which one strand is acetan and the other is glucomannan is stereochemically feasible. The authors suggested that this molecular model could be generalized to the other binary systems.

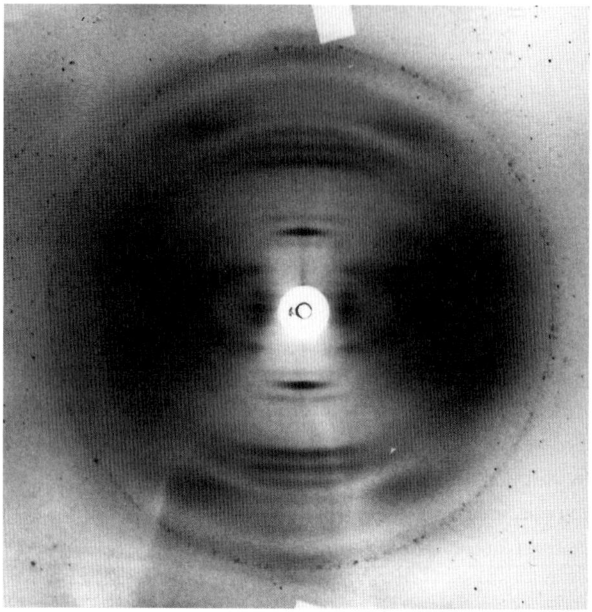

Figure 10.19 X-ray fibre diffraction pattern for 100% stretched deacetylated acetan–KGM (1:1) mixed gel, X-ray wavelength 0.154 nm. Reprinted with permission from Ridout *et al.* (1998) © 1998 Elsevier.

10.7 Conclusions

As we have shown, studies on mixed gels cover a wide range of techniques and systems. For the segregative gels, microscopy and scattering techniques have been applied, whereas for the 'molecular' systems, perhaps understandably, studies have also included a number of more specialized molecular techniques. Much of the work on the mixed polysaccharide systems seems to be driven by efforts to establish the rather idealized models. Mixed-gel formation occurs through various mechanisms including segregation, aggregation and synergy; a precise knowledge of both the structure of junctions and the kinetics of the phase separation at larger scales is necessary in order to classify the underlying processes. Multi-distance scale studies such as rheology, microscopy and scattering on the same samples are required. Despite this difficulty, many of the systems discussed here have proved of great value commercially and, we have no doubt, will continue so to do.

References

Amici, E., Clark, A. H., Normand, V., Johnson, N. B., 2000. *Biomacromolecules* **1**, 721–729.
Amici, E., Clark, A. H., Normand, V., Johnson, N. B., 2001. *Carbohydr. Polym.* **46**, 383–391.
Amici, E., Clark, A. H., Normand, V., Johnson, N. B., 2002. *Biomacromolecules* **3**, 466–474.

Annable, P., Williams, P. A., Nishinari, K., 1994. *Macromolecules* **27**, 4204–4211.
Antonov, Y. A., Grinberg, V. Y., Tolstoguzov, V., 1977. *Colloid Polym. Sci.* **255**, 937–947.
Bordi, F., Paradossi, G., Rinaldi, C., Ruzicka, B., 2002. *Phys. A (Amsterdam)* **304**, 119–128.
Bresolin, T. M., Milas, M., Rinaudo, M., Ganter, J. L. M., 1998. *Int. J. Biol. Macromol.* **23**, 263–275.
Brownsey, G. J., Cairns, P., Miles, M. J., Morris, V. J., 1988. *Carbohydr. Res.* **176**, 329–334.
Cahn, J. W., 1965. *J. Chem. Phys.* **42**, 93–99.
Cahn, J. W., Hilliard, J. E., 1958. *J. Chem. Phys.* **28**, 258–267.
Cahn, J. W., Hilliard, J. E., 1959. *J. Chem. Phys.* **31**, 688–699.
Cahn, R. W., Haasen, P., Kramer, E. J., 1991. *Materials Science and Technology: A Comprehensive Treatment*. Vol 5: *Phase Transformations in Materials*. VCH, Weinheim.
Cairns, P., Miles, M. J., Morris, V. J., 1986. *Nature* **322**, 89–90.
Cairns, P., Miles, M. J., Morris, V. J., Brownsey, G. J., 1987. *Carbohydr. Res.* **160**, 411–423.
Chandrasekaran, R., Janaswamy, S., Morris, V. J., 2003. *Carbohydr. Res.* **338**, 2889–2898.
Cheetham, N. W. H., Mashimba, E. N. M., 1990. *Carbohydr. Polym.* **14**, 17–27.
Cheetham, N. W. H., McCleary, B. V., Teng, G., Lum, F., Maryanto, 1986. *Carbohydr. Polym.* **6**, 257–268.
Clark, A. H., Richardson, R. K., Robinson, G., Ross-Murphy, S. B., Weaver, A. C., 1982. *Prog. Food Nutr. Sci.* **6**, 149–160.
Clark, A. H., Richardson, R. K., Ross-Murphy, S. B., Stubbs, J. M., 1983. *Macromolecules* **16**, 1367–1374.
Comfort, S., Howell, N. K., 2002. *Food Hydrocolloids* **16**, 661–672.
Cross, M. M., 1965. *J. Colloid Sci.* **20**, 417–437.
Dea, I. C. M., 1981. Specificity of interactions between polysaccharide helices and β-1,4-linked polysaccharides. In Brant, D. A. (ed), *Solution Properties of Polysaccharides*, ACS Symposium Series 150. American Chemical Society, Washington, DC, pp. 439–454.
Dea, I. C. M., Morrison, A., 1975. *Adv. Carbohydr. Chem. Biochem.* **31**, 241–312.
Dea, I. C. M., McKinnon, A. A., Rees, D. A., 1972. *J. Mol. Biol.* **68**, 153–172.
Dea, I. C. M., Morris, E. R., Rees, D. A. *et al.*, 1977. *Carbohydr. Res.* **57**, 249–272.
de Gennes, P.-G., 1980. *J. Chem. Phys.* **72**, 4756–4763.
Dickinson, E., 2012. *Food Hydrocolloids* **28**, 224–241.
Doublier, J. L., Garnier, C., Renard, D., Sanchez, C., 2000. *Curr. Opin. Colloid Interface Sci.* **5**, 202–214.
Fox, P. F., McSweeney, P. (eds), 2003. *Advanced Dairy Chemistry 1: Proteins*. Springer, New York.
Funami, T., Kataoka, Y., Omoto, T. *et al.*, 2005. *Food Hydrocolloids* **19**, 1–13.
Goycoolea, F. M., Richardson, R. K., Morris, E. R., Gidley, M. J., 1995. *Macromolecules* **28**, 8308–8320.
Harding, S. E., Hill, S. E., Mitchell, J. R., 1995. *Biopolymer Mixtures*. Nottingham University Press, Nottingham.
Hashimoto, T., Kumaki, J., Kawai, H., 1983. *Macromolecules* **16**, 641–648.
Hashimoto, T., Sasaki, K., Kawai, H., 1984. *Macromolecules* **17**, 2812–2818.
Ikeda, S., Nitta, Y., Kim, B. S. *et al.*, 2004. *Food Hydrocolloids* **18**, 669–675.
Kalichevsky, M. T., Ring, S. G., 1987. *Carbohydr. Res.* **162**, 323–328.
Kasapis, S., 2008. *Crit. Rev. Food Sci. Nutr.* **48**, 341–359.
Kavanagh, G. M., Clark, A. H., Gosal, W. S., Ross-Murphy, S. B., 2000. *Macromolecules* **33**, 7029–7037.
Kerner, E. H., 1956. *Proc. Phys. Soc. London, Sect. B* **69**, 808–813.
Kim, B.-S., Takemasa, M., Nishinari, K., 2006. *Biomacromolecules* **7**, 1223–1230.

Kitano, T., Kataoka, K., Shirota, T., 1981. *Rheol. Acta* **20**, 207–209.
Koningsveld, R., Stockmayer, W. H., Nies, E., 2001. *Polymer Phase Diagrams: A Textbook*. Oxford University Press, Oxford.
Langton, M., Hermansson, A. M., 1992. *Food Hydrocolloids* **5**, 523–539.
Lazaridou, A., Biliaderis, C. G., Izydorczyk, M. S., 2001. *J. Sci. Food Agric.* **81**, 68–75.
Lipatov, Y. S., Alekseeva, T., 2007. Phase-separated interpenetrating polymer networks. In *Advances in Polymer Science* **208**. Springer, Berlin, p. 6.
Lorén, N., Altskär, A., Hermansson, A.-M., 2001. *Macromolecules* **34**, 8117–8128.
Lorén, N., Langton, M., Hermansson, A.-M., 2002. *J. Chem. Phys.* **116**, 10536–10546.
Loveday, S. M., Rao, M. A., Creamer, L. K., Singh, H., 2009. *J. Food Sci.* **74**, R47–55.
Matsumura, Y., Kang, I.-J., Sakamoto, H., Motoki, M., Mori, T., 1993. *Food Hydrocolloids* **7**, 227–240.
McCleary, B. V., 1979. *Carbohydr. Res.* **71**, 205–230.
McEvoy, H., Ross-Murphy, S. B., Clark, A. H., 1985. *Polymer* **26**, 1493–1500.
Miles, M. J., Morris, V. J., Ring, S. G., 1985. *Carbohydr. Res.* **135**, 257–269.
Miyoshi, E., Takaya, T., Williams, P. A., Nishinari, K., 1996. *J. Agric. Food Chem.* **44**, 2486–2495.
Mo, Y., Kubota, K., Nishinari, K., 2000. *Biorheology* **37**, 401–408.
Moritaka, H., Nishinari, K., Taki, M., Fukuba, H., 1995. *J. Agric. Food Chem.* **43**, 1685–1689.
Morris, A. M., Watzky, M. A., Finke, R. G., 2009. *Biochim. Biophys. Acta, Proteins Proteomics* **1794**, 375–397.
Morris, E. R., 1990. Mixed polymer gels. In Harris, P. (ed.), *Food Gels*. Elsevier Applied Science, London, pp. 291–359.
Morris, E. R., 1992. *Carbohydr. Polym.* **17**, 56–70.
Morris, E. R., Rees, D. A., Young, G., Walkinshaw, M. D., Darke, A., 1977. *J. Mol. Biol.* **110**, 1–16.
Morris, V. J., 1998. Gelation of polysaccharides. In Hill, S. E., Ledward, D. A., Mitchell, J. R. (eds), *Functional Properties of Food Macromolecules*. Aspen Publishers, Gaithersburg, MD, pp. 143–226.
Nishinari, K., Koide, S., Ogino, K., 1985. *J. Phys. France* **46**, 793–797.
Nishinari, K., Takaya, T., Watase, M., 1993. Rheology and DSC of gellan-agarose mixed gels. In Nishinari, K., Doi, E. (eds), *Food Hydrocolloids: Structures, Properties, and Functions*. Plenum Press, New York, pp. 473–476.
Nishinari, K., Miyoshi, E., Takaya, T., Williams, P. A., 1996. *Carbohydr. Polym.* **30**, 193–207.
Nishinari, K., Takemasa, M., Zhang, H., Takahashi, R., 2007. Storage plant polysaccharides: Xyloglucans, galactomannans, glucomannans. In Kamerling, J. P. *et al.* (eds), *Comprehensive Glycoscience*. Elsevier, Amsterdam, pp. 614–652.
Nishinari, K., Takemasa, M., Yamatoya, K., Shirakawa, M., 2009. Xyloglucan. In Phillips, G. O., Williams, P. (eds), *Handbook of Hydrocolloids*. CRC Press, Cambridge, UK, pp. 247–267.
Nitta, Y., Kim, B. S., Nishinari, K. *et al.*, 2003. *Biomacromolecules* **4**, 1654–1660.
Okano, K., 1962. *Rep. Prog. Polym. Phys. Jpn.* **5**, 79–82.
Pai, V. B., Khan, S. A., 2002. *Carbohydr. Polym.* **49**, 207–216.
Papon, P., Leblond, J., Meijer, P. H. E., 2002. *The Physics of Phase Transitions*. Springer, Berlin.
Piculell, L., Nilsson, S., Muhrbeck, P., 1992. *Carbohydr. Polym.* **18**, 199–208.
Piculell, L., Iliopoulos, I., Linse, P. *et al.*, 1994. Association and segregation in ternary polymer solutions and gels. In Phillips, G. O., Williams, P. A., Wedlock, D. J. (eds), *Gums and Stabilisers for the Food Industry* **3**. IRL Press, Oxford, pp. 309–322.
Ragone, D. V., 1995. *Thermodynamics of Materials*. John Wiley & Sons, NY.
Rayment, P., Ross-Murphy, S. B., Ellis, P. R., 1995. *Carbohydr. Polym.* **28**, 121–130.
Rayment, P., Ross-Murphy, S. B., Ellis, P. R., 2000. *Carbohydr. Polym.* **43**, 1–9.

Richardson, P. H., Norton, I. T., 1998. *Macromolecules* **31**, 1575–1583.
Richardson, R. K., Robinson, G., Ross-Murphy, S. B., Todd, S., 1981. *Polym. Bull.* **4**, 541–546.
Ridout, M. J., Cairns, P., Brownsey, G. J., Morris, V. J., 1998. *Carbohydr. Res.* **309**, 375–379.
Ross-Murphy, S. B., Todd, S., 1983. *Polymer* **24**, 481–486.
Sasaki, K., Hashimoto, T., 1984. *Macromolecules* **17**, 2818–2825.
Scholten, E., Tuinier, R., Tromp, R. H., Lekkerkerker, H. N. W., 2002. *Langmuir* **18**, 2234–2238.
Shatwell, K. P., Sutherland, I. W., Ross-Murphy, S. B., Dea, I. C. M., 1990a. *Carbohydr. Polym.* **14**, 29–51.
Shatwell, K. P., Sutherland, I. W., Ross-Murphy, S. B., Dea, I. C. M., 1990b. *Carbohydr. Polym.* **14**, 115–130.
Shatwell, K. P., Sutherland, I. W., Ross-Murphy, S. B., Dea, I. C. M., 1990c. *Carbohydr. Polym.* **14**, 131–147.
Smith, T. L., 1963. *J. Polym. Sci., Part A: Gen. Pap.* **1**, 3597–3615.
Takayanagi, M., Harima, H., Iwata, Y., 1963. *Mem. Fac. Eng., Kyushu Univ.* **23**, 1.
Tolstoguzov, V. B., 2006. Phase behavior in mixed polysaccharide systems. In Stephen, A., Phillips, G., Williams, P. (eds), *Food Polysaccharides and Their Applications*. CRC Press, Boca Raton, FL, pp. 589–627.
Tolstoguzov, V. B., 1998. Functional properties of protein-polysaccharide mixtures. In Hill, S. E., Ledward, D. A., Mitchell, J. R. (eds), *Functional Properties of Food Macromolecules*. Aspen Publishers, Gaithersburg, MD, , pp. 252–277.
Tolstoguzov, V. B., Braudo, E. E., Grinberg, V. Y., Gurov, A. N., 1985. *Usp. Khim.* **54**, 1738–1759.
Tromp, R. H., Rennie, A. R., Jones, R. A. L., 1995. *Macromolecules* **28**, 4129–4138.
Tuinier, R., Dhont, J. K. G., de Kruif, C. G., 2000. *Langmuir* **16**, 1497–1507.
Vittadini, E., Dickinson, L. C., Chinachoti, P., 2002. *Carbohydr. Polym.* **49**, 261–269.
Walkenstrom, P., Hermansson, A. M., 1994. *Food Hydrocolloids* **8**, 589–607.
Wang, Q., Ellis, P. R., Ross-Murphy, S. B., Reid, J. S. G., 1996. *Carbohydr. Res.* **284**, 229–239.
Watase, M., Nishinari, K., 1980. *Rheol. Acta* **19**, 220–225.
Watase, M., Nishinari, K., Clark, A. H., Ross-Murphy, S. B., 1989. *Macromolecules* **22**, 1196–1201.
Williams, P. A., Clegg, S. M., Day, D. H., Phillips, G. O., Nishinari, K., 1991. Mixed gels formed with konjac mannan and xanthan gum. In Dickinson, E. (ed.), *Food Polymers, Gels and Colloids*. Royal Society of Chemistry, Cambridge, UK, pp. 339–348.
Williams, P. A., Clegg, S. M., Langdon, M. J., Nishinari, K., Piculell, L., 1993. *Macromolecules* **26**, 5441–5446.
Zasypkin, D. V., Braudo, E. E., Tolstoguzov, V. B., 1997. *Food Hydrocolloids* **11**, 159–170.

11 Innovative systems and applications

Our final chapter is intended to serve as a guide not only to novel systems but also to recent and potential future applications. The majority of exciting developments are in the pharmaceutical and biomedical areas, including scaffolds for tissue engineering, so biomedical applications make up, by some way, the largest section here. Moreover, for practical reasons this chapter is shorter than we might wish and covers only a limited selection of topics. We apologize in advance if this does not give appropriate weight to all relevant developments.

11.1 Innovative systems

11.1.1 Slide-ring gels

Perhaps the most intriguing of recent novel physical gel systems are the so-called slide-ring gels pioneered by Ito and co-workers (Okumura and Ito, 2001; Ito, 2007, 2010, 2012). Two polymer chains are 'cross-linked' – actually topologically interlocked – by a supramolecular closed figure-of-eight–shaped molecule. Each polymer chain threads through one of the two loops and can translate freely, but is always held adjacent to the other chain by the figure-of-eight (Figure 11.1). Such a structure was first proposed by de Gennes (1999), but only subsequently has supramolecular chemistry helped to realize the concept. As the cross-links can pass along the polymer chains freely, equalizing the tension of the threading polymer chains like a system of pulleys, their behaviour has been called the 'pulley effect'. The chemistry of synthesis is obviously testing, but first produces a polyrotaxane (PR). A rotaxane is a supramolecular structure consisting of one or more ring-shaped cyclic molecules and a linear molecule threaded through the rings and 'capped' at both ends, forming a dumbell shape so that the ring can move along the molecular axis but no further. If the linear molecule is polymerized it should give a necklace-like chain onto which cyclic molecules can be threaded.

In the work by Ito and co-workers, the polyrotaxane contained a small amount of an α-cyclodextrin ring (α-CD) on poly(ethylene glycol) (PEG). The slide-ring gel was then prepared by cross-linking α-CDs on different PR chains, though not all α-CD rings were so linked to give figures-of-eight. Since the chains are capped, the α-CDs cannot diffuse off of them. Although the precursor system was prepared in 2001 (Okumura and Ito,

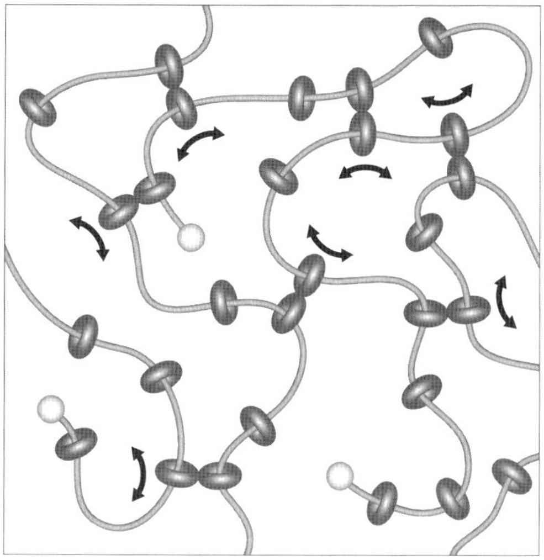

Figure 11.1 Schematic representation of a slide-ring gel with freely moveable figure-of-eight cross-links functioning as 'pulleys' (arrows). Free cyclic α-cyclodextrin molecules are threaded onto the linear polymer, forming a polyrotaxane (PR) chain. Reproduced from Ito (2012) with permission from Nature Publishing Group.

2001), only more recently has a fuller physical characterization involving rheology and scattering been carried out.

In his recent review, Ito has gone beyond the synthesis to describe the mechanical properties of his slide-ring system (Ito, 2012). The modulus does not increase proportionally to the amount of 'cross-linker' but shows a weaker dependence, although this may be because of wastage – it is obviously difficult to estimate the fraction of elastically effective slide rings. That said, given enough of them, there is evidence that the system does have an equilibrium Young's modulus, as if the cross-links were permanent. The overall modulus is quite low, but Ito has suggested that this reflects a new type of elastic response, which he terms 'sliding elasticity'. Here it is assumed that the pulley effect relaxes stress in a way similar to the reptation processes in an entangled polymer system. He also suggests that free rings would tend to cluster along the chain, giving rise to a heterogeneous density distribution, which then results in substantial entropy loss.

11.1.2 Polyelectrolyte complexes

A polyelectrolyte complex (PEC) involves a mixture of two polymeric components, one of which is positively charged (cationic) and the other negatively charged (anionic). As we have already noted (Chapter 10), when two polymers are mixed, the normal tendency is for phase separation. However, if the two systems are oppositely charged, it is enthalpically favourable for them to 'neutralize' so that the two macromolecules are bound together by electrostatic interaction (Figure 11.2). Although this method can be used to prepare small complexes, there is a greater interest in either building up multi-layer structures by adding

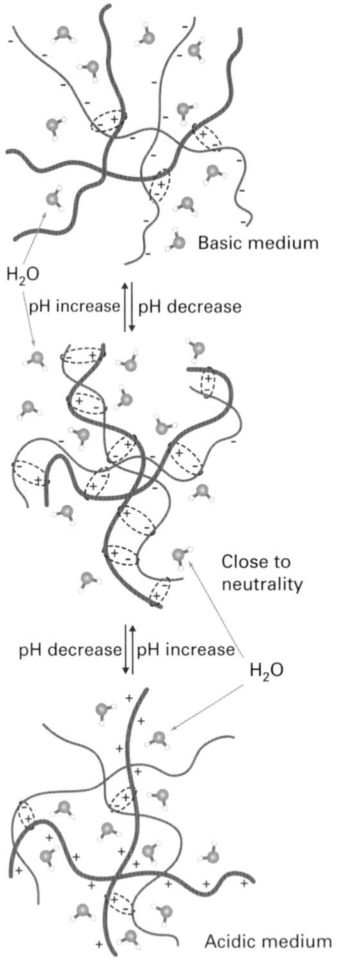

Figure 11.2 Structure and pH-sensitive swelling of a polysaccharide complex containing chitosan: – negative charge of the anionic polymer; + positive charge of chitosan; --- ionic interaction; **—** chitosan; – anionic polymer. Adapted with permission from Berger *et al.* (2004a) © 2004 Elsevier.

positively and negatively charged species sequentially or in 'fixing' the PEC by adding chemical or other non-covalent interactions.

Of course the number of positively charged, biocompatible polymers is quite limited, but here chitosan (Chapter 5) has proved of major interest, and, as we will see below, chitosan complexes have been suggested for a wide range of biomedical applications. The number of anionic polymers is less limited, and the most commonly used include alginates, pectins and xanthans (also introduced in Chapter 5). More recently there has been renewed interest in exploiting the sulphated glycosoaminoglycans (GAGs, sometimes referred to as mucopolysaccharides) hyaluronic acid and chondroitin sulphate. Proteins such as collagen and its derivative gelatin, synthetic polymers such as PAA, and even DNA have also been investigated. A list of polyelectrolytes forming complexes with chitosan is given by Berger *et al.* (2004a).

11.1.3 Gel micro- and nanoparticles

As mentioned in Chapter 4, the rate of swelling (and de-swelling) of a spherical particle varies approximately inversely with the square of its radius, and this general behaviour holds even for arbitrary shapes. The release of an entrapped material is controlled by similar factors – the surface area to volume ratio being much larger, the smaller the particle. For that reason there has always been an advantage in producing enhanced swelling rate micro- and nanogel particles. It was realized in the 1950s that intramolecularly (chemically) cross-linked macromolecules (ICMs), particularly based on methacrylates, could have useful properties in the paint industry. According to Graham and Cameron (1998), the term 'microgel' goes back to Baker in 1949, who from his work on emulsion polymerizations recognized that his systems were indeed intramolecularly cross-linked. In 'solution' these microgels behaved more like Einstein spheres than dissolved linear polymer coils, and exhibited much lower than expected solution viscosities. Subsequently ICM microgels became widely used in paints. By controlling molecular mass by change of initiator concentration, for example, species could be produced of radius <10 nm, referred to as nanogels.

Arguably the first commercial physical gel 'spheres' were those artificial fruit products produced in the 1960s by adding Na^+ alginate plus fruit puree drop-wise into a Ca^{2+} bath (Chapter 5). This technique is still used, although interest now is more in micro- and nanogels, particularly for controlled release of active ingredients in the biomedical area.

The Trondheim group (Thu *et al.*, 2000) has extended its work on alginates to investigate the homogeneity of such gel particles, and small spherical alginate beads (1.0–0.7 mm diameter) were obtained under various conditions. Micro-images of these obtained by magnetic resonance imaging (MRI) (Figure 11.3) illustrate the polymer concentration gradient inside the beads, and reveal differences which depend on the exact gelling procedure. For example, low calcium concentration during gelation gave rise to more inhomogeneous beads. The images show that there is no smooth polymer gradient throughout the gel but that micro-heterogeneities exist, although, because of resolution limitations, their sizes are difficult to obtain. That said, for controlled-release properties, such heterogeneities are valuable, and the ability to design them becomes very useful.

Traditional ionic or thermally induced gelation of polysaccharide drops is a reasonably efficient way to prepare simple 'monodisperse' gel microspheres for application as drug delivery systems, enzyme carriers for detergents or supports for biocatalysts, but the characteristics can be improved. Several routes have been tried, with different emphases for specific applications. These include so-called 'core-shell' designs, in which the microsphere has an outer layer with various properties, either for mechanical/chemical stability or to alter release characteristics, and some of these are described below. The texture of the polysaccharide gel plays a key role in the kinetics of drug release or in support for catalysts in fine chemistry reactions or live tissue, and relevant properties can be optimized using core-shell structures.

Another approach that has been used – particularly for certain physical gel systems such as the carrageenans and agarose (Chapters 5 and 7) – is simply to agitate a fast-gelling system, producing arbitrarily shaped gel pieces. Contrary to some

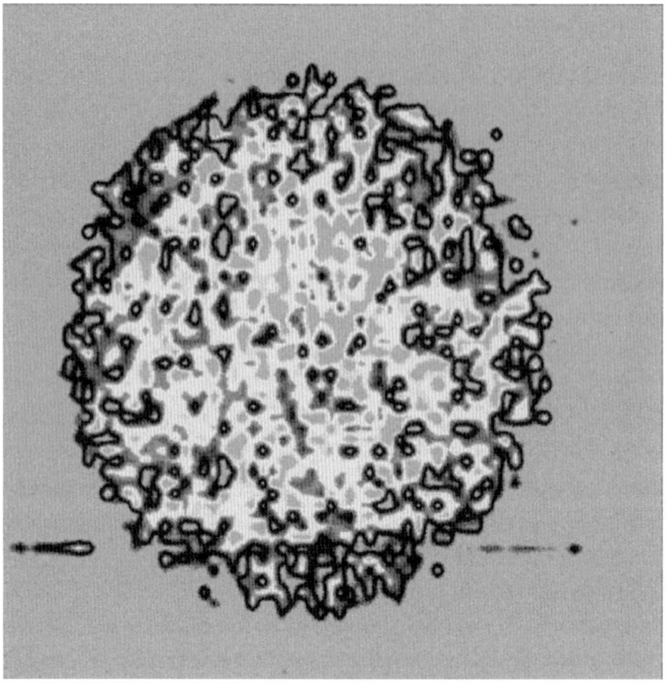

Figure 11.3 Magnetic resonance imaging (MRI) of an alginate bead (1.8% w/v alginate in 50 mM $CaCl_2$). Polymer concentration decreases from dark grey to light grey. Images obtained with Bruker DMX 200 spectrometer. Adapted from Thu *et al.* (2000) with permission from John Wiley & Sons.

expectations, this does not disrupt gelation *per se* but simply produces a system where intramolecular cross-links become more favoured, just as in the original microgel concept. An alternative is, of course, to prepare the gels as emulsions, and this does have the advantage of maintaining more homogeneous sizes and shapes. Norton and co-workers (Norton *et al.*, 1999; Adams *et al.*, 2004) have exploited both approaches to produce so-called 'fluid gels'. Here the gelling system is allowed to cool quite rapidly through the disorder–order transition while being agitated (in their terms 'sheared', although during the process analysis must be difficult, so other more complex deformation fields are probably involved). The gel particles produced lie typically in the range 5–50 μm and appear to be roughly ellipsoidal, although with many surface defects. The resultant properties of these systems are not those of typical physical gels, but are described as 'fluid or paste-like'. The exact properties are dependent upon the choice and concentration of polymer, and on ionic composition if gelation is ion-mediated.

Adams *et al.* (2004) have examined agar microgels produced by first preparing a classical water-in-oil emulsion and subsequently cooling it to gel the internal phase. The gelled particles (generally close to spherical) were then separated and re-suspended in water. Depending upon the initial agar concentration, particles sizes are typically of the order of 25 μm. They then studied the rheological response of these high volume fraction (>80%, ~ϕ_m, the maximum packing fraction) soft spherical particle suspensions. Above

ϕ_m there is a dramatic increase in modulus, and an approximate fit found to the classical Hertz elastic equation, while at lower volume fractions steady shear measurements could be made. Results from these suggest an apparent yield stress, and were related to classical hard-sphere repulsion models. The 'fluid gel' method has been reviewed by Frith (2010). Since gelling under shear flows produces quite complex microstructures, such properties as the interfacial tension(s) become increasingly important. By controlling these, more precisely ellipsoidal microgel particles can be produced. The phase diagram (strictly, the state diagram) of a mixture which is both phase separating and gelling while being deformed, for example in a shear or extension field, may be perturbed. This route can help to produce novel 'textures' of use in particular applications.

11.1.4 Multi-membrane hydrogels

Ladet *et al.* (2008) have described the formation of so-called multi-membrane structured materials based on purely physical gels. They applied this approach to both chitosan and alginate (Chapter 5), but they claim the method can be generalized to other natural polyelectrolytes such as hyaluronic acid. In the multi-stage process for chitosan, an 'alcohol gel' is first formed by evaporating water at 55°C from a 50:50 water:alcohol solution of the polymer, and with the concentration above that for chain entanglements. When the water is almost fully removed, a gel is formed. The system is then neutralized with NaOH, and subsequent washings in water yield a material that contains only water (over 95 wt%) and chitosan in the free amine form. When gelation is achieved, almost 40% NH_3^+ groups are still present and have to be neutralized to favour inter-chain interactions allowing the formation of a stable material (Figure 11.4). The neutralization of NH_3^+ sites removes ionic repulsion between polymer chains and, they suggest, produces physical cross-links from hydrogen bonding, hydrophobic interactions and crystallite formation. There appear to be parallels between these structures and the freeze/thaw gels described in Chapter 8.

Modification of the balance between hydrophobic and hydrophilic interactions de-swells the neutralized gel, and helps create a separate gel membrane structure with what the authors term an interphase of lower polymer concentration between the neutralized and alcohol gels. The concentration of NaOH determines the kinetics of neutralization. At high NaOH, chains are rapidly and completely neutralized, and their aggregation is

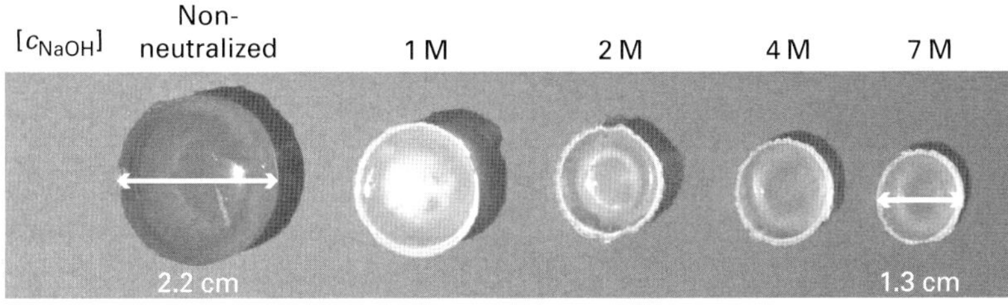

Figure 11.4 Variation of hydrogel shrinkage during neutralization as a function of concentration of NaOH. The initial chitosan concentration in the non-neutralized alcohol gel was constant and close to 4.5% w/w. Adapted from Ladet *et al.* (2008) with permission from Nature Publishing Group.

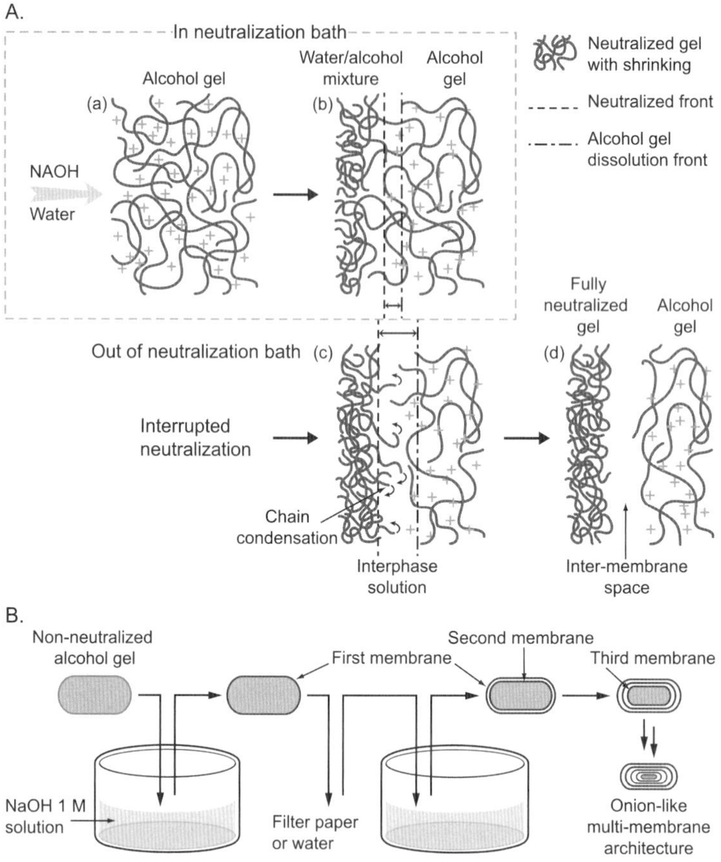

Figure 11.5 (A) Schematic representation of the neutralization steps for a polyelectrolyte alcohol gel and the methodology for building a multi-membrane structure: (a) structure of alcohol polyelectrolyte gel; (b) the neutralization step: chain condensation and shrinkage of the alcohol gel with the disappearance of ionic repulsions; (c) formation of the interphase in solution; (d) formation of an inter-membrane space outside the neutralization bath and complete retraction of residual polymer chains from the interphase solution. (B) Overview of the multi-step neutralization process. Adapted from Ladet *et al.* (2008) with permission from Nature Publishing Group.

a maximum, leading to more homogeneous gels. In contrast, for low NaOH, water diffusion within the alcohol gel must be considered, and the neutralization route used to generate novel concentric (onion-) shelled multi-membrane architectures.

During the neutralization step, if water diffusion is fast enough compared to ionic neutralization kinetics, an interfacial solution is formed in which the polymer mobility is high enough to allow disentanglement of the polymer chains and their ‘condensation’ on to the neutralized gel (Figure 11.5). In this way, formation of an inter-membrane space was promoted by slowing down the neutralization, simply by removing the gel from the neutralization bath and washing with water. According to Ladet *et al.*, membrane formation occurs in a three-step sequence: (i) generation of a water–alcohol interphase solution between the neutralized and alcohol gels, (ii) disentanglement of chains located

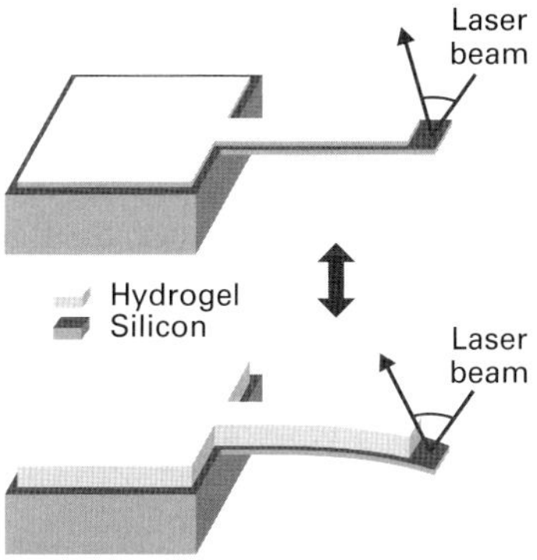

Figure 11.6 Schematic view of the MEMS sensor platform based on a microcantilever patterned with environmentally responsive hydrogel: at rest (top) and after bending by swelling (bottom). Adapted with permission from Hilt *et al.* (2003) © 2003 Springer.

within the interphase and (iii) their condensation and contraction on to the neutralized gel. This sequence could be repeated up to 20 times, to give a macroscopic gel of a few cubic centimetres with an onion-like structure composed of independent gel membranes.

11.1.5 Hydrogels as ultra-sensitive sensors

Another active research area is the development of micro-electromechanical systems (MEMS) based on the ultra-sensitivity of microcantilevers and the various responsive abilities of environmentally sensitive hydrogels. For example, pH-sensitive hydrogels can be laid down in a precise pattern on microcantilevers to create an ultra-sensitive pH microsensor, as shown schematically in Figure 11.6 (Hilt *et al.*, 2003). The chemically cross-linked hydrogel film is capable of sensing a change in environmental pH and equilibrates within a few minutes. Photolithography is used to pattern the polymer networks – for example of poly(methacrylic acid) (PMMA), produced by reacting methacrylic acid monomer with poly(ethylene glycol) dimethacrylate (UV free radical copolymerization) – and adhesion is generated between the silicon substrate and the polymer using an organosilane coupling agent.

In the anionic network of the hydrogel, ionization of the acid groups occurs when the pH is raised above the pK_a. With deprotonation of the acid groups, electrostatic repulsion between chains increases the hydrophilicity of the network, so water is absorbed. Swelling of the hydrogel induces a deflection of the cantilever. The dynamic and equilibrium deflection characteristics of such microcantilevers were examined in buffered solutions of various pH.

An example of data from such a sensor is shown in Figure 11.7. A value of 20 μm per pH unit was found for the highest sensitivity sample, which gives a resolution for optical-based laser detection set-ups of 1 nm deflection for $\Delta pH = 5\times10^{-5}$. It is to be expected that the swelling properties of the polymer hydrogels, and therefore the sensitivity of such devices, can be tailored by changing the amount of cross-linking, by increasing the porosity of the network or by varying the polymer film thickness.

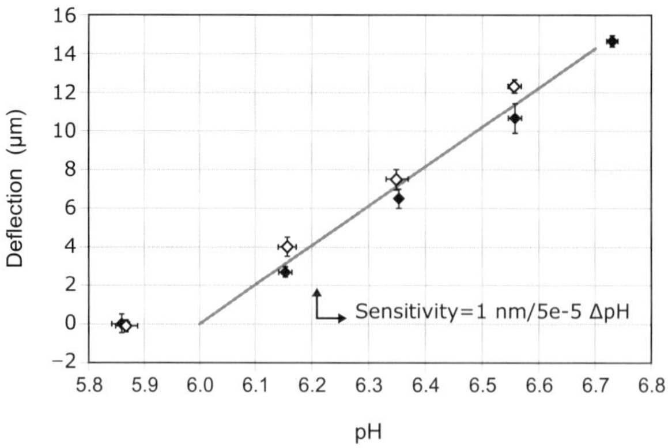

Figure 11.7 Equilibrium deflection versus pH for increasing pH (solid symbols) and decreasing pH (open symbols) at constant ionic strength (0.5 M). The solid line is a linear fit. Adapted with permission from Hilt *et al.* (2003) © 2003 Springer.

11.2 Food and cosmetic applications

The truly innovative work in this area tends to be limited by statutory requirements. That said, as far as food applications are concerned, the major new potential use of physical gels is in low-calorie products. We stress 'potential' since, starting in the 1980s, the food patent literature seems large but the number of real low-calorie products based on gels appears limited. The principle of this 'materials science' approach is quite sound. Taking the case of full-fat (*c*.80%) spreads such as butter and margarine, if the solid fat is removed from a product, something else is needed to maintain its solid character, and a gelling biopolymer component appears to be very appropriate. Unfortunately there are problems. First, the taste and 'mouth-feel' of some polysaccharide gels are not attractive. Here gelatin has a number of advantages: it has little flavour and 'melts' around body (i.e. mouth) temperature. However, there are ethical issues in using mammalian gelatin, particularly for vegetarians. There is less of a problem with piscine gelatins, but, as noted in Chapter 7, these generally melt at lower temperatures. The external literature includes a number of proposals, including work by Kasapis and co-workers (Roberts *et al.*, 2000) on the respective binary systems of starch, caseinate (from milk) and alginate. By changing the compositions, and the concentration of added Ca^{2+}, they claim to have matched at least some of the mechanical properties of traditional spreads. More recently, Cheng and co-workers (Cheng *et al.*, 2008) have used a mixed gel of piscine gelatin and pectin, and prepared model 'zero-fat' spread systems with rheological properties that begin to mimic those of fat spreads.

In Japan and elsewhere in the Far East there is an extended culture of food gels, particularly from polysaccharides (Nishinari, 1988). More novel applications include artificial salmon roe (widely used as an important ingredient in sushi), made from salad oil and seaweed extract, with the 'skin' made from carrageenan or sodium alginate, again gelled by the addition of Ca^{2+}. Apparently its appearance, texture and taste cannot be distinguished from real salmon roe, until the artificial roe is placed in hot water. Under these conditions, real salmon roe forms an egg-white colour and texture. That said, some consumers prefer artificial salmon roe because of its lower cholesterol content. Crab sticks (imitation crab-meat sticks) are a form of kamaboko, a processed seafood made of finely pulverized white fish flesh (surimi), shaped and cured to resemble crab leg meat, and a gelling agent such as egg white (albumen). Crab flavouring and colouring is added. The fibrous texture is also appreciated in commercialized 'string cheese', which can be torn into thin strings just like dried squid, a favourite Japanese snack.

Clearly another way to produce low-calorie products, many of which are sold by volume rather than mass, is simply to aerate them. To stabilize the resultant air bubbles, an appropriate gel can be added. The gas phase in a food matrix not only makes the product softer but changes its appearance, colour and mouth-feel in a controllable manner. As Campbell and Mougeot (1999) have pointed out, aeration has been used for millennia in beer and bread products, and most commercial 'soft scoop' ice creams contain up to 50% air, together with a mixture of gelling/stabilizing agents including guar, LBG, pectin, alginate and gelatin (Clarke, 2004). Commercial marshmallows are another example of aerated gels: they usually contain gelatin and sometimes also albumin.

Both the science and further applications of aerated gelatin gels have been discussed by Zuniga and co-workers (Zuniga and Aguilera, 2008, 2009; Zuniga *et al.*, 2011). They prepared a series of what they term gas-filled gelatin gels (GGGs) and compared the results with control samples. Originally they 'aerated' with air, nitrogen and helium using a syringe pump and gas cylinder assembly to bubble gas through a pre-gelled solution, which they then allowed to cool. In the gel, gas bubble sizes lay mainly in the range 40–120 μm and volume fractions were *c.*70%. They then made a number of measurements of both gas-filled and control gelatin gels, and subjected them to uniaxial compression. As in Chapter 7, the stress–strain 'curves' were almost linear up to failure, which occurred, in this case, at strains of around 60% for the control samples and around 40% for the aerated samples. In particular, the GGGs did not show the sigmoidal stress–strain behaviour expected for cellular foams, an observation which may reflect the small gas bubble sizes and the thick interface walls. In their most recent paper (Zuniga *et al.*, 2011), they generated the gas-filled gels ultrasonically and studied the interface stabilized with partially denatured β-lactoglobulin samples. The latter samples failed at lower strains than the control aerated samples, although the effect of the extent of denaturation is limited.

There is interest in so-called 'nutraceutical microgels'. A nutraceutical is defined as a food material containing a component with a specific physiological benefit beyond inherent general nutrition, thus preventing or alleviating specific diseases (Zuniga and Aguilera, 2008). Such a microgel includes those made from proteins and polysaccharides. Proteins in particular have prospective tuneable release properties because of the dependence of their charge on pH. In other work (Remondetto *et al.*, 2004), in the presence of proteolytic enzymes such as pepsin at pH 1.2 or pancreatin at pH 7.5, particulate gels of β-lactoglobulin, typically made at pH > 6, have been shown to release more iron than fibrillar gels made at pH <3, so gel microstructure is also significant. According to Zuniga and co-workers (Zuniga and Aguilera, 2008), tailoring the porous structure of an aerated gel would also offer opportunities for protection or release of a physiologically bioactive component (or a nutrient).

The personal products and cosmetic industries have also made use of what they term 'gels', but in practice most of these seem to fall outside our remit: they are often entirely surfactant-based or they are simply viscous liquids, and commonly both. A typical example is a branded commercial 'shower gel', which consists of worm-like micelles which naturally entangle. Nonetheless the work by Miyazawa and co-workers (Miyazawa *et al.*, 2000) is of relevance. They prepared novel soft microcapsules for cosmetics, consisting of an oil–water–oil emulsion in which the water phase was then gelled with a commercial agar. These microcapsules were typically around 300–400 μm in diameter, and the concentration of agar in the water phase was typically 2–4% w/w. Various agar preparations, differing in composition and molecular mass, were measured in compression, and it was established that the best stabilized microcapsules were those where the original agar gels had the highest modulus.

11.3 Biomedical applications

It was recently estimated (Peppas *et al.*, 2006) that there were over 8000 medical devices and 2500 diagnostic products employing biomaterials in various medical applications. Hydrophilic polymers in particular have demonstrated great potential for biological and medical applications, and the ability to engineer these with specific properties is the source of new developments in tissue engineering, drug delivery and bio-nanotechnology. The design and synthesis of 'smart' hydrophilic polymers are of interest in biomedical and nanotechnology applications. Many such 'hydrogels' are particularly appealing for biological applications because of their high water content and biocompatibility. In this volume we use the term 'hydrogel' sparingly since, as mentioned in Chapter 1, it is sometimes used simply for viscous, entangled solutions.

11.3.1 Applications of chitosan

Chitosan has been studied very widely over the last decade, and here we give only a brief summary since detailed mechanical measurements seem scarce. There are a number of excellent reviews, including those by Berger, Dash and their respective co-workers (Berger *et al.*, 2004a, 2004b; Dash *et al.*, 2011). The use of 'entangled chitosan hydrogels' is limited by their lack of mechanical strength and their tendency to dissolve but, as noted, these will not be discussed here. 'Chemical' hydrogels are formed by irreversible covalent links, for instance using glutaraldehyde. Unfortunately, for biomedical applications the addition of covalent cross-linkers may decrease biocompatibility (Berger *et al.*, 2004b).

Physical chitosan hydrogels can be formulated with various reversible links: ionic interactions in ionically cross-linked hydrogels and in polyelectrolyte complexes (PEC), or secondary (physical) interactions as in grafted chitosan hydrogels and in chitosan–poly(vinyl alcohol) (PVA) hydrogels.

11.3.1.1 Ionic interactions

Ionic interactions between the negative charges of the cross-linker and positively charged groups of chitosan depend on the type of cross-linker. Metallic ions induce formation of coordinate-covalent (dipolar) bonds between the positively charged ammonium groups of chitosan. Among anionic molecules, phosphate-bearing groups such as β-glycerophosphate and particularly tripolyphosphate can be employed.

PECs are generally biocompatible networks and exhibit interesting swelling characteristics, and these can be reinforced by the addition of ions. Typically Ca^{2+} can be added to alginate or pectin and Al^{3+} to sodium carboxymethylcellulose (NaCMC), distinct from ionically cross-linked chitosan gels since then chitosan is not cross-linked but plays the role of the additional polymer. For example, as well as forming a PEC with the anionic mucopolysaccharide chondroitin sulphate, chitosan can also be ionically cross-linked, although control of pH during the cross-linking reaction is critical to generating a reproducible product.

During complexation, polyelectrolytes can either coacervate, i.e. form dispersed emulsion-like droplets, or form a more or less compact gel, whereas if ionic interactions are too strong, precipitation rather than gel formation occurs. Such precipitation can be avoided if electrostatic attractions are weakened by the addition of salts such as NaCl. The presence of salts reduces the attraction between oppositely charged polyelectrolytes by contributing to the counterion environment. To overcome the difficulties encountered with large-scale processing during PEC preparation, and so obtain a homogeneous mixture, mixing can be carried out at a pH value where complexation does not occur. This pH can then be adjusted to a value where interactions are favoured.

PEC gels exhibit pH-sensitive, and to a minor extent ion-sensitive, swelling. As pH changes, the swelling rate is controlled by the diffusion of mobile ions and changes in the degree of ionization. As swelling of PECs is influenced by many factors, it can be used for fine modulation of drug release. Chitosan gels formed by a PEC can be used as scaffolds in cell culture and enzyme immobilization. As no potentially toxic molecules or covalent cross-linkers are added to these gels, they represent a better medium for cell culture than covalently cross-linked systems. The GAGs, chondroitin sulphate and hyaluronic acid are of interest here, since they can be used in cartilage reconstruction and wound-healing, as carrier materials and as scaffolds for tissue engineering.

11.3.1.2 Grafted chitosan hydrogels

Grafting (functionalizing) chitosan is a common way to improve properties such as chelating and complexation capability, solubility in water or organic solvents, bacteriostatic effects and absorption properties. Depending on the nature of the graft, the secondary interactions between grafted groups can be hydrogen bonds or hydrophobic interactions. Hydrogen bonds occur, for example, with poly(ethylene glycol) (PEG). Hydrophobic interactions are responsible for the formation of a network when alkyl chains are grafted to chitosan by an acid or an aldehyde. Moreover, these interactions can arise from intrinsic properties of the graft. For instance, functionalization with Pluronics (Chapter 6) allows the formation of thermally gelling systems due to graft dehydration, leading to increased inter-chain association. Grafting with NIPAm (Chapter 4) allows the formation of a temperature-sensitive system.

As swelling depends on the electrostatic repulsion of the free ammonium groups, it only occurs in acidic conditions, which limits its potential applications. Grafted chitosan presents interesting properties, for example, in wound-healing, where biocompatible chitosan derivatives can exhibit enhanced bacteriostatic activity. Grafting of hydrophobic groups onto chitosan also allows the incorporation of lipophilic drugs. That said, such grafted chitosan gels do not have significant advantages over covalently cross-linked gels, since they may require the use of potentially toxic auxiliary molecules for their preparation, and this preparation is usually more difficult.

11.3.1.3 Chitosan–PVA complexes

Chitosan–PVA complexes can be prepared with less difficulty than PECs, but not as easily as ionically cross-linked gels. They do not exhibit pH-sensitive swelling, and PVA does not have any specifically exploitable properties, as do some polyelectrolytes. Therefore, chitosan–PVA complexes have a relatively narrow field of application. However, as the use of PVA in medical and pharmaceutical applications is well documented, chitosan–PVA complexes represent an interesting alternative to PEC and ionically cross-linked gels for the preparation of biocompatible drug delivery systems, if a pH-controlled drug delivery is not required. Like chitosan, PVA is non-toxic, biodegradable and highly biocompatible.

The structure of chitosan–PVA hydrogels (Figure 11.8) can be considered as intermediate between PEC hydrogels and networks formed by grafted chitosan, discussed in

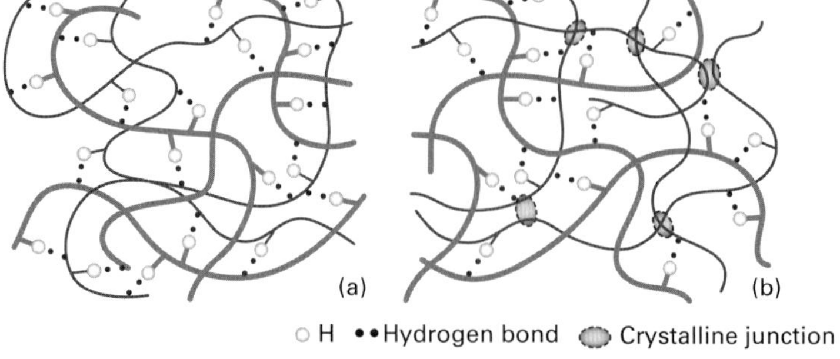

Figure 11.8 Proposed structure of chitosan–PVA hydrogel (– chitosan; ▬ PVA): (a) prepared by autoclaving method; (b) prepared by freeze/thaw method, showing hydrogen bonds and crystalline junctions. Adapted with permission from Berger *et al.* (2004a) © 2004 Elsevier.

the previous section. They can be prepared either by autoclaving or by the freeze/thaw method (Chapter 8), although the resultant structures are different.

Complexes formed by autoclaving can be used as scaffolds in cell culture. In Figure 11.8, hydrogen bonds between the hydroxyl groups of PVA and the hydroxyl or amino groups of chitosan are thought to be the main interactions inside the complex. Chitosan–PVA blended gel membranes have been prepared by this procedure (Koyano *et al.*, 2000), but observations of the membrane by microscopy suggest that the chitosan components for certain compositions are localized in islands in the air-surface side of the membrane, because chitosan is less hydrophilic than PVA. Complexes formed by autoclaving are readily soluble under acidic conditions and therefore not suitable as drug delivery systems, although they do favour cell attachment: the higher the chitosan ratio, the higher the water retention, which enhances the cell growth rate. Such complexes can be used, for example, in fibroblast cultures, where they provide a better scaffold than collagen.

The chitosan–PVA blends formed by freeze/thaw methods have been used as drug delivery systems. In addition to the interactions between the PVA hydroxyl groups and the hydroxyl or amino groups on chitosan, crystallite junction zones between PVA polymeric chains are formed (see Chapter 8). This complex with a less regular structure forms gels with a high capacity to swell, which is an important property for drug delivery systems. Since there is no pH-sensitive swelling, drug release is diffusion-controlled and modulated by the chitosan ratio. Overall, chitosan–PVA hydrogels represent an interesting biocompatible alternative to other types of gels, but they are not as versatile. Because of the intrinsic properties of chitosan, its inclusion favours the adhesion of the system at the site of administration, and the bioavailability of poorly absorbable drugs is enhanced. Such complexes can be used, for example, for the controlled release of growth hormones and for oral administration of certain drugs.

11.3.2 Applications of microspheres

11.3.2.1 Inhomogeneous alginate beads

Alginate gels are particularly versatile, especially for immobilization of living cells. The most exciting prospect for such immobilized cells is their potential use in cell transplantation, where the main goal of the gel is to act as a barrier between the transplant and the immune system of the host. We have already mentioned in Chapter 5 the work on encapsulation, and the implantation of alginate beads containing pancreatic cells into insulin-dependent diabetics, where the alginate gel surrounds serve to protect from immune reactive responses and so preserve cell viability (King *et al.*, 2003). However, there are problems with this approach, including the low biocompatibility and low stability of such systems. By forming capsules with a strong polyanion–polycation membrane on their surface and a low-swelling gel network in the core, one can obtain systems that are stable for extended periods under physiological conditions. Such conditions are achieved by forming inhomogeneous gels, where the highest polymer concentration is close to the surface and the lowest is in the centre of the capsule. Such inhomogeneities may be preferred in microcapsules since they help create low porosities and higher stabilities.

11.3.2.2 Beads with hierarchical macroporosity

As noted above, the formation of a shell-structured microsphere with higher density at the surface considerably improves the retention of cells in biocatalysts. Heterogeneities inside the gel microspheres are also expected to affect diffusion and swelling phenomena. One mechanism of generation of heterogeneous structures is described by Thumbs and Kohler (1996). When a solution of Cu^{2+} salt (instead of the more usual Ca^{2+}) is placed on top of a solution of Na^{+} alginate, the chains cross-link to form a 'capillary gel'. These capillaries (8–300 μm diameter) are examples of dissipative structures, and a periodic pattern, due to convective motion of the aqueous solution in the neighbourhood of the gelation front, appears. The complete theory was proposed by Treml *et al.* (2003), and involves a counter-current diffusion between the polysaccharide solution and the gelling agent.

For appropriate values of the ratio between diffusivity and gelling rate, convection cells are formed, through which the solution of gelling cations can feed the advancing gelation front. In alginate solutions at the interface with a Cu^{2+} solution, the rising channels of the convection cells may remain in the final gel, as a system of parallel void channels normal to the gelation front.

Another strategy is to form cavities in the core of polysaccharide beads by dehydration treatment. Cavities can be generated by controlled syneresis of the core, in the presence of a shell able to prevent isotropic shrinking of the bead. Such structures can lead to improved diffusivity inside the macroporous core, and present intermediate properties between core-shell systems and hollow-shell microcapsules, extending the range of materials available for biocatalysis and drug delivery.

Di Renzo *et al.* (2005) investigated the influence of a rigid shell on the formation of channel structures in the polysaccharide core. Core-shell microspheres were prepared with chitosan–silica composites. Beads of nearly 3 mm diameter were obtained from drops of chitosan solution gelling in an alkaline solution. Mixing was achieved by rotating the flask around its horizontal axis, and the chitosan hydrogel could easily be impregnated with a silica sol, allowing a shell of silica gel to be deposited on the outer surface of the beads. The beads were then washed with water. Supercritical drying of the gel beads was preceded by dehydration in a series of ethanol–water baths with increasing alcohol concentration. The gel beads were dried under supercritical CO_2 conditions. Silica-coated chitosan beads show an outer shell, 200 μm thick, which contains only silica, and a distribution of silica inside the beads below this depth, the core of the bead being composed of 80% silica and 20% polysaccharide. A radial pattern of channels with an average diameter 10 μm can be observed on micrographs. The shafts do not reach the outer surface of the sphere, but appear to run from the inner surface of the silica shell toward the centre of the bead.

The presence of a similar structure is illustrated in Figure 11.9, which shows, at various levels of magnification Cu–alginate gels (Di Renzo *et al.*, 2005) before and after drying,

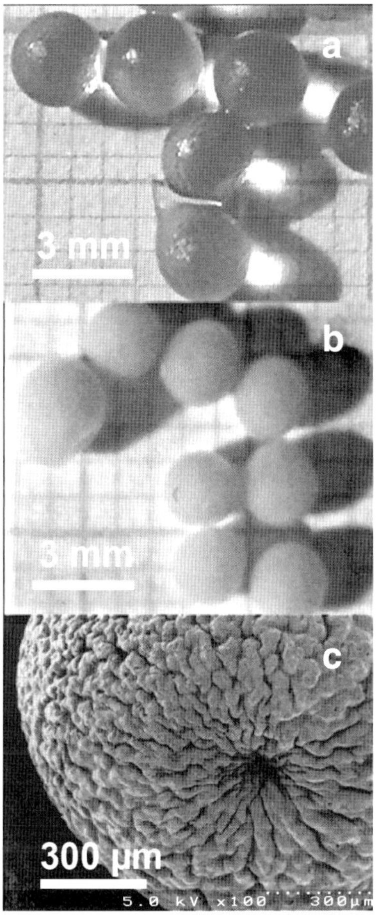

Figure 11.9 Cu^{2+}alginate beads: (a) hydrogel beads; (b) so-called aerogel after 22 h ageing; (c) SEM image of aerogel. Adapted with permission from Di Renzo *et al.* (2005) © 2005 American Chemical Society.

and the core shell structure of Cu-alginate beads and a section of this crust. The outer shell is formed by a flattened layer of fibrils, 'densified' by the radial retraction of the gel. Below the dense shell, the core presents a thick radial pattern of shafts. In Figure 11.10, a cross-section normal to the shaft axis is shown. The average diameter of the shafts is about 10 μm. After taking into account the shrinking of the bead and the formation of the shafts, the dry gel volume is nearly half the overall volume and the void fraction of the dried gel is still higher than 45%.

In these examples, the drying process and gelation conditions created the core-shell microspheres containing radial shafts in the depth of the beads. The transport properties of such core-shell systems with radial shafts are improved relative to untreated microcapsules, with the additional advantage of a large internal surface for stabilizing active groups.

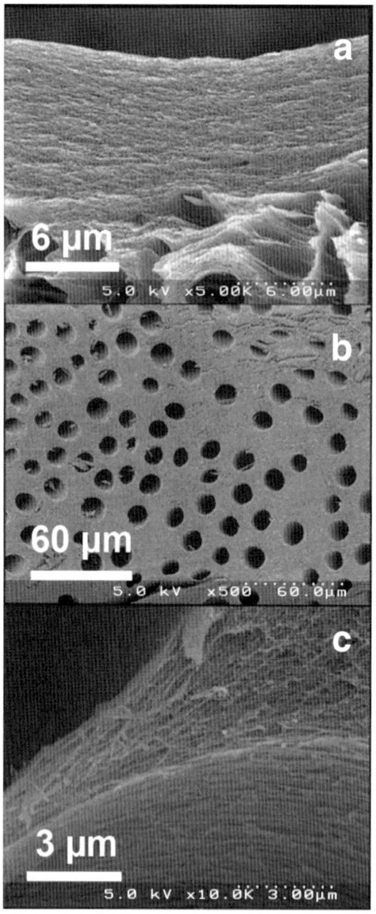

Figure 11.10 SEM micrographs of the cross-section of Cu^{2+}–alginate aerogel dried after 22 h of ageing of the hydrogel: (a) section of the crust; (b) cross-section normal to the shaft; (c) detail of the shaft wall indicating an anisotropic contraction of the gel away from the shaft axis. The aerogel volume is nearly half of the volume of the hydrogel. Adapted with permission from Di Renzo *et al.* (2005) © 2005 American Chemical Society.

11.3.2.3 Amphiphilic alginate beads

Controlled-release properties are of considerable interest for numerous pharmaceutical and medical applications (Leonard *et al.*, 2004; Rastello de Boisseson *et al.*, 2004). In the selection of a gel network for a given application, properties such as modulus, swelling ratio and porosity needed to be matched via appropriate formulations. In addition, knowledge of gel stability is essential as it is directly related to gel performance with time. A novel way to gel alginates is to transform them into amphiphilic derivatives that then form networks by associating in water (Chapter 6).

Such amphiphilic Na^{+} or Ca^{2+} alginate gels were prepared by Rastello de Boisseson *et al.* (2004). Their objective was to control properties such as mechanical strength, porosity or stability through the introduction of covalently bound hydrophobic alkyl chains. The solubility of such hydrophobically modified alginates is strongly reduced as ionic strength increases. Particles can then be formed by dropping a solution containing the polysaccharide into a Na^+ solution or into a solution containing Na^+ and Ca^{2+}. Appropriate conditions to induce gel particle formation, and avoid polymer precipitation, were determined by adjusting concentrations of NaCl, $CaCl_2$ and amphiphilic alginates. Amphiphilic derivatives of Na^+ alginate were prepared by esterifying long alkyl chains (12 or 18 carbon atoms) on to the hydrophilic backbone. Highly viscous solutions and strong gels were obtained in water, within a narrow range of substitution degree, i.e. 3–5% and 13–16% for C_{18}- and C_{12}-modified alginates, respectively, since at higher degree of substitution the polymers were no longer water-soluble. The microstructure of Ca^{2+} modified alginate gels was more heterogeneous than that of Ca^{2+} unmodified alginate gels, because the dense hydrophobic domains restrict egg-box formation (Chapter 5).

Another method of preparing microparticles from hydrophobically modified alginate derivatives uses an 'all-aqueous' procedure (Leonard *et al.*, 2004); see Figure 11.11. The main problem arising from the use of amphiphilic alginates is the high pre-gel viscosity of the solution, so dispersion in NaCl solution was difficult and particle sizes much higher than required.

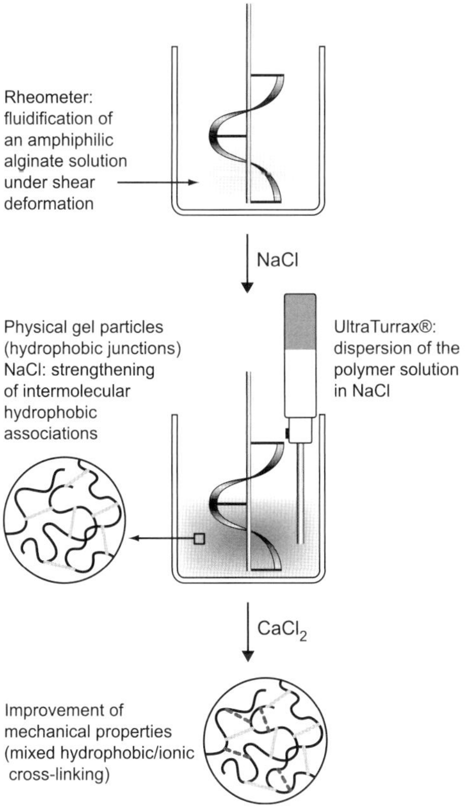

Figure 11.11 Schematic representation of the process used for preparation of microparticles of amphiphilic alginate. Adapted from Leonard *et al.* (2004) with permission from John Wiley & Sons.

Consequently Leonard *et al.* used a three-stage process: first agitating the partially dodecyl-substituted alginate derivative (alginate substituted with low amounts of dodecyl chains) under mechanical stress, creating a shear thinning water solution without NaCl; then repeating this in the presence of NaCl, resulting in physical gelation and particle formation via hydrophobic interactions; and finally adding small amounts of $CaCl_2$ to improve the mechanical properties of the beads. The average size (10–100 μm) and distribution of the resulting particles depended on the hydrophobic character of the solution and on dispersion conditions, but because of the absence of NaCl in the initial stages, this procedure can be used for the mild entrapment of biological substances such as proteins.

11.3.3 Nanoparticles for protein delivery: production, assembly and structure

New approaches for water-based drug delivery vehicles, and strategies to optimize nanoparticle properties such as size, charge and hydrophilicity, are receiving increasing attention. Prokop *et al.* (2001) have proposed a novel production method for nano-particles of multi-component polymeric complexes that are candidates as delivery vehicles for biological molecules such as proteins and drugs. Using either batch or continuous processing, and in the absence of organic solvents, stable nanoparticles that are insoluble in water or buffered media are produced using biocompatible and/or natural polymers. In the batch system an anionic polymer solution is introduced into a syringe and slowly extruded (by means of controlled air pressure) through a needle and into an ultrasonic nebulizer. A fine mist of anionic solution forms at the probe tip and falls into a cationic solution placed in an agitated receiving bath (Figure 11.12). Nanoparticles are produced ‘instantly’ on contact of the extruded liquid with the receiving bath. The

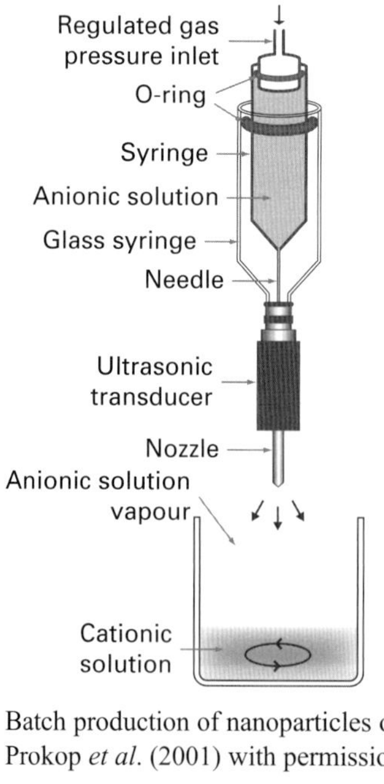

Figure 11.12 Batch production of nanoparticles of multi-component polymeric complexes. Adapted from Prokop *et al.* (2001) with permission from John Wiley & Sons.

continuous technology is a modification of the batch system made by introducing two inflow lines (anionic and cationic polymer solutions) and one overflow line to keep the receiving bath volume constant. The anionic solution is again dispersed in the form of a spray.

These nanoparticles consist of an inner core which holds the desired drug or protein, and a semi-permeable membrane shell (corona). Both positively and negatively charged nanoparticles can be produced, depending on the sequence of polymer addition. Various chemical constituents have been tested, including a combination of *at least two anionic polymers* in the droplet-forming phase, and one cationic polymer plus a small inorganic ion in the corona-forming phase. Components used in the core of the particles include alginates, gellan gum, sodium cellulose sulphate, κ-carrageenan, pentasodiumtripoly (phosphate) and, for the shell chitosan glutamate, poly(methylene-co-guanidine) hydrochloride and calcium chloride. For loaded samples, the model protein chosen was ovalbumin, a prototype for vaccine delivery.

In general, such nanoparticles are created using a minimum of two interacting pairs. The mechanism of particle formation, according to Prokop *et al.* (2001), is that a core gelling polysaccharide (or polyanion blend) and a small inorganic ion form the first pair, reacting almost instantaneously and providing mechanical stability to the particles, as well as a template (preformed structure) for the second interacting pair. The second interaction develops more slowly when polymers (corona, shell) complex via mutual interpenetration (diffusion) since at least one component of the second pair is of high molecular mass. Nanostructures result primarily from PECs.

It is likely that thermodynamic rearrangements of the already formed aggregates proceed over rather longer times, and involve conformational changes, disentanglements etc., which are a source of complex instability. The corona-forming complex also provides some degree of permeability control. The corona contains a so-called poloxamer triblock copolymer type (PEO_n–PPO_m–PEO_n) dissolved into the batch. As the guest polyanion solution drops into the host polycation solution, cationic particles form and grow in size. If the particles' charge is neutralized (i.e. at a stoichiometric charge ratio), aggregation occurs and a bulky precipitate is formed, so useful nanoparticles are produced before the stoichiometric charge ratio is reached. Surface properties of nanoparticles are therefore charge, porosity, size and degree of hydrophilicity. The poloxamer type triblock copolymer acts as steric stabilizer, the corona reducing surface adhesion or interaction with various biological components. The entropic and osmotic barrier overcomes the protein or particle–particle interaction that causes aggregation. The amount of poloxamer is very low, but sufficient to provide adequate stability and prevent aggregation during a centrifugation step (Peppas, 1986). Encapsulation efficiency is defined as the percentage of protein mass recovered, and drug loading is defined as the percentage of protein mass in the total mass of the final product. Nanoparticles were assessed against these criteria.

The manner in which the protein is released from the nanoparticles is dependent on the chemical nature of the nanoparticles. The protein is released by diffusion if its liberation occurs faster than the degradation of the matrix. Particles also swell according to the matrix composition (ionizable groups, degree of cross-linking) and environmental conditions (pH, ionic strength). The release of a drug from a hydrogel that swells involves the dissociation of the complex drug-matrix through exchange with counterions penetrating into the particle from the dissolution medium. Such release is indeed enhanced when the salt concentration is lowered. In addition to interactions between the constituent polymers, the incorporation of a charged species (e.g. protein) into the particle core introduces additional electrostatic interactions, particularly where the particle loading is high. A short incubation of the nanoparticles at elevated pH, for instance, forms cross-links between the carboxyl groups of polysaccharides of the polymeric mixture and the amino groups of the protein (Schiff-base linkages) that can also occur *in vivo* (Prokop *et al.*, 2002). Such linkages are relatively labile. Charged entrapped proteins contribute to the structure of particles and enhance high encapsulation efficiency. When tested with animal tissue, this type of particle formulation shows reduced cytotoxic effects.

11.3.4 Application of multi-membrane systems

The interesting multi-membrane systems of Ladet *et al.* (2008) described in Section 11.1.4 have a number of potential applications, but their work was very much geared towards studies of cell growth and attenuation for tissue engineering. For example, to validate the usefulness of such systems as biomaterials, the authors carried out a chondrocyte culture within the multi-membrane onion-like hydrogels. Cell aggregates were observed in several inter-membrane spaces, showing that cells can be introduced and cultured within these new systems for tissue growth.

11.3.5 Superporous hydrogels

A superporous hydrogel (SPH) differs from a superabsorbent polymer (SAP) or a normal gel in that SPHs swell fast, within minutes, to the equilibrium (swollen) state, regardless of their size. Their fast-swelling property is based on water absorption by capillary forces through an open porous structure. This unique property of SPHs – size-independent fast-swelling kinetics – follows from their interconnected, open cellular structure; using specially designed syntheses, pore sizes of up to 100 μm can be formed (Figure 11.13).

11.3.5.1 Harmonized foaming and gelation

The reactions involved in the preparation of SPHs are combinations of cross-linking polymerization (gelation) and foaming, just as in the traditional RIM process (Stanford, 1998) for production of urethane foams, but on a microscale. In the synthesis of SAP or SPH, the following general procedure is applied (Figure 11.14). (1) the monomer, such as acrylic acid salt or hydrophilic acrylamide, is first diluted with water to reach the desired concentration. (2) The monomer is partially neutralized, and (3) this is followed by addition of an appropriate cross-linker. (4) Foaming aids and stabilizers (PEO_n–PPO_m–PEO_n triblock copolymers) are then added, since to produce homogeneous SPHs foam stability is essential. To promote polymerization, (5) oxidant and (6) reductant are added to the monomer solution. Finally, to produce large numbers of pores, (7) $NaHCO_3$ is

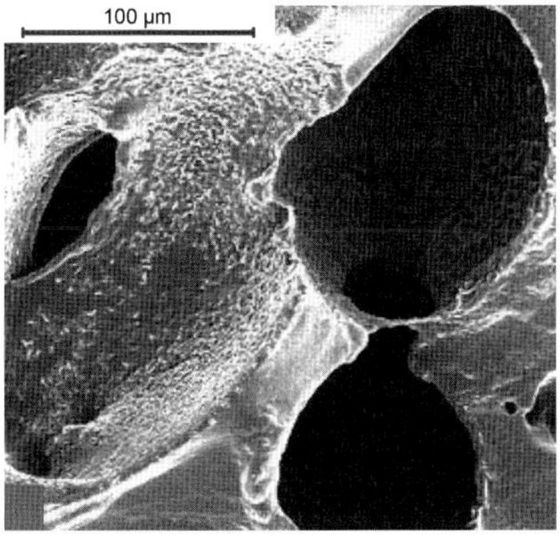

Figure 11.13 Scanning electron micrograph of a superporous hydrogel (SPH). Reproduced with permission from Omidian *et al.* (2005) © 2005 Elsevier.

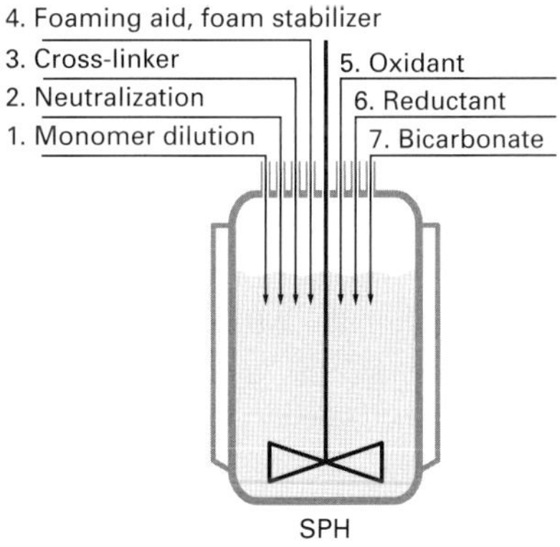

Figure 11.14 Steps in the production of a superporous hydrogel (SPH). Reproduced with permission from Omidian *et al.* (2005) © 2005 Elsevier.

added to generate gas bubbles. Dispersion and dissolution of HCO_3^- increases the pH of the reaction medium to a level at which the initiator decomposes faster. When the formation of initiator radicals reaches a certain level, polymerization proceeds rapidly and the reacting mixture becomes viscous over time. At the same time, HCO_3^- reacts with the acid component to produce the CO_2 required for the blowing process.

The two processes, i.e. the cross-linking reaction (gelation) and foaming processes, need to be controlled so that 80% of the foaming develops before gelation. Then the rest of the foaming occurs, the cross-linking reaction takes place and temperature increases by 5–10°C. As cross-linking proceeds further and the temperature reaches its maximum, the foam turns into a flexible, rubber-like material. Then the synthesized foam is added to a non-solvent, usually ethanol, in several batches until it is completely dehydrated, to stabilize the product and prevent it from shrinking. This results in a solid, brittle, porous product which can be ground into particles (like SAP particles), sliced into absorbent sheets or machined to a desired shape and size. Dried SPHs swell fast in water and to a size a few hundred times their initial volume. In this highly swollen state, they are sometimes difficult to handle without fracturing, and a second generation of SPH composites has been developed which are characterized by fast swelling, but a medium swelling ratio and improved mechanical properties.

11.3.5.2 Superporous 'hybrid' gels

These third-generation SPH 'hybrids' are designed to exhibit improved elastic properties, valuable for gastrointestinal devices as well as in other pharmaceutical and biomedical applications. To synthesize these, a so-called hybrid component that can be cross-linked after the SPH is formed is also included. This is a water-soluble or water-dispersible polymer that can form structures with either chemical or physical cross-linking. Examples of hybrid components are polysaccharides including sodium alginate, pectin and chitosan, and synthetic water-soluble polymers such as PVA. The overall system is structurally similar to an interpenetrating polymer network (IPN). An example of the process is the synthesis of acrylamide-based SPH in the presence of sodium alginate, followed by the cross-linking of alginate chains by Ca^{2+}.

One of the unique properties of SPH hybrids is that the gels are highly elastic in the swollen state and, compared to conventional SPHs, do not easily fail when subjected to large tensile strains. This property makes SPH hybrids a better choice in applications where resilient gels are preferred. They can resist various types of deformation, including tension, compression, bending and twisting. Such a hybrid hydrogel can also swell by up to 40 to 50 times, is very elastic in its swollen state and in this state can be extended by up to 3 times its original length, and the loading/unloading cycle can be repeated many times. This property could be exploited in the development of fast- and high-swelling elastic hydrogels for a variety of pharmaceutical, biomedical and industrial applications.

11.3.6 Superdisintegrant materials

Superdisintegrants are another class of superabsorbing materials with tailor-made swelling properties (Omidian and Park, 2008). They are not intended to absorb significant amounts of water or aqueous fluids, but to swell very fast. Such superdisintegrants are used to accelerate disintegration of solid samples (dosage forms) by being physically dispersed within the matrix of the dosage form, and then to allow the whole system to expand when exposed to a wet environment. The swelling pressure and the isotropic swelling of the particles create high stresses in localized areas, so the whole structure breaks apart, as shown in Figure 11.15.

To ensure that the tablet or solid dosage form breaks apart at an appropriate rate, the disintegrant should be evenly distributed within the matrix and should swell very fast to a size typically 10–40 times its original volume, in water or other appropriate aqueous medium. It should also be compatible with the other components (excipients).

In recent years there has been considerable growth in the number of orally soluble and chewable tablet products. These disintegrate rapidly in contact with saliva, thus

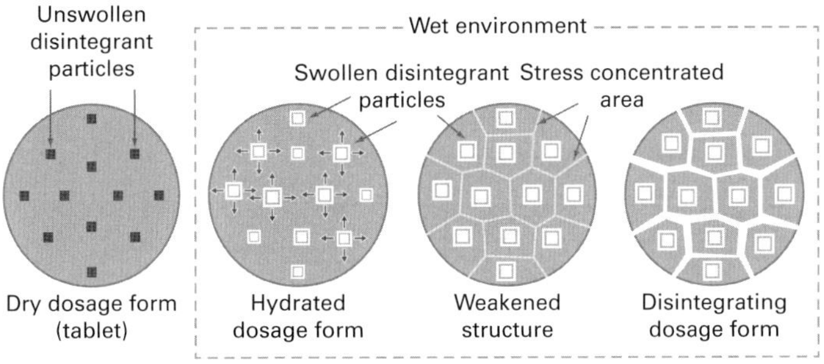

Figure 11.15 Disintegration mechanism for superdisintegrant tablets. From Omidian and Park (2008) with permission from Editions de Santé.

eliminating the need for water, and are adapted especially for children and the elderly, who may have difficulty swallowing the whole tablet. Initially the polymeric material should be in the glassy state, and form the matrix containing the drug particles. When water comes into contact with the polymer, swelling will begin as water penetrates between these chains. The process is self-accelerating, and a sheath of the swollen material is formed. In the swollen state, drug release is usually diffusion-controlled for water-soluble drugs, while water-insoluble drugs are released through a different, erosion-controlled mechanism. This so-called 'pseudo-swelling' is the major mechanism of controlled drug delivery in tablet matrices.

Such anisotropic swelling behaviour can be designed into tablets. These are generally manufactured under compression, and the polymer used in the tablet formulation is compressed uniaxially. In the presence of water it will swell to a greater extent axially than radially (Palleschi *et al.*, 2006). This is generally observed in sustained-release tablets where hydroxypropyl methylcellulose is used as a swelling agent (Papadimitriou *et al.*, 1993). Anisotropic swelling with scleroglucan–borax (Coviello *et al.*, 2005) and xanthan gum tablets (Talukdar and Kinget, 1995) has also been reported. Tablet disintegration times are typically found to be inversely proportional to the swelling ability of the superdisintegrant. The disintegrant particles are generally small and porous, which allows for rapid tablet disintegration in the mouth without undesirable mouth-feel from either large particles or the gel texture. Such tablets prepared by direct compression in the presence of gel-forming microparticles disintegrate in less than 10 s.

11.3.7 Further applications in tissue engineering

Tissue engineering has been defined in a number of ways, but in practice it covers the whole area pertinent to designing and growing cells on substrates of appropriate materials. In the present volume we are only interested in cases where physical gels have been used as the material support. Some cases, including the artificial pancreas using alginate gels as cell supports, have already been described (sections 5.5.4 and 11.3.2). This is a vast and very rapidly growing field, so we can only produce a very selective summary. Many papers on physical gels now tend to stress potential applications in this field, even though no tissue work has been carried out. Others deal with chemically cross-linked materials, for example glutaraldehyde as a biocompatible cross-linker. Again, space does not allow us to include these. Instead we will concentrate on a limited range of work using physical gels as 'scaffolds' – materials which allow, indeed encourage, *in vitro* cell growth to proceed unhindered, and generally in three dimensions. The material itself has to allow initial cell attachment and then transport appropriate biochemical factors and nutrients, eliminate any waste products and ideally encourage successful development, most often in a specific geometric fashion – for example to encourage growth of body part substitutes such as artificial bladders.

A typical study is that by Stevens and co-workers (Stevens *et al.*, 2004), in which the Ca^{2+} plus Na^{+} alginate system is employed as a scaffold for articular cartilage tissue. As already mentioned, this forms a very rapid and essentially irreversible gel. In this study a range of alginate types and concentrations was employed, and both pre-gel viscosity and post-gel Young's modulus were measured. Subsequent *in vitro* cell culture of chondrocytes in the gel and on the gel surface yielded so-called 'alginate/cell constructs'. These were composed of cartilage that contained appropriate cartilage-specific proteoglycans and collagen. The authors suggested that such 'explants' could be either transplanted or regenerated within the alginate matrix. In this way, partial or full-thickness defects in articular cartilage could be treated. They also commented that, because of the speed of gelation (in this case, <1 min), injectable delivery of the gel, a fairly minor medical procedure, could be used to fill defects in the articular surface of a patient.

Another paper, with several authors in common (Stevens *et al.*, 2005), used agarose, but fabricated into a 'stamp'. First, a silicone mould was made of a biochemical well plate. This was peeled away and a silicone replica of this produced by moulding, but this time of the original silicone construct. Finally, hot agarose solution was poured onto the second replica – now a direct silicone copy of the well plate – and allowed to cool, and the gel was then peeled away. This consisted of a series of 'posts' – an inversion of the original wells – which could be used as surfaces for subsequent cell growth. This procedure seems indirect, but presumably it was too difficult to produce direct agarose moulds of the cell plate without employing cytotoxic release agents.

The posts were then 'inked' with a suspension of human osteoblasts in a bone mineral (hydroxyapatite) scaffold, and their growth studied as a function of treatment and time. The authors argued that the technique has the advantage that it produces patterns of osteoblasts with control over the geometry, size and spacing between patterns of cells. They asserted that the procedure of stamping cells directly on tissue engineering scaffolds may have many future uses in controlling the spatial incursion on scaffolds, promoting cell hierarchical organization and controlling cell–cell interactions.

Finally, in the context of scaffold materials, Hyland *et al.* (2011) examined changes in the bulk material properties of various concentrations of chitosan–alginate mixtures on addition of the mucopolysaccharide chondroitin, and how this was altered by the order of addition. They found that chondroitin increased the mechanical strength of chitosan–alginate networks, and the highest modulus was obtained from samples made with chitosan and alginate modified first by chondroitin and then later by Ca^{2+}. Average pore size was slightly larger in samples modified by Ca^{2+} first and then chondroitin. In this study, small-angle neutron scattering (SANS) was employed to examine the mesh size of the fibrous networks and of the fibre dimensions prior to freeze-drying. These studies revealed that addition of Ca^{2+} and chondroitin modifiers increased fibre compactness and thickness, respectively.

11.4 Conclusions

Future development of 'smart' physical gels will continue, although we anticipate that the majority of novel applications will be in the biomedical area, not least because here the 'added value' is highest. Exciting new 'architectures' – as described above – will no doubt continue to be fashioned because of the potential of these gels in all aspects of nanotechnology. A considerable variety of physical and chemical cross-linking methods have been evaluated for application in products employed in various environments, and in both traditional and novel pharmaceutical areas. Innovative products have resulted from the molecular and supramolecular properties of networks of associating biological and synthetic polymers, and, in parallel with improvements in processing routes, exploit the various cross-linking mechanisms. For example, scaffold applications have already made quite dramatic progress, and a number of whole-body organs have been prepared using this approach. Physical gels have already contributed to numerous applications in everyday life so that, in our view, innovative products are already more advanced than aspects of understanding. However, we believe that this will stimulate further debate and initiate more thorough investigation of the underlying mechanisms of gelation. Finally, we expect the interplay between the system creators (in most cases synthetic chemists or biochemists) and concerned theorists to continue. We conclude with the hope that our book will help to stimulate further such cross-disciplinary interactions.

References

Adams, S., Frith, W. J., Stokes, J. R., 2004. *J. Rheol.* **48**, 1195–1213.

Berger, J., Reist, M., Mayer, J. M., Felt, O., Gurny, R., 2004a. *Eur. J. Pharm. Biopharm.* **57**, 35–52.

Berger, J., Reist, M., Mayer, J. M. *et al.*, 2004b. *Eur. J. Pharm. Biopharm.* **57**, 19–34.

Campbell, G. M., Mougeot, E., 1999. *Trends Food Sci. Technol.* **10**, 283–296.

Cheng, L. H., Lim, B. L., Chow, K. H., Chong, S. M., Chang, Y. C., 2008. *Food Hydrocolloids* **22**, 1637–1640.

Clarke, C., 2004. *The Science of Ice Cream*. Royal Society of Chemistry, Cambridge, UK.

Coviello, T., Grassi, M., Palleschi, A. *et al.*, 2005. *Int. J. Pharm.* **289**, 97–107.

Dash, M., Chiellini, F., Ottenbrite, R. M., Chiellini, E., 2011. *Prog. Polym. Sci.* **36**, 981–1014.

de Gennes, P.-G., 1999. *Phys. A (Amsterdam)* **271**, 231–237.

Di Renzo, F., Valentin, R., Boissière, M. *et al.*, 2005. *Chem. Mater.* **17**, 4693–4699.

Frith, W. J., 2010. *Adv. Colloid Interface Sci.* **161**, 48–60.

Graham, N. B., Cameron, A., 1998. *Pure Appl. Chem.* **70**, 1271–1275.

Hilt, J. Z., Gupta, A. K., Bashir, R., Peppas, N. A., 2003. *Biomed. Microdevices* **5**, 177–184.

Hyland, L. L., Taraban, M. B., Hammouda, B., Yu, Y. B., 2011. *Biopolymers* **95**, 840–851.

Ito, K., 2007. *Polym. J.* **39**, 489–499.
Ito, K., 2010. *Curr. Opin. Solid State Mater. Sci.* **14**, 28–34.
Ito, K., 2012. *Polym. J.* **44**, 38–41.
King, A., Andersson, A., Strand, B. L. *et al.*, 2003. *Transplantation* **76**, 275–279.
Koyano, T., Koshizaki, N., Umehara, H., Nagura, M., Minoura, N., 2000. *Polymer* **41**, 4461–4465.
Ladet, S., David, L., Domard, A., 2008. *Nature* **452**, 76–79.
Leonard, M., Rastello de Boisseson, M., Hubert, P., Dellacherie, E., 2004. *J. Biomed. Mater. Res., Part A* **68**, 335–342.
Miyazawa, K., Yajima, I., Kaneda, I., Yanaki, T., 2000. *J. Cosmet. Sci.* **51**, 239–252.
Norton, I. T., Jarvis, D. A., Foster, T. J., 1999. *Int. J. Biol. Macromol.* **26**, 255–261.
Okumura, Y., Ito, K., 2001. *Adv. Mater.* **13**, 485–487.
Omidian, H., Park, K., 2008. *J. Drug Delivery Sci. Technol.* **18**, 83–93.
Omidian, H., Rocca, J. G., Park, K., 2005. *J. Controlled Release* **102**, 3–12.
Palleschi, A., Coviello, T., Bocchinfuso, G., Alhaique, F., 2006. *Int. J. Pharm.* **322**, 13–21.
Papadimitriou, E., Buckton, G., Efentakis, M., 1993. *Int. J. Pharm.* **98**, 57–62.
Peppas, N. A. 1986. *Hydrogels in Medicine and Pharmacy, Vol 1: Fundamentals*. CRC Press, Boca Raton, FL.
Peppas, N. A., Hilt, J. Z., Khademhosseini, A., Langer, R., 2006. *Adv. Mater.* **18**, 1345–1360.
Prokop, A., Holland, C. A., Kozlov, E., Moore, B., Tanner, R. D., 2001. *Biotechnol. Bioeng.* **75**, 228–232.
Prokop, A., Kozlov, E., Newman, G. W., Newman, M. J., 2002. *Biotechnol. Bioeng.* **78**, 459–466.
Rastello de Boisseson, M., Leonard, M., Hubert, P. *et al.*, 2004. *J. Colloid Interface Sci.* **273**, 131–139.
Remondetto, G. E., Beyssac, E., Subirade, M., 2004. *J. Agric. Food Chem.* **52**, 8137–8143.
Roberts, S. A., Kasapis, S., Lopez, I. D., 2000. *Int. J. Food Sci. Technol.* **35**, 227–234.
Stanford, J. L., 1998. Reactive processing of polymer networks. In Stepto, R. F. T. (ed.), *Polymer Networks: Principles of Their Formation, Structure and Properties*. Chapman and Hall, Glasgow, pp. 125–186.
Stevens, M. M., Qanadilo, H. F., Langer, R., Shastri, V. P., 2004. *Biomaterials* **25**, 887–894.
Stevens, M. M., Mayer, M., Anderson, D. G. *et al.*, 2005. *Biomaterials* **26**, 7636–7641.
Talukdar, M. M., Kinget, R., 1995. *Int. J. Pharm.* **120**, 63–72.
Thu, B., Gåserød, O., Paus, D. *et al.*, 2000. *Biopolymers* **53**, 60–71.
Thumbs, J., Kohler, H.-H., 1996. *Chem. Phys.* **208**, 9–24.
Treml, H., Woelki, S., Kohler, H.-H., 2003. *Chem. Phys.* **293**, 341–353.
Zuniga, R., Aguilera, J., 2008. *Trends Food Sci. Technol.* **19**, 176–187.
Zuniga, R., Aguilera, J., 2009. *Food Hydrocolloids* **23**, 1351–1357.
Zuniga, R., Kulozik, U., Aguilera, J., 2011. *Food Hydrocolloids* **25**, 958–967.

Index